PHOTOSTABILITY OF DRUGS AND DRUG FORMULATIONS

Second Edition

PHOTOSTABILITY OF DRUGS AND DRUG FORMULATIONS

Second Edition

Edited by
Hanne Hjorth Tønnesen

CRC Press
Taylor & Francis Group
Boca Raton London New York

CRC Press is an imprint of the
Taylor & Francis Group, an **informa** business

CRC Press
Taylor & Francis Group
6000 Broken Sound Parkway NW, Suite 300
Boca Raton, FL 33487-2742

First issued in paperback 2019

© 2004 by Taylor & Francis Group, LLC
CRC Press is an imprint of Taylor & Francis Group, an Informa business

No claim to original U.S. Government works

ISBN-13: 978-0-415-30323-1 (hbk)
ISBN-13: 978-0-367-39410-3 (pbk)
Library of Congress Card Number 2004043590

Library of Congress Cataloging-in-Publication Data

Photostability of drugs and drug formulations / edited by Hanne Hjorth Tønnesen. -- 2nd ed.
        p. cm.
Includes bibliographical references and index.
    ISBN 0-415-30323-0 (alk. paper)
    1. Drug stability. I. Tønnesen, H. Hjorth (Hanne Hjorth). II. Title.

RS424.P48 2004
615′.19--dc22   2004043590

Visit the Taylor & Francis Web site at
http://www.taylorandfrancis.com

and the CRC Press Web site at
http://www.crcpress.com

Preface

This is the second edition of *Photostability of Drugs and Drug Formulations*; the first was published in 1996. The philosophy of this new edition remains unchanged, i.e., to provide the background and overview necessary for planning standardized photochemical stability studies; evaluating drug photoreactivity as a part of drug development and formulation work; and safely handling photolabile drug products. Thus, the book covers *in vitro* and *in vivo* aspects of drug photoreactivity. It should thereby provide a useful reference for R&D scientists in the pharmaceutical and healthcare industries, hospital pharmacists, and pharmaceutical regulatory bodies.

The content structure of this edition has changed somewhat to focus more clearly on standardized photostability testing of drug substances and products; *in vitro* photoreactivity screening of drugs; and various aspects of the formulation of photoreactive substances. The *ICH Harmonized Tripartite Guideline on Photostability Testing of New Drug Substances and Products* (ICH Q1B) was implemented by January 1, 1998. Despite this implementation, issues not specifically covered in the documents have been left to the applicant's discretion. Photochemical reactions are far more complex than thermal processes and many questions must be considered in addressing photostability testing. It is hoped that this book will assist nonexperts in this field in designing test protocols and interpreting the results by discussing kinetic and chemical aspects of drug photodecomposition, as well as practical problems frequently encountered in photochemical stability testing.

Recent documents issued by the FDA (FDA,* CDER May 2003) and CPMP** (December 2002) provide guidance on how to address photosafety assessments and labeling requirements of potentially photoreactive drugs. A key issue in these documents is the physicochemical assessment of a compound's photoreactivity potential. The regulatory guidance documents, however, do not define how these assessments are made or interpreted. Although the interpretation of these particular guidelines is not a specific issue in this book, the chapters on *in vitro* screening of drug photoreactivity should be of great relevance for developing and validating a set of testing protocols to address photosafety. Practical implications of drug photodecomposition (e.g., formulation problems; photoactivation for triggering drug delivery; storage and handling of products) are covered from a pharmaceutical formulation viewpoint. This information should be useful for those in charge of drug development and for hospital pharmacists preparing and handling infusion solutions.

The involvement of a wide range of authors continues in this edition; each is an accepted expert in the field on which he or she has written. Many contributions of authors from the first edition remain. Other authors (who have retired or, sadly, died) have been replaced by a new generation of experts. Some material has been removed

---

* Guidance for industry: photosafety testing.
** Note for the guidance on photosafety testing (CPMP/SWP/398/01, December 2002).

and new material added. The new authorship should further emphasize the significance of information and knowledge on drug photoreactivity in a pharmaceutical context. I hope that the book will be helpful in understanding interactions between drugs and light.

Hanne Hjorth Tønnesen
June 2004
Oslo

Acknowledgments

I am extremely indebted to the authors for the work and time they put into their texts. I know that they were under pressure from numerous other commitments during this work. The time spent in contributing to the present book is warmly appreciated. Special thanks to Dr. D.E. Moore for guidance, valuable discussions, and critical comments throughout the preparation of this new edition.

Hanne Hjorth Tønnesen

Contributors

Angelo Albini
Department of Organic Chemistry
University of Pavia
Pavia, Italy

N.H. Anderson
Retired from Sanofi-Synthelabo
  Research
Alnwick Research Centre
Alnwick, Northumberland, U.K.

Steve W. Baertschi
Eli Lilly and Company
Lilly Corporate Center
Indianapolis, IN

Jörg Boxhammer
Atlas Material Testing Technology
  GmbH
Linsengericht-Altenhasslau, Germany

Stephen J. Byard
Sanofi-Synthelabo Research
Alnwick Research Centre
Alnwick, Northumberland, U.K.

P.L. Carter
Pharmaceutical Technologies
GlaxoSmithKline Pharmaceuticals
Harlow, Essex, U.K.

D. Clapham
Pharmaceutical Technologies
GlaxoSmithKline Pharmaceuticals
Harlow, Essex, U.K.

Elisa Fasani
Department of Organic Chemistry
University of Pavia
Pavia, Italy

Joseph C. Hung
Department of Radiology
Mayo Clinic
Rochester, MN

Petras Juzenas
Department of Biophysics
Institute for Cancer Research
Oslo, Norway

Jan Karlsen
Department of Pharmaceutics
School of Pharmacy
University of Oslo
Oslo, Norway

Solveig Kristensen
Department of Pharmaceutics
School of Pharmacy
University of Oslo
Oslo, Norway

D.R. Merrifield
Pharmaceutical Technologies
GlaxoSmithKline Pharmaceuticals
Harlow, Essex, U.K.

Johan Moan
Department of Biophysics
Institute for Cancer Research
Oslo, Norway

Douglas E. Moore
Department of Pharmacy
The University of Sydney
New South Wales, Australia

Suppiah Navaratnam
Bioscience Research Institute
University of Salford
Salford, U.K.
and
FRRF
CLRC Daresbury Laboratory
Warrington, U.K.

Joan E. Roberts
Fordham University
New York, NY

F.D. Sanderson
Pharmaceutical Technologies
GlaxoSmithKline Pharmaceuticals
Harlow, Essex, U.K.

Artur Schönlein
Atlas Material Testing Technology
  GmbH
Linsengericht-Altenhasslau, Germany

Hanne Hjorth Tønnesen
Department of Pharmaceutics
School of Pharmacy
University of Oslo
Oslo, Norway

Table of Contents

Introduction: Photostability Testing of Drugs and Drug Formulations — Why and How?

Hanne Hjorth Tønnesen

CONTENTS

It is well known that light can change the properties of different materials and products. This is often observed as bleaching of colored compounds like paint and textiles or as a discoloration of colorless products. Photostability has for many years been a main concern within several fields of industry, e.g., the textile, paint, food, cosmetic, and agricultural industries. In the field of pharmacy, photostability has played a less important role. Meanwhile, the number of drugs found to be photo-chemically unstable is steadily increasing. The European pharmacopoeia prescribes light protection for more than 250 medical drugs and a number of adjuvants. New compounds are frequently added to this list, although the justification of light protection required for certain compounds has been questioned (Reisch and Zappel, 1993).

Several points need to be clarified before developing and adopting a protocol for photostability testing:

- What is the rationale for evaluating drug photostability?
- What can be achieved by evaluating drug photostability?
- How can adequate information about drug photostability be obtained?

In this context the term "photostability" is used to describe how a compound responds to light exposure and includes not only degradation reactions but also other processes such as formation of radicals, energy transfer, and luminescence.

1.1 RATIONALE FOR EVALUATION OF DRUG PHOTOSTABILITY

The most obvious result of drug photodecomposition is a loss of potency of the product. In the final consequence, this can result in a therapeutically inactive drug product. Although this is not often the case, even less severe degradation can lead to problems. Adverse effects due to the formation of minor degradation products during storage and administration have been reported (de Vries et al., 1984). The drug substance can also cause light-induced side effects after administration to the patient by interaction with endogenous substances. Therefore, two aspects of drug photostability must be considered: *in vitro* and *in vivo* stability. The possible consequences of drug photoinstability are illustrated in Figure 1.1. Independent of concern about *in vitro* stability or *in vivo* effects, characterization of the photochemical properties of drug substances and drug formulations is a part of the formulation work and cannot be ignored.

Many drug substances and drug products are found to decompose *in vitro* under exposure to light, but the practical consequences will not necessarily be the same

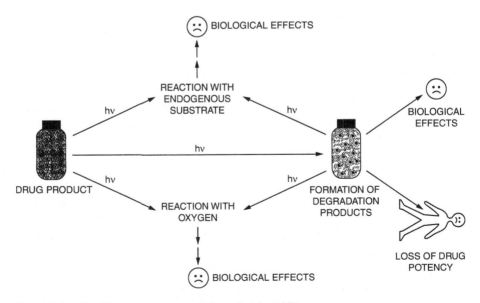

Figure 1.1 Possible consequences of drug photoinstability.

in all cases. Derivatives of the drug nifedipine have a photochemical half-life of only a few minutes, while other drugs may decompose only a few percent after several weeks' exposure (Squella et al., 1990). All are sensitive to light, but the same precautions are not required in handling these compounds. Knowledge about the photostability of drug substances and drug products is important in order to evaluate:

- Handling, packaging, and labeling
- Adverse effects
- Therapeutic aspects and new drug delivery systems

1.1.1 Handling, Packaging, and Labeling

The ability of a drug substance to degrade or undergo a gradual change in color upon light exposure is not an uncommon property. Polymorphs of drug substances can even exhibit different sensitivity to light (Nyqvist and Wadsten, 1986). In practice, the drug substance would mainly experience exposure to visible light (i.e., cool white fluorescent tubes) during storage and production. Many drug substances are white; essentially, no visible light will be absorbed by these compounds. It is, however, important to know that all lamps, even incandescent ones, emit some radiation in the UV region of the spectrum. Light protection of the drug substance during storage and production must therefore be recommended in many cases. Solid-state photostability is not fully evaluated. A change in color upon exposure is necessarily not correlated with the extent of chemical degradation of the material (Matsuda and Tatsumi, 1990).

A change in the selection of packing materials combined with a change in storage conditions or conditions during administration of the drug products seems to generate new stability problems *in vitro*. Most people are familiar with the traditional brown medicinal flask or the white pill box. These containers offer adequate protection of most drug products during storage and distribution. In modern hospital pharmacies, drugs are often stored in unit-dose containers on an open shelf. In many cases, the protective market pack is removed; the inner container can be made of transparent plastic materials that offer little if any protection toward UV and visible radiation (Tønnesen, 1989; Tønnesen and Karlsen, 1987). The unprotected drug product can then be exposed to fluorescent tubes and/or filtered daylight for several days or weeks (Tønnesen and Karlsen, 1995).

Infusion solutions should be stored in transparent infusion bottles or infusion bags. Long-term infusions can lead to exposure of the drug to filtered daylight for hours. During intravenous medication of premature babies under treatment for hyperbilirubinemia, the drug can experience radiation of high intensity. Portable drug delivery devices are often used to treat patients with severe pain and various types of plastic materials are used in the drug reservoirs for these pumps.

The precautions taken in handling these drugs, including adequate labeling and selection of packaging, will in each case depend on the photochemical half-life of the drug substance in the formulation. Because basic information about the photostability of the compounds is needed, evaluation of *in vitro* stability is essential to ensure good quality over the entire life span of the drug.

1.1.2   Adverse Effects

Although a drug product is shown to be photochemically inert in the sense that it does not decompose during exposure to light, it can still act as a source of free radicals or form phototoxic metabolites *in vivo* (Beijersbergen van Henegouwen, 1981). The drug will then be photoreactive after administration if the patient is exposed to light, causing light-induced adverse effects (Epstein and Wintroub, 1985). This emphasizes the importance of including studies of reaction mechanisms and sensitizing properties of the parent compound and its degradation products and *in vivo* metabolites in the evaluation of photostability of drugs. The increase in number of reported adverse effects that can be ascribed to the combination of drugs and light is due to an increase in exposure to artificial light sources such as daylight lamps and solaria; a change in human leisure habits (more time spent outdoors); and a widespread use of drugs.

Several requirements are to be met if a drug is to cause phototoxic reactions. First, the drug or metabolites of the compound must be distributed to tissues near the body surface, e.g., skin, eye, and hair, that are exposed to light. Then the absorption spectrum of the drug must overlap with the transmission spectrum of light through the tissue. Recent documents issued by the FDA (FDA, CDER, May 2003) and EMEA (CPMP, December 2002) provide guidance on how to address photosafety assessments and labeling requirements for potentially photoreactive drugs.

1.1.3   Therapeutic Aspects and New Drug Delivery Systems

In vivo photodecomposition and radical formation should not always be avoided because these properties can be advantageous from a therapeutic viewpoint. More than 3000 years ago, the Egyptians, Chinese, and Indians were using photosensitization in attempts to cure such disorders as vitiligo, rickets, psoriasis, skin cancer, and psychosis (Harber et al., 1982; Spikes, 1985). Treatment of psoriasis by combination of psoralens and UV-A light (PUVA therapy) is now well established. Alternative photosensitizers are certainly in demand. The potential for new drug delivery systems such as light-activated liposomes or prodrugs should not be ignored. Fiber optics can be used for activation of therapeutic compounds in drug targeting (Bayley et al., 1987). New developments in the field of topical preparations and other novel drug delivery systems activated by radiation have advantages, especially in the treatment of localized tumors and resistant bacterial infections.

1.2   WHAT CAN BE ACHIEVED BY EVALUATION OF DRUG PHOTOREACTIVITY?

Great effort is taken to stabilize a formulation in such a way that the shelf-life becomes independent of the storage conditions. Photostability of drugs and excipients should be evaluated at the formulation development stage in order to assess the effects of formulation and packaging on the stability of the final product. The information obtained should also result in label storage recommendations. The purpose of these recommendations is to guarantee maintenance of product quality in relation to its safety, efficacy, and acceptability throughout the proposed shelf-life, i.e., during storage, distribution, and use (including reconstitution or dilution as recommended in the labeling). Details on photostability will also be helpful for advising the patient to avoid direct sun, wear sunglasses, and use sun-protective creams in order to minimize side effects.

1.3 HOW TO OBTAIN INFORMATION ON DRUG PHOTOREACTIVITY

The ICH Harmonized Tripartite Guideline on Stability Testing of New Drug Substances and Products (ICH Q1B) was implemented in Europe in 1996 and in the U.S. and Japan in 1997. Since January 1, 1998 it has been obligatory to provide photostability data for all new drug license applications to the markets in the U.S., the European Union, Japan, and Canada. Photostability testing of drugs may be considered as consisting of two parts. Stress testing is undertaken to evaluate the overall photosensitivity of the drug substance. Such evaluation is not mandatory but should be established as a part of the preformulation work. Left to the applicant's discretion, the design of the photoassay may include a variety of exposure conditions. The photoassay should lead to determination of degradation pathways; identification of degradation products; and evaluation of sensitizing properties of the parent compound, its degradation products, impurities, or *in vivo* metabolites. Accelerated testing includes a standardized photostability test for drug substances and drug products in order to determine the need for a label warning according to regulatory requirements. This test is designed as a simple pass/fail test.

Knowledge about the photochemical and photophysical properties of the compound is essential for handling, packaging, and labeling the drug substance and drug product; however, it is also needed in order to predict drug phototoxicity. Several *in vitro* methods for phototoxicity studies have been previously described (Valenzeno et al., 1991), but in many cases *in vivo* test methods will also be required (Oppenländer, 1988). A complete assay for photostability/phototoxicity, however, consumes time and money and requires a broad spectrum of techniques. A selection of the drug compounds to undergo a full screening can be made based on certain criteria:

- The drug or metabolites of the drug accumulate in tissues frequently exposed to light (skin, eye, hair).
- The drug is administered at a high accumulative dosage.
- The drug is photolabile *in vitro*.
- The drug forms photolabile degradation products or *in vivo* metabolites.
- The drug is administered topically.
- The drug molecule contains essential functionalities known to induce phototoxicity reactions.

Large structural variations are found among molecules that can act as photosensitizers in biological systems, so photostability is difficult to predict (Greenhill and McLelland, 1990). It is also important to be aware that the photostability of a pure compound can change when the sample is introduced into a biological system. Interactions between the drug substance and excipients in the actual formulation can further influence the photostability. Tests on the final product should therefore be included in the total evaluation of photostability.

1.4 CONCLUSION

Photostability testing of the drug substance is undertaken to evaluate the overall photosensitivity of the material for development and validation purposes and to provide information necessary for handling, packaging, and labeling. A photostability assay for pharmaceutical products should provide information related to the practical use of the product, i.e., light-exposure conditions that the product will experience under its normal applications. Well-designed photostability studies ensure the quality of the product throughout shelf-life and guarantee its safety, efficacy, and acceptability to the patient. Standardized experimental conditions must be applied in stability testing. Demand is also increasing for photoreactivity data in order to address photosafety assessments and labeling requirements for potentially photoreactive drugs. The evaluation of interactions between drugs and light forms a natural part of the research and development work for new drug substances and drug products.

REFERENCES

Bayley, H., Gasparro, F. and Edelson, R., 1987, Photoactivable drugs, *TIPS*, 8, 138–143.
Beijersbergen van Henegouwen, G.M.J., 1981, Photochemistry of drugs *in vitro* and *in vivo*, in Breimer, D.D. and Speiser, D. (Eds.), *Topics in Pharmaceutical Sciences*, pp. 233–256. Holland: Elsevier/North-Holland Biomedical Press.
CPMP (December 2002) Note for the guidance on photosafety testing, CPMP/SWP/398/01, http://www.emea.eu.int/pdfs/human/swp/039801en.pdf.
de Vries, H., Beijersbergen van Henegouwen, G.M.J. and Huf, F.A., 1984, Photochemical decomposition of chloramphenicol in a 0.25% eyedrop and in a therapeutic intraocular concentration, *Int. J. Pharm.*, 20, 265–271.
Epstein, J.H. and Wintroub, B.U., 1985, Photosensitivity due to drugs, *Drugs*, 30, 42–57.
FDA, CDER (May 2003) Guidance for industry: photosafety testing, http://www.fda.gov/cder/guidance/3640fnl.pdf.
Greenhill, J.V. and McLelland, M.A., 1990, Photodecomposition of drugs, in Ellis, G.P. and West, G.B. (Eds.), *Progress in Medicinal Chemistry*, pp. 51–121. Holland: Elsevier Science Publishers, B.V.
Harber, L.C. Kochevar, I.E. and Shalita, A.R., 1982, Mechanisms of photosensitization to drugs in humans, in Regan, J.D. and Parrish, J.A. (Eds.), *The Science of Photomedicine*, pp. 323–347. New York: Plenum Press.
ICH Q1B (1997) Photostability testing of new drug substances and products, *Fed. Reg.*, 62, 27115–27122.
Matsuda, Y. and Tatsumi, E., 1990. Physiochemical characterization of frusemide modifications, *Int. J. Pharm.*, 60, 11–26.

Nyqvist, H. and Wadsten, T., 1986, Preformulation of solid dosage forms: light stability testing of polymorphs as a part of a preformulation program, *Acta Pharm. Technol.,* 32, 130–132.

Oppenländer, T., 1988, A comprehensive photochemical and photophysical assay exploring the photoreactivity of drugs. *CHIMIA,* 42, 331–342.

Reisch, J. and Zappel, J., 1993, Photostabilitätsuntersuchungen an Natrium-Warfarin in kristallinem Zustand, *Sci. Pharm.,* 61, 283–286.

Spikes, J.D., 1985, The historical development of ideas on applications of photosensitized reactions in the health sciences, in Bensasson, R.V., Jori, G., Land, E.J., and Truscott, T.G. (Eds.), *Primary Photo-Processes in Biology and Medicine,* pp. 209–277. New York: Plenum Press.

Squella, J.A., Zanocco, A., Perna, S. and Nuñez–Vergara, L.J., 1990, A polarographic study of the photodegradation of nitrendipine, *J. Pharm. Biomed. Anal.,* 8, 43–47.

Tønnesen, H.H., 1989, Emballasjens betydning ved formulering av fotokjemisk ustabile legemidler, *Norg. Apot. Tidsskr.,* 97, 79–85.

Tønnesen, H.H. and Karlsen, J., 1987, Studies on curcumin and curcuminoids. X. The use of curcumin as a formulation aid to protect light-sensitive drugs in soft gelatin capsules, *Int. J. Pharm.,* 38, 247–249.

Tønnesen, H.H. and Karlsen, J., 1995, Photochemical degradation of components in drug formulations. III. A discussion of experimental conditions, *PharmEuropa,* 7, 137–141.

Valenzeno, D.P., Pottier, R.H., Mathis, P. and Douglas, R.H., 1991, *Photobiological Techniques,* pp. 85–120, 165–178, 347–349. London: Plenum Press.

CHAPTER 2

Photophysical and Photochemical Aspects of Drug Stability

Douglas E. Moore

CONTENTS

2.1 ABSORPTION SPECTRA OF DRUGS

A photon corresponding to the ultraviolet wavelength 300 nm has energy of 400 kJ mol^{-1}, which is of comparable magnitude to the bonding energy of organic compounds. The fact that a drug absorbs radiation in the ultraviolet or visible region of the electromagnetic spectrum means that it is absorbing energy sufficient to break a bond in the molecule. Thus, the property of absorption is a first indication that the drug *may* be capable of participating in a photochemical process leading to its own decomposition or that of other components of the formulation. The statement is a qualified one because a number of processes may occur following absorption of UV or visible light, some of which lead to no net change to the absorbing molecule or the system. The photochemical reaction must follow the basic law of photochemical absorption, established by Grotthus and Draper in 1818, that no photochemical (or subsequent photobiological) reaction can occur unless electromagnetic radiation is absorbed. The absorption spectrum of a compound is therefore an immediate way of determining the wavelength range to which the drug may be sensitive.

Some drug substances and formulation excipients are colored, meaning that they absorb light in the visible region. The color they display is complementary to the light they absorb, e.g., a red powder is absorbing blue light. The great majority of therapeutic substances are white in appearance, meaning that they do not absorb light in the visible region, but they may absorb in the UV region as a consequence of their chemical structure. The presence of aromatic residues and conjugated double bonds containing N, S, or O in the structure is usually associated with the ability of the molecule to absorb light.

Two contrasting examples, ibuprofen and sulindac, chosen from the wide range of anti-inflammatory drugs, are given in Figure 2.1. Ibuprofen is a white powder with a weak absorption centered on 265 nm, due to the aromatic chromophore, unaffected by substituents. On the other hand, sulindac is yellow and absorbs strongly across the UV and into the visible regions. Its absorption maxima occur at 280 and 327 nm due to the extended chromophore. When each of these compounds is irradiated with wavelengths corresponding to their maximum absorption, photodegradation occurs. However, if the extent of decomposition is equated with the amount of radiation absorbed, it transpires that ibuprofen is significantly more photoreactive than sulindac (unpublished results). The difference in the way these two substances need to be stored stems, of course, from the fact that ibuprofen would experience exposure to UV radiation of around 265 nm only under the most unusual storage conditions involving germicidal lamps that emit at 254 nm. On the other hand, sulindac can absorb the output from regular room lighting as well as sunlight. Thus, sulindac must be packaged so that it is protected from light, e.g., with amber glass, but that precaution is not deemed necessary for ibuprofen.

Two important factors should be pondered in relation to the potential of a drug to be degraded following absorption of electromagnetic radiation. First, the absorption spectrum is normally described by the maximum absorption wavelength and the molar absorptivity at that wavelength; however, the spectrum of a drug molecule is usually broad, and *any overlap* of the absorption spectrum with the output of the photon source impinging upon it has the potential to lead to photochemical change.

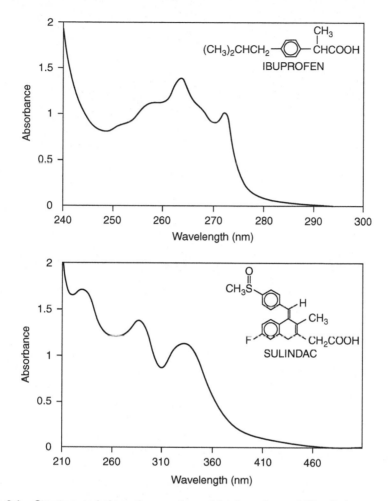

Figure 2.1   Structure and absorption spectrum of (a) ibuprofen and (b) sulindac.

Second, the decomposition may be initiated by another component of the formulation that has absorption characteristics that overlap with the incident radiation while the therapeutic component does not. In this case, the process is called photosensitization and the absorbing component, or photosensitizer, may transfer the absorbed energy completely and not be altered in the process (although it is more likely that it will undergo some degradation).

2.2 SPECTRAL CHARACTERISTICS OF SUNLIGHT AND ARTIFICIAL LIGHT SOURCES

In the course of manufacture and storage through ultimate use by the consumer, pharmaceuticals may be exposed to light from a number of sources, ranging from direct sunlight through filtered sunlight to a variety of artificial light conditions. In

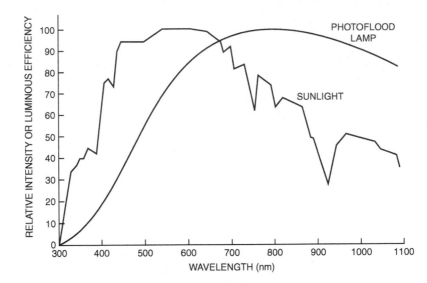

Figure 2.2   Spectral power distribution of sunlight compared with an incandescent lamp (the curves are normalized).

terms of the possibility of photochemical reaction, the UV component of sunlight is the most potentially damaging. However, there may be long periods of exposure to fluorescent and incandescent lighting during the various stages of manufacture, storage, and use, so it is important to consider their spectral distribution as well. Relative spectral intensity curves are shown in Figure 2.2 for sunlight and an incandescent (filament) lamp. Each of these spectra extends from near 300 nm in the ultraviolet region to beyond 3000 nm in the infrared, but with differing intensity distribution.

Ultraviolet radiation (UVR) has been divided into three sub-bands: UVC, UVB, and UVA (Grossweiner, 1989). The UVC band ranges from 200 to 280 nm and is often called shortwave or far-UV because the wavelengths in this region are the shortest UVR transmitted through air. Although most drugs and all cellular constituents absorb UVC, sunlight at the Earth's surface contains no UVC because of efficient absorption by molecular oxygen and ozone in the upper atmosphere (Frederick et al., 1989). Despite its absence from natural sunlight at the Earth's surface, UVC is present in artificial radiation sources, such as discharge and germicidal lamps and welding arcs, and can cause rapid photochemical degradation, as well as serious damage to the skin and cornea following exposure. The determination of chemical and biological effects of UVC is still receiving much attention today, partly because of the increasing knowledge of far-UV photochemistry and the specificity of the damage generated (Cadet et al., 1992).

The UVB spectral region is currently defined as encompassing wavelengths from 280 to 320 nm (Grossweiner, 1989). However, no solar radiation penetrates to the ground at wavelengths between 280 and 290 nm, and this remains true even in the case of a large reduction in atmospheric ozone such as occurs over Antarctica in springtime. Therefore, it has been suggested that the interval from 290 to 320 nm be adopted as a practical definition of the UVB (Frederick, 1993), but this has not received official sanction. The purine and pyrimidine bases of DNA and the aromatic amino acids are the major cellular absorbers of UVB. Although the intensity of UVB in the solar UVR reaching the Earth's surface is relatively small (Thorington, 1985), it is abundantly clear that UVB is the most important band because it causes sunburn, skin cancer, and other biological effects and is responsible for the direct photoreaction of many chemicals in natural sunlight (Epstein, 1989). The UVB intensity at a particular latitude varies greatly with time of day and the season of the year, as the variation of the solar azimuthal angle varies the path length of the Sun's rays through the stratospheric ozone layer.

UVA is the long wavelength UV region from 320 to 400 nm. It is also called near-UV because of its proximity to the visible spectrum. In total energy the amount of solar UVA reaching the Earth's surface is enormously greater than that of UVB (Gates, 1966). Chemical and biological effects induced by UVA may involve direct energy absorption, e.g., in the long wavelength absorption tail of proteins and DNA, or photosensitization by endogenous or exogenous substances.

Sunlight has a very high output in the visible (400 to 800 nm) and infrared (800 to 3200 nm) regions, while the incandescent lamp typifies black body radiation with a higher relative infrared output. The only importance that infrared radiation can accrue in the context of photodegradation is that the sample can be heated, thereby activating thermal decomposition. The visible region is relevant when a colored substance is present in the formulation.

Artificial light sources can have varying spectral characteristics depending on the particular construction. The key component of a fluorescent light is the low-pressure mercury discharge at 254 nm within a glass tube coated internally with a phosphor having specific emission characteristics. The spectral power distribution of several commercial fluorescent artificial light sources is shown in Figure 2.3. Although the principal output is in the visible region, there is a significant UV component. Note also the discontinuous line spectrum superimposed upon the background of continuous radiation.

It has been estimated that at least 90% of all lighting in the business and manufacturing sectors in the U.S. is achieved by "cool-white" fluorescent tubes, while the domestic sector uses incandescent lighting for 80% of its artificial light needs (Thorington, 1985). The glass bulb or tube in an artificial light source can be said to act as the ozone layer does with respect to natural sunlight, limiting the UVR component to about 300 nm, depending on the glass used. According to Thorington (1985), no criteria exist for the UVR component in most commercial artificial light sources because the sole function is to provide light in the narrow definition of illuminating engineering (i.e., visible light).

2.3 ACTION SPECTRUM AND OVERLAP INTEGRAL

The term *action spectrum* has been used in two rather different ways. The first usage is strictly incorrect in that it relates to the overlap integral of the spectra for the

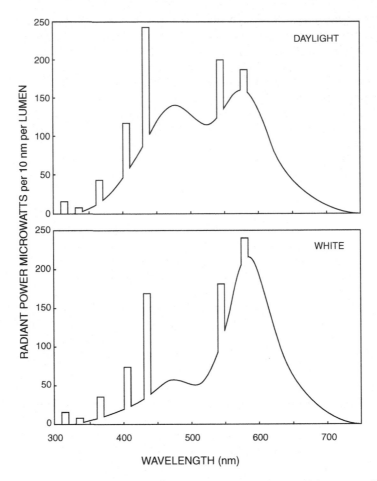

Figure 2.3   Spectral power distribution of daylight and white fluorescent light sources. (Redrawn from Thorington, L., *Ann. N. Y. Acad. Sci.,* 453: 28–54,1985.)

particular combination of photon source and absorbing substance. A familiar example of this definition of an action spectrum is the sunburn or *erythemal effectiveness spectrum*, which is the overlap of the sunlight UV spectrum and the absorption spectra of proteins and nucleic acids as shown in Figure 2.4 (Parrish et al., 1978). The sunburn response (erythema) can be elicited in human skin by an artificial light source emitting any of the wavelengths corresponding to the absorption spectra of protein and nucleic acid.

Sunburn (caused by the Sun) occurs only for the narrow range of wavelength for which the overlap with the solar emission is finite. This type of overlap integral would be found for quite a large number of drugs whose absorption maxima occur at around 270 to 280 nm with a broad tail extending into the UVB region. For examples one need look no further than the sulfonamide group of antibacterials. Sulfamethoxazole has its absorption maximum at 268 nm but is decomposed on exposure to sunlight. Indeed, such is its change in decomposition rate with time of

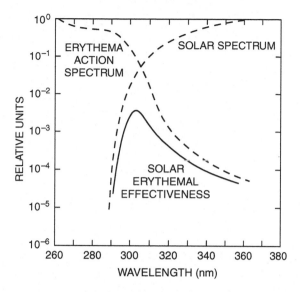

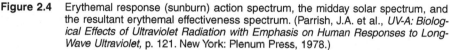

Figure 2.4   Erythemal response (sunburn) action spectrum, the midday solar spectrum, and the resultant erythemal effectiveness spectrum. (Parrish, J.A. et al., *UV-A: Biological Effects of Ultraviolet Radiation with Emphasis on Human Responses to Long-Wave Ultraviolet,* p. 121. New York: Plenum Press, 1978.)

day and season of the year that its use has been suggested for an absolute chemical reaction standard for measuring the seasonal variation in UVB intensity (Moore and Zhou, 1994).

The second usage of the term *action spectrum* is the more correct, according to photochemists. The action spectrum is obtained by measuring the radiation dose required to evoke the same chemical or biological response at different wavelengths. It will usually coincide with the absorption spectrum of the compound when the irradiation source variation with wavelength is corrected. To parallel the absorption spectrum, an action spectrum should meet the following conditions:

1. The action mechanism is the same at all wavelengths.
2. The quantum efficiency is the same at all wavelengths.
3. Absorption of radiation by inactive chromophores and radiation scattering is negligible.
4. Only a small fraction of the incident radiation is absorbed by the sample in the wavelength range of interest.
5. The exposure time is inversely proportional to the fluence rate for the same effect.

The action spectrum for any specific photosensitized reaction would normally overlay the absorption spectrum of the sensitizer (Grossweiner, 1989). The erythemal response (sunburn) spectrum in Figure 2.4 is an example.

In the context of drug photostability, the action spectrum is less important, in that the formulation developer is concerned with the overlap of the drug's spectrum with the spectral output of the incident radiation. In order to avoid confusion, the term *overlap integral* is recommended for this situation.

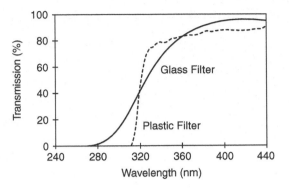

Figure 2.5   UV transmission curves of Corning O-53 Pyrex glass filter and a plastic filter made from overhead transparency film.

2.4 PENETRATION OF UV

The extent to which UV radiation (UVR) is able to provoke reactions is obviously dependent on its penetration of the system. For pharmaceutical formulations, this will depend on the degree of transparency of the packaging material. The most frequently used materials for which this is an issue are glass and plastic, but a variation in light transmission characteristics is caused by different compositions. The transmission cutoff can only be clearly delineated in terms of a filter of defined composition and thickness. Thus, the Corning glass filter O-53 in Figure 2.5 corresponds to standard Pyrex glass and can be characterized as giving 50% transmission at 310 nm for a sheet of 2 mm thickness.

Note, however, that it still transmits 1% at 280 nm; this means that glass will cut down the incident radiation in the UVB region by a significant proportion but not completely. Thus, for a substance that absorbs in the UVB region but whose absorption spectrum does not extend above 310 nm, storage in glass containers is not sufficient to protect it. If the substance is exposed long enough, the possibility of photochemical reaction remains. Also shown in Figure 2.5 is the transmission spectrum of a plastic film used for overhead transparencies. For experimental purposes, this film provides a much sharper cutoff than does glass, although it does not completely exclude the UVB region. The transmission characteristics of plastics vary according to their composition.

With respect to human response to UVR, the transmission of Caucasian skin is such that most of the UVR shorter than 320 nm is absorbed in the stratum corneum (Epstein, 1989). To evoke a photochemical reaction in the skin, UVR must penetrate to the site of the absorbing molecule in the peripheral blood capillaries. The penetration is governed by the optical properties of the skin and modified by absorption by melanin and scattering processes, which vary dramatically with wavelength (Diffey, 1983). The transmission of radiation through the human stratum corneum, for example, was estimated to vary from 15% at 297 nm, through 33% at 313 nm and 50% at 365 nm, to 72% at 546 nm (Bruls et al., 1984). It is widely accepted that UVA can penetrate nonmelanized skin and has the potential to cause photo-reactions in the skin at a greater depth than UVB, which can only reach the viable layers of the epidermis (Lovell, 1993).

2.5 EXCITED STATES, RADIATIVE, AND NONRADIATIVE PROCESSES

Photochemical damage to a substance is initiated by the compound's or a photosensitizer's absorption of energy. Many photochemical reactions are complex and may involve a series of competing reaction pathways in which oxygen may play a significant role. In fact, the great majority of photoreactions in biological systems involve the consumption of molecular oxygen and are photosensitized oxidation processes (Spikes, 1989).

Consider first the photophysical processes, which can be best described by an energy-level diagram (Figure 2.6) and Equation 2.1 to Equation 2.7:

$$\text{Absorption: } D_0 + h\nu \rightarrow {}^1D \tag{2.1}$$

$$\text{Internal conversion: } {}^1D \rightarrow D_0 \tag{2.2}$$

$$\text{Fluorescence: } {}^1D \rightarrow D_0 + h\nu' \tag{2.3}$$

$$\text{Photoionization: } {}^1D \rightarrow D^+ + e^- \tag{2.4}$$

$$\text{Intersystem crossing: } {}^1D \rightarrow {}^3D \tag{2.5}$$

$$\text{Intersystem crossing: } {}^3D \rightarrow D_0 \tag{2.6}$$

$$\text{Phosphorescence: } {}^3D \rightarrow D_0 + h\nu'' \tag{2.7}$$

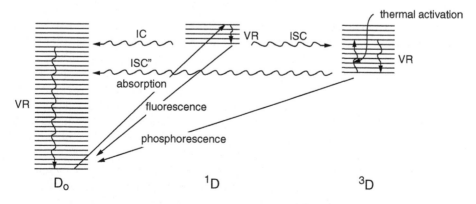

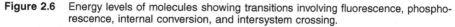

Figure 2.6 Energy levels of molecules showing transitions involving fluorescence, phosphorescence, internal conversion, and intersystem crossing.

Any UVR- or visible light-induced process begins with the excitation of drug molecules or sensitizers, from their ground state (D_0) to reactive excited states, by absorption of photons of certain wavelengths. As shown in Equation 2.1, upon absorption of radiation, the drug molecule, D_0, in the ground state in which the valence electrons are paired or antiparallel (a spin singlet state) is raised to a higher energy level, as a valence electron moves to the first available outer shell corresponding to the first excited singlet state, ^1D (the electron spins remain antiparallel). When the absorption spectrum shows more than one absorption band, this indicates a corresponding number of excited states that can be reached by irradiation with the appropriate excitation wavelength. For example, sulindac can be raised to the second excited state when irradiated with UVR in the wavelength range around 280 nm; longer wavelengths around 327 nm yield the first excited state only (see Figure 2.1).

The molecule cannot persist in an excited state indefinitely because it represents a less stable situation with respect to the ground state. A variety of competing physical processes involves energy dissipation and result in deactivation of the excited states. The energy dissipation may be via internal conversion (IC) (Equation 2.2), which is a nonradiative transition between states of like multiplicity, or via photon emission (fluorescence) resulting in return to D_0 (Equation 2.3). Even if excitation occurs to an excited state higher than the first, IC will always bring the molecule to the ^1D level (within a picosecond) before fluorescence occurs. Thus, the fluorescence emission wavelength is the same, irrespective of the irradiating wavelength. Any excess energy within a particular electronic state is dissipated as heat by collision with neighboring molecules — referred to as vibrational relaxation (VR).

Because the lifetime of the excited singlet state of a molecule is generally of the order of nanoseconds (but up to microseconds for rigid molecular structures), the possibility of interaction with neighboring molecules leading to chemical change is limited at this stage. However, in the excited singlet state, the ionization potential of the molecule is reduced; the excited electron is more easily removed than it is from the ground state molecule, but requires an appropriate acceptor to be present. This process of photoionization (Equation 2.4) is also more likely to occur if higher energy UVR is used (i.e., wavelengths less than 300 nm) and if the molecule is in the anionic state.

Alternatively, intersystem crossing (ISC) may occur from the excited singlet state to a metastable excited triplet state ^3D (electron spins parallel) (Equation 2.5). Despite the low probability in general for transfer between states of differing multiplicity, ISC occurs with relatively high efficiency for most photochemically active molecules. Because of its longer lifetime (microseconds to seconds, or even longer), the excited triplet state may diffuse a significant distance in fluid media and therefore has a much higher probability of interaction with other molecules. If no interaction occurs, it decays back to the ground state by another ISC event (Equation 2.6), or by phosphorescence emission (Equation 2.7).

The nature of the excited state decay processes is studied by the technique of laser flash photolysis, a description of which has been given by Bensasson et al. (1983). Briefly, flash photolysis involves irradiating a sample with a short (nanosecond) intense pulse from a laser, then observing by rapid response spectrophotometry the spectral changes that occur on the time scale nanoseconds to milliseconds.

Several standard tests have been established to aid in the identification of the transient species. Solvated electrons generated by photoionization in a nitrogen-gassed solution have a characteristic broad structureless absorption peak at about 700 nm, depending on the solvent (720 nm in aqueous solution). Oxygen quenches this absorption and also quenches the triplet state, while nitrous oxide gassing can be used to quench the solvated electron only, thereby gaining an indication of any transient absorption that arises from the triplet state.

One difficulty with flash photolysis experiments at present lies with the laser-exciting source. To achieve the required pulse intensity, the source usually employed is an Nd-YAG laser emitting at 1064 nm, with frequency doubling to produce 532 nm, tripling to 355 nm, and quadrupling to 266 nm. As far as the majority of drugs are concerned, this provides excitation at very specific wavelengths of 266 or 355 nm, leaving an unfortunate gap in the 280- to 340-nm region. Thus, for many drug molecules whose absorption does not extend to 355 nm, one is forced to use the high-energy 266-nm excitation, which may produce upper excited states and lead to events such as photoionization. In the context of photodegradation initiated by UVR greater than 300 nm, some of these events may not be relevant. The efficiency, or quantum yield, of each of the processes described by Equation 2.2 to Equation 2.7 is defined as the fraction of the molecules excited by absorption (Equation 2.1), which then undergo that particular mechanism of deactivation.

Although the quantum yield of fluorescence is readily determined by reference to quinine fluorescence as described by Calvert and Pitts (1966), those of the other processes can only be obtained by difference. Phosphorescence is usually too weak to be observed in solution at room temperature, but can be measured if the drug is held in a glassy matrix at low temperature. The usual procedure is to dissolve the drug in ethanol and immerse in liquid nitrogen. The phosphorescence accessory of the fluorimeter incorporates a mechanical chopper enabling the phosphorescence to be observed free of interference from any fluorescence. Because of the difference in temperature and matrix, it is not possible to compare the phosphorescence yield with that of fluorescence. Nevertheless, phosphorescence is worth measuring because it is an important indicator of the capacity of a molecule to populate its triplet state.

2.6 DIRECT REACTIONS FROM EXCITED STATES OF THE DRUG

The excited molecule has a different electronic character compared to the ground state and is often able to form a complex (called an exciplex) with another species (designated as Q), i.e., the complex is D^*Q. The symbol Q is used because, in effect, the interacting molecule is a quencher of the native fluorescence of D. Sometimes, at a high concentration of the absorbing molecule, this occurs with the ground state (in which case the D^*D species formed is called an excimer). The formation of the exciplex or excimer is observed as a shift in the fluorescence emission to longer wavelength — the difference in energy between exciplex and normal fluorescence reflecting the stability of the exciplex. More details of this type of interaction can be found in Gilbert and Baggott (1991).

The substances for which this phenomenon has been observed are invariably polycyclic aromatic hydrocarbon structures. No exciplex formation has been reported in the literature to involve drug molecules, but this remains a possibility in concentrated solution or perhaps in solid-state mixtures. The consequences of exciplex formation are a radiative or nonradiative return to the ground state without chemical change, or electron transfer leading to chemical reaction of the drug, the quencher, or both. Many photoaddition processes are postulated to proceed via exciplex formation with the quencher molecule becoming chemically bound.

The electronically excited state of a molecule will act as a more powerful electron donor or acceptor than the ground state. The reactions that can occur are, respectively, oxidative or reductive quenching:

$$\text{Oxidative quenching: } D^* + Q \rightarrow D^* Q \rightarrow D^{+\cdot} + Q^{-\cdot} \qquad (2.8)$$

$$\text{Reductive quenching: } D^*Q \rightarrow D^{-\cdot} + Q^{+\cdot} \qquad (2.9)$$

The exact nature of the reaction (oxidative vs. reductive) will depend on the redox properties of D^* and Q. The electron transfer process is a special case of exciplex formation favored in the strongly polar solvents, such as water. The involvement of an exciplex in a photochemical reaction is generally established by studying the effects of known exciplex quenchers such as amines on the exciplex fluorescence and the product formation. The heavy atom effect, due to the presence of substituents such as bromine or iodine intra- or intermolecularly, causes an exciplex to move to the triplet state preferentially, with a quenching of fluorescence.

2.6.1 Photodehalogenation Reactions

In regard to drug photodegradation reactions that appear to involve exciplex formation, the most frequently observed are those in which an aromatic chlorine substituent is lost in the photoreaction. Examples of drugs that lose their chlorine substituent are chlorpromazine (Davies et al., 1976); hydrochlorothiazide (Tamat and Moore, 1983); chloroquine (Moore and Hemmens, 1982); frusemide (Moore and Sithipitaks, 1983); diclofenac (Moore et al., 1990); and amiloride (Li et al., 1999). In each case, when the drug (Aryl–Cl) is photolyzed in aqueous or alcoholic (ROH) solution, HCl is liberated and a mixture of reduction (Aryl–H) and substitution (Aryl–OR) products is obtained. This is exemplified by the photodegradation of diclofenac shown in Figure 2.7. The photodechlorination occurs for these compounds more strongly in deoxygenated solution. When oxygen is present, it promotes ISC to the triplet state and the production of singlet oxygen (see below).

The mechanism is by no means completely clear, but the photodehalogenation reaction is postulated to occur through the formation of a pair of radical ions from an exciplex resulting in the excited state (Grimshaw and de Silva, 1981). The precursor of the reduction product (Aryl–H) is suggested to be a radical anion (Aryl–Cl$^{-\cdot}$), while a radical cation (Aryl–Cl$^{+\cdot}$) is postulated as the precursor of the substitution product (Aryl–OR). In a less polar solvent, e.g., iso-propanol, direct homolysis of the C–Cl bond occurring from the triplet state has been suggested,

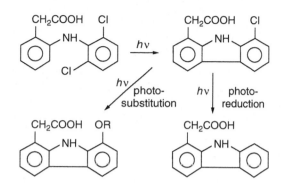

Figure 2.7 Photodegradation of diclofenac in aqueous solution at pH 7. (From Moore, D.E. et al., *Photochem. Photobiol.*, 52: 685–690, 1990.)

based on flash photolysis experiments with chlorpromazine (Davies et al., 1976). 3,3′,4′,5-Tetrachlorosalicylanilide represents a class of antibacterial agents formerly used in cosmetics and soaps. These compounds were found to undergo sequential photodehalogenation that was presumed to be related to their capacity to induce skin rashes upon sunlight exposure (Davies et al., 1975).

Not all chloroaromatic drugs appear to follow this type of reaction. For example, free chloride ion is not formed on irradiation of chlordiazepoxide for which an oxaziridine is the major photoproduct (Cornelissen et al., 1979). Reports on other drugs that contain chlorine substituents vary. This can arise due to differences in the irradiation conditions. If an unfiltered mercury arc source is used, the sample will receive 254-nm irradiation and the C–Cl bond will certainly break, while under longer wavelength irradiation (>300 nm), the bond may be stable.

2.7 PHOTOSENSITIZED REACTIONS

Any photochemical process in which there is a transfer of reactivity to a species other than that absorbing the radiation initially is called a photosensitization reaction. As a result of the long lifetime and the biradical nature with unpaired electron spins, the excited triplet states can mediate photosensitized reactions, the most common of which are photosensitized oxidations. Due to the triplet spin nature of its ground state, oxygen is spin matched with the drug triplet state and also is a very good scavenger of free radicals. These characteristics lead to two distinct mechanisms of photo-oxidation, as shown in Scheme 2.1 using AH to refer to an oxidizable substrate.

The excited triplet sensitizer can undergo its primary reaction with molecules in its vicinity by (Spikes, 1989):

1. Electron transfer, including simultaneous transfer of a proton corresponding to the transfer of a hydrogen atom, resulting in free radical reactions (Equation 2.10 to Equation 2.12) — termed type I, or free radical, reaction
2. Energy transfer, with spin conservation, to ground state molecular oxygen ($^3O_2$) to form singlet oxygen (Equation 2.14) — termed type II reaction

Free radical formation: $^3D + AH \rightarrow D_0 + A^{\cdot} + H^{\cdot}$ (2.10)

or $^3D + AH \rightarrow DH\cdot + A\cdot$ (2.11)

or $^3D + AH \rightarrow D_0 + A\cdot^- + H^+$ (2.12)

(Type I photo-oxidation): $A\cdot \text{ (or } A\cdot^-) + O_2 \rightarrow AO_2\cdot \text{ (or } AO_2\cdot^-)$ (2.13)

Energy transfer to oxygen: $^3D + {}^3O_2 \rightarrow D_0 + {}^1O_2$ (2.14)

Oxidation by singlet oxygen: $^1O_2 + AH \rightarrow AOOH$ (2.15)

(Type II photo-oxidation)

Scheme 2.1 Photosensitized Oxidation Reactions, Types I and II.

Type I and type II processes can take place simultaneously in a competitive fashion, as in the cases of thionine (Kramer and Maute, 1973) and chlorpromazine (Moore and Burt, 1981). The distribution between the two processes depends on the sensitizer; substrate; solvent; oxygen concentration; and affinity of sensitizer and substrate (Henderson and Dougherty, 1992). One of the processes may be dominant in a specific system. For example, in an air-saturated aqueous solution at neutral pH, the excited triplet of the dye Rose Bengal reacts overwhelmingly with oxygen rather than directly with DNA (Lee and Rodgers, 1987). For 2-methyl-1,4-naphtho-quinone, however, a similar study revealed that the one-electron transfer to thymine can effectively compete with singlet oxygen formation (Fisher and Land, 1983).

2.7.1 Type I Photosensitization of Chain Reactions

The type I mechanism of photosensitization commonly proceeds through the transfer of electrons or protons, depending on the polarity of the medium (Foote, 1968). The formed cation or neutral radical is expected to undergo further reactions, which, in the absence of oxygen, means recombination, dimerization, or disproportionation. When oxygen is available in sufficient concentration, molecular oxygen is rapidly added to the radical. The peroxy radical formed is also reactive and will seek to stabilize itself by proton abstraction from neighboring molecules. If the sample consists of a high concentration of the drug, the extent to which the reaction continues will depend on the reactivity of the drug.

This sequence may be thought of as a chain reaction because the radical activity is continually transferred and kept "alive." Except in very unusual structures, free radicals are considered high reactivity species, but a suitable donor or acceptor in the near vicinity is needed. Secondary alcohols are examples of molecules with readily abstractable hydrogens. Thus, *iso*-propanol, mannitol, and ascorbic acid are very good scavengers of free radicals and can be used to protect the therapeutic substance while they undergo oxidation.

Initiation: Ph–CHO + hv → Ph–CO· + H·

Propagation: Ph–CO· + O_2 → Ph–CO–OO·

peroxy radical

Ph–CO–OO· + Ph–CHO → Ph–CO–OOH + Ph–CO·

peroxybenzoic acid

Termination: Ph–CO–OO· + Ph–CO–OO· → dimeric peroxy compounds

Scheme 2.2 Chain Reaction Mechanism for Benzaldhyde Degradation.

The chain reaction mechanism is frequently referred to as autocatalytic because it starts slowly, but the rate becomes faster as the reaction proceeds. Not many examples of drug substances that decompose by a free radical chain mechanism are known because the process requires participation of a very reactive (i.e., unstable) compound. This usually means a compound susceptible to oxidation and is illustrated by the photo-oxidation of benzaldehyde, as shown in Scheme 2.2 (Moore, 1976):

Although the peroxy products are unstable and will break down, thus potentially generating new free radical species, the faster processes are those given as the propagation steps in Scheme 2.2. Although benzaldehyde has only a weak n → π^* absorption at 320 nm, it is only necessary to generate one radical by dissociation of an excited state molecule. This is quite sufficient to set off a chain reaction that results in the oxidation of thousands of benzaldehyde molecules (depending on temperature). The free radical chain reaction is categorized in terms of the *chain length*, which means the number of propagation steps occurring for every initiation event. In this case, the chain length and also the quantum yield for the overall photochemical process will be in the thousands.

The limit to the chain reaction is determined by the relative values of the rate constants for the propagation step and the branching or transfer reactions involving solvent or inhibitor molecules. As the concentration of the oxidizable molecule falls in the solution, the reaction rate also falls. The reaction is characterized by a "steady-state" or maximum rate represented by the linear portion of the sigmoidal reaction progress curve. This is achieved when the rate of generation of new initiating radicals is equal to their termination rate. Here, the kinetics is simplified by the steady-state approximation, and the maximum rate is first order with respect to the benzaldehyde concentration.

Inhibition of chain processes is achieved by the addition of free radical scavengers, which react by chain transfer more rapidly than the propagation step. The product of chain transfer is also a free radical, but the key to the transfer agent being a good inhibitor is that it must be a very unreactive radical, e.g., sterically hindered radicals formed from the widely used antioxidants BHT (2,6-di-*t*-butyl-hydroxy-toluene) and BHA (2,6-di-*t*-butyl-hydroxy-anisole).

Chain reactions are the major pathway by which hydrocarbon polymers as used in packaging are broken down, with the radicals for initiation arising from photoinduced

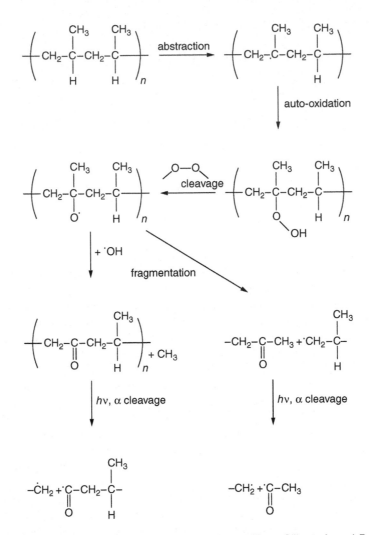

Figure 2.8   Photodegradation of a hydrocarbon polymer. (From Gilbert, A. and Baggott, J., *Essentials of Molecular Photochemistry*, pp. 145–228. Oxford: Blackwell, 1991.)

decomposition of trace amounts of peroxide or hydroperoxide impurities. Indeed, the development of "biodegradable packaging" is an application of this principle. Figure 2.8 shows an example of the chain reaction process leading to the breakdown of a hydrocarbon polymer backbone.

In biological systems, free radicals can react with cellular macromolecules in a variety of ways, the most important of which is hydrogen abstraction from DNA leading to chain scission or cross-linking. In proteins, tryptophan is the amino acid residue most susceptible to free radical attack. Lipid peroxidation by free radicals in turn is liable to cause alteration in cell membranes.

2.7.2   Electron Transfer-Sensitized Photo-oxidation

As mentioned earlier in the discussion of exciplex formation, electron transfer between an excited state species and a ground state molecule (Equation 2.8 and Equation 2.9) is frequently observed in the photochemistry of systems containing an electron donor–acceptor combination. As a result, a pair of radical ions is formed that react with oxygen but with different rates. The reaction of ground state oxygen with radical anions occurs rapidly and yields superoxide anion (Equation 2.16). The superoxide then adds to the radical cation forming DO_2 (Equation 2.17). When D is an olefin, DO_2 is a dioxetan that is liable to cleave to yield ketones as products.

$$Q^{-\cdot} + O_2 \rightarrow Q + O_2^{-\cdot} \tag{2.16}$$

$$D^{+\cdot} + O_2^{-\cdot} \rightarrow DO_2 \tag{2.17}$$

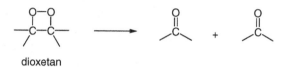

dioxetan

2.7.3   Detection of Free Radicals

The preceding discussion is a simplified view of some of the processes that may occur involving free radicals generated from the excited state. Determination of the detailed reaction mechanism is a difficult task and requires knowledge of the quenching efficiency of the sensitizer excited state by the substrate; the ability of the radical anion to transfer an electron to oxygen; and the rate of reaction of the substrate radical cation with ground state oxygen. A number of techniques have been developed to enable the detection of free radical intermediates in photochemical reactions, including electron paramagnetic resonance spectroscopy (EPR). EPR is useful for radicals formed in relatively high concentration that persist for relatively long times. Unfortunately, that is not the case for the great majority of photochemical reactions, and special procedures such as rigid solution matrix isolation are necessary. Addition of free radical trapping compounds to the system (spin traps) is an alternative (Mason and Chignell, 1982; Chignell et al., 1985). The superoxide anion is also readily trapped and identified by this technique.

An extremely sensitive technique to detect the nature of radical pairs in a photochemical reaction, called chemically induced dynamic nuclear polarization (CIDNP), depends on the observation of an enhanced absorption in a nuclear magnetic resonance (NMR) spectrum of the sample irradiated *in situ* in the cavity of the NMR spectrometer. The background to and interpretation of CIDNP are discussed by Gilbert and Baggott (1991).

Probably the main technique that has been used to detect free radical intermediates in photochemical reactions is the competitive reaction rate study in which various free radical scavengers are added to the sample during irradiation, and the rate of disappearance of drug and appearance of particular products is compared to that occurring without the scavenger. Typical scavengers include ascorbic acid and glutathione for aqueous systems, and BHT and α-tocopherol for lipophilic systems. However, interpreting the results of such a study is somewhat difficult because the relative reactivity of radicals and scavengers is determining the outcome, so the product profile will invariably change. If the radical intermediates are extremely reactive, they may react with the solvent before they encounter a scavenger molecule, and no change will be observed.

2.7.4   Polymerization as Detector of Free Radicals

The chain reaction process can be used as a diagnostic aid to determine whether free radicals are generated from a drug when irradiated. Acrylamide is an acrylic monomer widely used in gel electrophoresis as a polymer formed *in situ* by peroxide or UV-initiated polymerization. This monomer is a water-soluble solid more easily handled than most other vinyl monomers; the progress of its polymerization is readily followed by measuring the contraction in volume by dilatometry or the increase in viscosity in a viscometer. Details of the dilatometry technique applied to photochemical reactions are found in Moore and Burt (1981). Although this approach does not give any information as to the identity of the free radical generated by irradiation of the drug, it is a chemical amplification process in which very small concentrations of free radicals can be detected. The rate of polymerization caused by free radicals generated by the UV irradiation of a drug solution containing acrylamide is a reflection not only of the rate of radical generation, but also of their lifetime. Note that oxygen must be excluded from the system so that the polymer radicals are not scavenged and the reaction inhibited.

2.8 SINGLET OXYGEN AND ITS REACTIVITY

The type II reaction involves electronic energy transfer from the triplet-excited photosensitizer to ground state molecular oxygen that is spin-matched, thereby forming excited singlet molecular oxygen while the photosensitizer is regenerated (Equation 2.14). The two types of singlet oxygen with different spectroscopic symmetry notations are $^1\Delta_g$ and $^1\Sigma_g^+$. Their energies are, respectively, 92 and 155 kJ/mol higher than that of ground state oxygen $^3\Sigma_g^+$. The $^1\Delta_g$ state possesses a much longer lifetime and normally has a higher yield in biological systems than does $^1\Sigma_g^+$. Consequently, the $^1\Delta_g$ state is the main consideration here. Because of the relatively small energy difference from the ground state, many compounds are capable of acting as sensitizers for singlet oxygen formation. For example, the dyes methylene blue and Rose Bengal have a triplet state energy of about 140 and 170 kJ/mol, respectively.

The production of $^1O_2$ has been reported to occur by energy transfer from the singlet- and triplet-excited states of the sensitizer, but that from the triplet excited state is highly preferred because singlet–triplet interaction is of very low probability. The lifetime of $^1O_2$ is highly dependent on the solvent medium and the presence of scavengers or oxidizable acceptors; it was determined to be about 3.1×10^{-6} s in water (Rodgers and Snowden, 1982) and 50 to 100×10^{-6} s in lipid (Henderson and

Dougherty, 1992). A half-life in tissue was estimated to be less than 5×10^{-7} s (Patterson et al., 1990). Singlet oxygen might diffuse about 0.01 to 0.02 μm in a cellular environment (Moan et al., 1979).

Although the energy of $^1O_2$ is only 92 kJ/mol higher than that of ground state oxygen, its chemical reactivity is completely different because it is now spin matched with ground state molecules susceptible to oxidation. Thus, $^1O_2$ is capable of oxidizing a large variety of substances including biological cell components such as DNA, protein, and lipids. Because many sensitizers are in a reduced form, they also may act as substrates, giving fully oxidized products. As a consequence, many preparative organic chemical processes are carried out photochemically, with $^1O_2$ the mediator.

2.8.1 Quenchers of Singlet Oxygen

Singlet molecular oxygen is deactivated by physical or chemical quenching agents. The two physical mechanisms are energy-transfer and charge-transfer quenching. The carotenoid pigments play an important role in the protection of biological systems, apparently because they are particularly efficient energy-transfer quenchers. Beta-carotene is the most studied member of this group. The extended conjugated π-system of β-carotene has triplet energies close to or below that of singlet oxygen so that collisional energy transfer occurs. Subsequently, the excited β-carotene decays by vibrational relaxation and no net chemical change is observed (Gorman and Rodgers, 1981).

Amines generally are capable of quenching singlet oxygen via a charge-transfer process, but may react chemically as well. The primary process is envisaged as formation of a complex between the electron-donating quencher and the electron-deficient oxygen species; the quenching rate constants correlate with the amine ionization potential. The resulting triplet complex dissociates with loss of energy by vibrational relaxation or forms oxidation products. Formation of products requires an abstractable hydrogen α to the nitrogen; N-methyl groups are particularly susceptible. Diazabicyclo-octane (DABCO) is unable to react chemically, presumably on steric grounds, but is an efficient physical quencher. Some phenols are also able to quench singlet oxygen by a mixture of physical and chemical processes, e.g., the 2,4,6-trisubstituted phenols used as antioxidants, BHT, and α-tocopherol.

Other chemical reactions or quenching of singlet oxygen rely on the fact that singlet oxygen is more electrophilic than ground state oxygen and therefore can react selectively with electron-rich regions of many molecules, e.g., olefins and aromatics. Some examples of the addition of singlet oxygen are given in Figure 2.9, including the ene-reaction in which an olefin possessing an allylic hydrogen atom forms allylic hydroperoxides, and endoperoxide formation by 1,4-addition to π-systems such as furan and anthracene derivatives. As with other oxidation reactions, the initial products are metastable and secondary reactions will occur, but on a slower time scale. Dioxetan formation occurs by singlet oxygen addition to olefins in which the double bond possesses an electron-donating heteroatom, generally N, O, or S; this leads ultimately to cleavage of the double bond in a way similar to the reaction of superoxide in Equation 2.17. The similarity leads to some controversy as to the mechanism of dioxetan formation (Gorman and Rodgers, 1981).

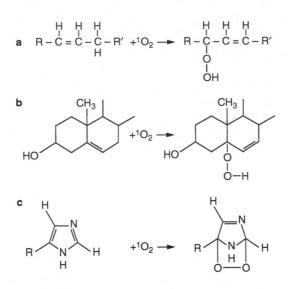

Figure 2.9   Chemical quenching of singlet oxygen. (a) The *ene* reaction — addition of singlet oxygen to an olefin with allylic hydrogen; (b) the *ene* reaction of cholesterol; (c) endoperoxide formation by singlet oxygen to imidazole residue as in histidine.

2.8.2 Detection of Singlet Oxygen

There are several methods for the detection of $^1O_2$ generated in an irradiated solution. A characteristic luminescence at 1270 nm, corresponding to the return of singlet oxygen to the ground state, can be detected with the appropriate equipment (Hall et al., 1987) The alternative is to measure the rates of reaction in the presence of molecules that react readily with or quench singlet oxygen. Here the choice depends on the solvent used, with sodium azide, 2,5-dimethylfuran, and the amino acid histidine suitably soluble for use in aqueous systems; β-carotene, DABCO, and diphenylisobenzofuran (DPBF) are more appropriate for organic solvents.

Analysis of the reaction rates is achieved in terms of oxygen uptake measured with an oxygen electrode, or by product separation and quantification. DPBF absorbs intensely at 415 nm and reacts rapidly with singlet oxygen to form a colorless intermediate endoperoxide. The DPBF reaction can be used as a benchmark against which the effect of an added quencher is compared. A note of caution must be applied: the use of inhibitors and quenchers alone is not unambiguous in its outcome and should be strictly supplemented with flash photolysis experiments. Thus, if a photosensitized reaction is quenched by millimolar concentrations of azide ion, it should also be established that azide does not quench the triplet state of the sensitizer directly because that would also affect the reaction rate.

It has also been reported that the furans and histidine can be oxidized to the same products by free radical processes. Nevertheless, these compounds have such a high reactivity with singlet oxygen that they are very rarely wrong as indicators of its generation by a photosensitizer. Cholesterol is regarded as an unambiguous trapping compound because singlet oxygen reacts with it to form a single product, the 5-α-hydroperoxide, whereas reaction with radicals gives a mixture of other

products (Spikes, 1989). The analytical procedure involved in isolating the cholesterol product is more technically demanding than that required when histidine or
the furans are used as the substrate.

Another kinetic technique is to compare the rates in heavy water (D_2O) with
normal water because, as noted earlier, the lifetime is about 10 times greater in D_2O.
This will only achieve a meaningful result when singlet oxygen deactivation by the
solvent is the rate-determining process. Frequently, other species in the solution are
capable of reacting with singlet oxygen, and the effect of the longer lifetime is not
manifest. Typical photosensitizers that generate singlet oxygen include dyes such as
methylene blue, Rose Bengal, and rhodamine. Many drug molecules such as phenothiazines; quinine and other antimalarials; thiazides; naproxen and other anti-inflam-
matories; and psoralens have been demonstrated to generate singlet oxygen under
the influence of UV-R or visible light. Environmental contaminants such as the
polycyclic aromatic hydrocarbons also are very efficient $^1O_2$ generators.

2.9 ACTIVE FORMS OF OXYGEN AND OXIDANT SPECIES

As noted previously, formation of free radicals or singlet oxygen is very often
accompanied by the generation of various other short-lived species (such as hydroxyl
radicals, superoxide radicals, and peroxyl radicals) that, together with singlet oxygen,
are termed reactive oxygen species (Pryor, 1986). For example, superoxide radicals
can be generated following photoionization (Equation 2.4) from singlet oxygen by
electron transfer between $^1O_2$ and the ground state sensitizer (Equation 2.18) or the
appropriate substrates (Equation 2.19). In some cases, the subsequent reactions may
result in formation of toxic hydrogen peroxide (Equation 2.20), which in turn
decomposes to produce hydroxyl radicals (Equation 2.21) (Proctor and Reynolds,
1984).

$$^1O_2 + D_0 \rightarrow O_2^{\cdot-} + D^{\cdot+} \tag{2.18}$$

$$^1O_2 + AH \rightarrow O_2^{\cdot-} + A\cdot + H^+ \tag{2.19}$$

$$O_2^{\cdot-} + H^+ \rightarrow HO_2^{\cdot} \rightarrow H_2O_2 + O_2 \tag{2.20}$$

$$H_2O_2 \rightarrow 2\cdot OH \tag{2.21}$$

Apart from the photodynamic reactions, a photosensitized reaction may proceed
through the direct photoionization of the sensitizer in which oxygen is not required
(Equation 2.4). Because photoionization is found to occur particularly from molecules containing one or more heteroatoms, a significant number of drugs undergo
photoionization, although, in general, higher energy radiation (<300 nm) is required.
Drugs said to photoionize include chlorpromazine (Navaratnam et al., 1978);
4-hydroxybenzothiazole (Chedekel et al., 1980); sulfacetamide (Land et al., 1982);
and hydrochlorothiazide (Tamat and Moore, 1983). The hydrated electron is one of

the strongest known reducing agents, reacting rapidly with oxygen to form the superoxide radical anion ($O_2\cdot^-$), the precursor of hydrogen peroxide (Buxton et al., 1988). Photoionization and subsequent superoxide formation are supported to a greater extent by an aqueous medium because the electron is readily stabilized by the solvent.

2.10 CONSEQUENCES OF EXCITED STATE PROCESSES TO DRUG STABILITY *IN VITRO*

The photodecomposition reactions of a large number of drug substances were reviewed by Greenhill and McLelland (1990) and Albini and Fasani (1998). A catalogue of the most common reaction types shows that, following light absorption, a drug might experience (see also Chapter 4)

Addition
Cyclization
N-Dealkylation
Decarbonylation
Decarboxylation
Dehalogenation
Dimerization
Oxidation
Reduction
Isomerization
Rearrangement
Hydrolysis

It is also possible that a drug sensitizes its own degradation. Elucidating the mechanism for some photodegradation reactions is very difficult because several pathways are reported for many drugs. Additionally, the irradiation conditions used by different laboratories have significant differences in terms of wavelength range, time of exposure, and drug concentration. Another factor that should be considered is the effect of oxygen on the product distribution. In many cases oxidation may be responsible for products that are secondary, or even tertiary, in the overall sequence when long exposures are used. Because of the broad array of chemical structures of drug molecules and the variety of processes that can occur following the absorption of light, one can predict the possible outcome for a new compound only if it is a close structural analogue of a previously studied compound.

Two case studies, naproxen and sulfamethoxazole, will be described briefly to illustrate how information concerning the excited state assists in elucidation of the mechanism of photodegradation. Their spectroscopic properties are detailed in Table 2.1, together with several other drugs given for comparative purposes. The state of ionization of the molecule can be a factor in the photochemistry, so it is relevant to show the information for neutral and anionic sulfamethoxazole (pK_a 5.6). Ionization of naproxen (pK_a 4.2) has no significant effect on its spectroscopic properties because the carboxyl group is distant from the naphthalene chromophore.

Table 2.1 Spectroscopic Data (Absorption, Fluorescence, and Phosphorescence) and Photosensitization Reaction Rates (Photo-oxidation and Photopolymerization) for Selected Drugs in Aqueous Solution

Drug (5.0 × 10^-5 M)	Buffer	λ_{max} (nm)	ε_{max} (M^{-1} cm^{-1})	λ_F (nm)	Φ_F	λ_P[a] (nm)	Relative Photoabsorption[b]	Oxygen Uptake Rate (mmol l^{-1} min^{-1}) Drug Alone	Oxygen Uptake Rate (mmol l^{-1} min^{-1}) Drug + DF	Polymerization Rate (mmol l^{-1} min^{-1})
Sulfamethoxazole	pH 9.0	258	17,000	342	0.0020	—	25.8	<0.3	8.2	0.7
Sulfamethoxazole	pH 3.0	268	17,360	342	0.0019	410	12.1	<0.4	9.6	1.3
Diclofenac	pH 7.0	285	8,220	—	0	460	100	<0.2	3.9	1.79
Chlorpromazine	pH 7.0	315	4,550	455	0.13	nd	860	3.1	32.8	3.42
Frusemide	pH 7.0	330	5,800	400	0.06	nd	630	0.7	2.8	0.59
Naproxen	pH 7.0	316; 331	1972; 1988	355	0.40	510	1500	5.3	25.1	0.24

Note: nd means not determined; DF is 2,5-dimethylfuran.

[a] Phosphorescence measured in an ethanol glass at 77 K (drug in neutral form).

[b] Relative photoabsorption is the overlap integral of the drug's absorption spectrum and the mercury arc output spectrum, normalized to diclofenac set to 100.

Sources: diclofenac data from Moore, D.E. et al., *Photochem. Photobiol.*, 52: 685–690, 1990; data for chlorpromazine and frusemide from Moore, D.E. and Burt, C.D., *Photochem. Photobiol.*, 34: 431–439, 1981; naproxen data from Moore, D.E. and Chappuis, P.P., *Photochem. Photobiol.*, 47: 173–181, 1988.

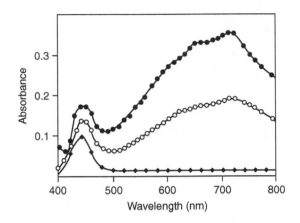

Figure 2.10   Transient absorption spectra obtained on laser flash photolysis at 266 nm of naproxen (0.1 mM; pH 7.0). The spectra were recorded 0.02 (•), 0.26 (○), and 1.29 μsec (♦) after the flash. (From Moore, D.E. and Chappuis, P.P., *Photochem. Photobiol.*, 47: 173–181, 1988.)

Both molecules have the capacity to emit fluorescence and phosphorescence, the latter suggesting an appreciable population and lifetime of their respective triplet states. The first photochemical tests involve using spectrophotometry and liquid chromatography to analyze the reaction mixtures after irradiation for varying times. It is important to follow the reaction from the very early stages to ensure that the products seen are those formed from the very beginning of the reaction. Some primary products may decompose in secondary thermal and/or photochemical reactions.

The data from flash photolysis experiments (266-nm excitation) on naproxen in aqueous solution at pH 7 are shown in Figure 2.10. The transient spectra were identified as due to the solvated electron (absorbing at 700 nm) and a triplet state (absorbing at 430 nm). Similar species were found for sulfamethoxazole. Under this pH condition, both molecules are present as the anion, so photoionization is a likely process with the high energy excitation. The solvated electron generation implied a radical production so EPR with spin trapping was performed.

Sulfamethoxazole failed to produce any trappable radicals with an array of different spin traps, but naproxen afforded the EPR spectrum shown in Figure 2.11 when irradiated with 330 nm UV-R in the instrument cavity in the presence of 2-methyl-2-nitroso-propane (MNP). The spectrum contains contributions from di-t-butyl nitroxide, a known photoproduct of MNP. The H-atom adduct MNP–H also evident can arise by several different mechanisms, including the trapping of an H atom by MNP; the reaction of MNP with an electron followed by protonation; and the direct reduction of MNP by an excited state species. In view of the flash photolysis results, it was concluded that photoionization was the major precursor of MNP–H. The third radical corresponded to a C-centered radical carrying a single H atom, leading to the postulate of a decarboxylation reaction as the primary photochemical step. Confirmation of the participation of free radical intermediates came from the initiation of the free radical polymerization of acrylamide with rates as shown in Table 2.1.

Photophysical and photochemical aspects

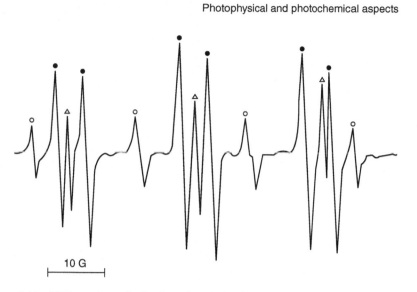

Figure 2.11 EPR spectrum obtained on photolysis of naproxen at 330 nm in the presence of 2-methyl-2-nitrosopropane (MNP). The spectrum contains contributions from DTBN (Λ); MNP-H (○); and decarboxy-naproxen (•). (From Moore, D.E. and Chappuis, P.P., *Photochem. Photobiol.*, 47: 173–181, 1988.)

Although sulfamethoxazole did not yield trappable radicals, there was evidence of free radical intermediates from polymerization experiments (Table 2.1) together with results from oxygen uptake experiments using 2,5-dimethylfuran as substrate. The latter measurements are an indication of the relative capability of the drugs to generate singlet oxygen, in confirmation of the population of the triplet state. All of these data, in combination with the identification of the isolated photoproducts, play a role in postulating intermediates and fully elucidating the mechanism of the photo-degradation, shown in Figure 2.12 for naproxen and Figure 2.13 for sulfamethoxazole.

2.11 CONSEQUENCES OF EXCITED STATE PROCESSES TO ADVERSE EFFECTS *IN VIVO*

A drug displaying photochemical reactivity *in vitro* may give rise to adverse photosensitivity effects in patients that manifest in responses labeled *phototoxicity* and *photoallergy*. Phototoxicity is defined as an alteration of cell function by an interaction between a chemical and nonionizing radiation; the response is likened to an exaggerated sunburn. The reaction occurs upon simultaneous exposure to a phototoxic chemical and radiation of the appropriate wavelength. The wavelength of radiation implicated in most phototoxic reactions ranges from 300 to 400 nm. Most drug-induced phototoxic reactions are acute, occurring within a few minutes to several hours after exposure. They reach a peak from several hours to several days later, and usually disappear within a short time after stopping the drug or the exposure to radiation.

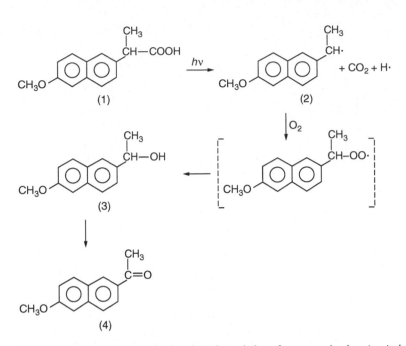

Figure 2.12 Reaction mechanism for the photodegradation of naproxen in air-saturated aqueous solution pH 7.0. (From Moore, D.E. and Chappuis, P.P., *Photochem. Photobiol.*, 47: 173–181, 1988.)

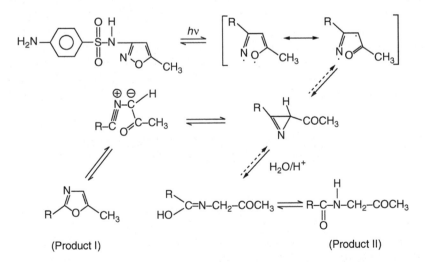

Figure 2.13 Reaction mechanism for the photodegradation of sulfamethoxazole in aqueous solution pH 4.0. (From Zhou, W. and Moore, D.E., *Int. J. Pharm.*, 110: 55–63, 1994.)

Photoallergy is an acquired altered capacity of the skin to react to light. Thus it is immune mediated while a drug molecule is stimulated by radiation to combine with a protein or other biomacromolecule in the skin to form an antigen. Photoallergy is believed to require several exposures and an induction period before the response is observed, and it has a histology different from that of phototoxicity. The incidence of drug-induced cutaneous photosensitivity and the mechanisms involved are given in a recent review (Moore, 2002).

It is believed that phototoxicity causes cellular damage by direct modification of certain target molecules, such as DNA, lipids, and/or amino acids and proteins. In principle, this can occur by the photosensitization reactions described earlier, and there is a reasonable correlation between the capacity of a drug to participate in type I and type II processes *in vitro* and the number of adverse photosensitivity reports registered against them. It is, of course, difficult to compare different drugs in this way because a number of extra variables enter — for example, the dose and pharmacokinetics of the drug's biodistribution.

Phototoxic reactions may be oxygen dependent (photodynamic) or oxygen independent (nonphotodynamic). In general, the capacity of the drug to generate free radicals has been regarded as the most potentially damaging characteristic because of the possibility of chain reactions subsequently occurring (Trush et al., 1982). Chlorpromazine is the classic example of this, with a record as the most photosensitive drug (Magnus, 1976) and a high yield of a long-lived free radical on irradiation (see Table 2.1, and Kochevar, 1981). Nonetheless, the importance of singlet oxygen can be seen by the fact that the polycyclic aromatic hydrocarbons are efficient producers of singlet oxygen and acute phototoxicity responses (Kochevar et al., 1982; Burt and Moore, 1987). On the basis that the demonstrated photochemical mechanisms involve free radical and singlet oxygen mediated damage, a logical approach to lessen the biological impact of the drug–sunlight interaction has been the implementation of diet supplementation with antioxidants (Moore, 2002).

Toxic photoproducts may also be formed by the action of sunlight on the drug in the epidermal layers of the patient's skin. These may cause adverse photosensitivity effects because they possess undesirable physiological properties or because they can easily transfer energy to body compounds (Rahn et al., 1974). For example, a stable photoproduct of chlorpromazine was cytotoxic to the plasma membrane so that it could result in haemolysis of erythrocytes in the dark (Kochevar and Lamola, 1979). Another example is the non-steroidal anti-inflammatory drug benoxaprofen, which was the subject of many adverse reports before its withdrawal in 1982. A two-step mechanism was proposed for benoxaprofen photosensitization. First, photoexcited benoxaprofen decarboxylates and thereby loses its ability to dissolve in aqueous media. The lipophilic photoproduct partitions into the lipid bilayer of the red blood cell membrane, where it acts as a singlet oxygen generator to lead to the oxidation of membrane lipids, thus ultimately disrupting the membrane structure (Reszka and Chignell, 1983; Kochevar et al., 1984).

In certain cases, a direct reaction between the excited state of a phototoxin and a biological target may result in the formation of a covalent photoaddition product comprising the compound and the biological target. Such a mechanism was demonstrated in the phototoxicity of psoralen, in which a direct photochemical reaction of psoralen with DNA occurred (Song and Tapley, 1979). For some photosensitizers, such as amiodarone and chlorpromazine, metabolites may play a significant role in their phototoxic reactions (Ljunggren and Möller, 1977; Ferguson et al., 1985).

The mechanism of photoallergy is postulated in a similar fashion to the nonphoto-dynamic processes. The drug becomes attached to a protein by a free radical process, thereby creating a hapten that can lead to an immune response. It is presumed that the reason for the low incidence of photoallergic reactions is that the body's immune response mechanisms generally deactivate the hapten, or the oxygen present in the blood is able to scavenge the free radical intermediates.

2.12 APPROACHES TO STABILIZATION OF FORMULATIONS AGAINST PHOTODEGRADATION

In principle, formulations containing drugs susceptible to photoreactions should be clearly marked and stored appropriately. However, in some situations the ideals are not maintained, and it is worth considering whether special procedures or additives should be included.

Using an inert atmosphere in the container is an important first approach that, clearly, will limit reactions involving oxygen. However, many examples have been found in which oxygen does not take part, e.g., sulfamethoxazole. The procedure would need to be considered on an individual basis following a full study of the role of oxygen in the photoreactions of a particular drug. On the other hand, it would not be advisable for intravenous fluids to be depleted of their oxygen prior to injection, so the application to which the formulation will be put is also a factor for consideration.

When photoreactions proceed through a type I or type II photosensitization mechanism, the possibility of adding quenchers to the formulation can be considered. If they can be clearly shown to be nontoxic and not to affect the therapeutic action, this is perhaps an option. The major contenders would be substances such as ascorbic acid, α-tocopherol, and BHT, which are capable of acting as free radical scavengers and weak singlet oxygen quenchers. They are already in use as food additive antioxidants. Beta-carotene is the only major singlet oxygen quencher that may be regarded as a possible food additive, but the trace concentration that can be used may not be effective. In view of the present regulatory environment, the use of these and other quenchers would involve a considerable amount of development work.

2.13 INFLUENCE OF EXCIPIENTS ON DRUG STABILITY

The compatibility of the drug with the excipients in a given formulation must be established early in the development program. For solutions and suspensions, this is usually a matter of choice of buffer and pH, determined from solution kinetic studies. For parenterals, compatibility with packaging components (plugs, plastics), as well as the possibility of metal ion contamination, is also investigated. For solid products, the investigation is more complex because of the greater range of excipients. The choice of excipient is company dictated and may include any of the following: lactose, dicalcium phosphate anhydrous, dicalcium phosphate dihydrate, corn starch, mannitol, terra alba, sugar, magnesium and calcium stearates, talc, stearic acid, polyvinylpyrrolidone (PVP), crosspovidone, and modified starches (Carstensen, 1995).

Photostability testing adds another dimension to the testing with excipients, and what is not mentioned in the preceding list are the coloring agents frequently used in elixirs, ointments, tablets, and capsules. To consider those first, it is obvious that such coloring agents should be of negligible reactivity when irradiated with light so that photosensitized decomposition of the therapeutic agent is not an issue. Many vegetable dyes are regarded as safe, although they may gradually fade. It is not always clear whether any drug decomposition is occurring under those circumstances because the product would be discarded if the color changed.

In relation to other excipients, one can see some components liable to participate in photochemical processes. For example, lactose, mannitol, sugar, starches, and PVP are susceptible to free radical attack in that they have abstractable hydrogens. If a drug substance acts as a photosensitizer and initiates a chain reaction in the product formulation, some of the excipient could be degraded to oxidation products while the drug is largely unchanged. In that sense, such excipients may serve to protect the drug, as long as their oxidation products are safe. Thus, testing the photostability of the final product should include a screening for impurities arising from the excipients in addition to the drug substance.

REFERENCES

Albini, A. and Fasani, E. (1998) Photochemistry of drugs: an overview and practical problems, in Albini, A. and Fasani, E. (Eds.), *Drugs: Photochemistry and Photostability*, pp. 1–73. Cambridge: Royal Society of Chemistry.

Bensasson, R.V., Land, E.J., and Truscott, T.G. (1983) *Flash Photolysis and Pulse Radiolysis*, pp. 1–19. Oxford: Pergamon Press.

Bruls, W.A.G., Slaper, H., van der Leun, J.C., and Berens, L. (1984) Transmission of human epidermis and stratum cornea as a function of thickness in the ultraviolet and visible wavelengths, *Photochem. Photobiol.*, 40: 485–494.

Burt, C.D. and Moore D.E. (1987) Photochemical sensitization by 7–methylbenz[c]acridine and related compounds, *Photochem. Photobiol.*, 45: 729–739.

Buxton, G.V., Greenstock, C.L., Helman, W.P., and Ross, A.B. (1988) Critical review of rate constants for reactions of hydrated electrons, hydrogen atoms and hydroxyl radicals ($\cdot$OH/$\cdot$O$^-$) in aqueous solution, *J. Phys. Chem. Ref. Data*, 17: 513–886.

Cadet, J., Anselmino, C., Douki, T., and Voituriez, L. (1992) Photochemistry of nucleic acids in cells, *J. Photochem. Photobiol. B: Biol.*, 15: 277–298.

Calvert, J.G. and Pitts, J.N. (1966) Experimental methods in photochemistry, in *Photochemistry*, pp. 783–804. New York: John Wiley & Sons.

Carstensen, J.T. (1995) *Drug Stability*, 2nd ed., pp. 486–537. New York: Marcel Dekker, Inc.

Chedekel, M.R., Land, E.J., Sinclair, R.S., Tait, D., and Truscott, T.G. (1980) Photochemistry of 4-hydroxybenzothiazole: a model for pheomelanin degradation, *J. Am. Chem. Soc.*, 102: 6587–6590.

Chignell, C.F., Motten, A.G., and Buettner, G.R. (1985) Photoinduced free radicals from chlorpromazine and related phenothiazines: relationship to phenothiazine-induced photosensitization, *Environ. Health Perspect.*, 64: 103–110.

Cornelissen, P.J.G., Beijersbergen van Henegouwen, G.M.J., and Gerritsma K.W. (1979) Photochemical decomposition of 1,4-benzodiazepines. Chlordiazepoxide, *Int. J. Pharm.*, 3: 205–220.

Davies, A.K., Hilal, N.S., McKellar, J.F., and Phillips, G.O. (1975) Photodegradation of salicylanilides, *Br. J. Dermatol.*, 92: 143–147.

Davies, A.K., Navaratnam, S., and Phillips, G.O. (1976) Photochemistry of chlorpromazine (2-chloro-*N*-(3-dimethyl-aminopropyl)phenothiazine) in propan-2-ol solution, *J. Chem. Soc. Perkin Trans.*, 2, 25–29.

Diffey, B.L. (1983) A mathematical model for ultraviolet optics in skin, *Phys. Med. Biol.*, 28: 647–657.

Epstein, J.H. (1989) Photomedicine, in Smith, K.C. (Ed.), *The Science of Photobiology*, pp. 155–192. New York: Plenum Press.

Ferguson, J., Addo, H.A., and Jones, S. (1985) A study of cutaneous of photosensitivity induced by amiodarone, *Br. J. Dermatol.*, 113: 537–549.

Fisher, G.J. and Land, E.L. (1983) Photosensitization of pyrimidines by 2-methyl-naphthoquinone in water: a laser flash photolysis study, *Photochem. Photobiol.*, 37: 27–32.

Foote, C.S. (1968) Mechanisms of photosensitized oxidation, *Science*, 162: 963–970.

Frederick, J.E. (1993) Ultraviolet sunlight reaching the Earth's surface: a review of recent research, *Photochem. Photobiol.*, 57: 175–178.

Frederick, J.E., Snell, H.E., and Haywood, E.K. (1989) Solar ultraviolet radiation at the Earth's surface, *Photochem. Photobiol.*, 50: 443–450.

Gates, D.M. (1966) Spectral distribution of solar radiation at Earth's surface, *Science*, 151: 523–529.

Gilbert, A. and Baggott, J. (1991) *Essentials of Molecular Photochemistry*, pp. 145–228. Oxford: Blackwell.

Gorman, A.A. and Rodgers, M.A.J. (1981) Singlet molecular oxygen, *Chem. Soc. Rev.*, 10: 205–231.

Greenhill, J.V. and McLelland, M.A. (1990) Photodecomposition of drugs, *Progr. Med. Chem.*, 27: 51–121.

Grimshaw, J. and de Silva, A.P. (1981) Photochemistry and photocyclization of aryl halides, *Chem. Soc. Rev.*, 10: 181–203.

Grossweiner, L.I. (1989) Photophysics, in Smith, K.C. (Ed.), *The Science of Photobiology*, pp. 1–45. New York: Plenum Press.

Hall, R.D., Buettner, G.R., Motten, A.G., and Chignell, C.F. (1987) Near-infrared detection of singlet molecular oxygen produced by photosensitization with promazine and chlorpromazine, *Photochem. Photobiol.*, 46: 295–300.

Henderson, B.W. and Dougherty, T.J. (1992) How does photodynamic therapy work? *Photochem. Photobiol.*, 55: 145–157.

Jagger, J (1985) *Solar-UV Actions on Living Cells*, p. 119. New York: Praeger Scientific.

Kochevar, I.E., (1981) Phototoxicity mechanisms: chlorpromazine photosensitized damage to DNA and cell membranes, *J. Invest. Dermatol.*, 77: 59–64.

Kochevar, I.E. and Lamola, A.A. (1979) Chlorpromazine and protriptyline phototoxicity: photosensitized oxygen-independent red cell hemolysis, *Photochem. Photobiol.*, 29: 1177–1197.

Kochevar, I.E., Armstrong, R.B., Einbinder, J., Walther, R.R., and Harber, L.C. (1982) Coal tar phototoxicity: active compounds and action spectra, *Photochem. Photobiol.*, 36: 65–69.

Kochevar, I.E., Hoover, K.W., and Gawienowski, M. (1984) Benoxaprofen photosensitization of cell membrane disruption, *J. Invest. Dermatol.*, 82: 214–218.

Kramer, H.E.A. and Maute, A. (1973) Sensitized photooxygenation: change from type I (radical) to type II (singlet oxygen) mechanisms, *Photochem. Photobiol.*, 17: 413–423.

Land, E.J., Navaratnam, S., Parsons, B.J., and Phillips, G.O. (1982) Primary processes in the photochemistry of aqueous sulfacetamide: a laser flash photolysis and radiolysis study, *Photochem. Photobiol.*, 35: 637–642.

Lee, P.C.C. and Rodgers, M.A.J. (1987) Laser flash photokinetic studies of Rose Bengal sensitized photodynamic interactions of nucleotides and DNA, *Photochem. Photobiol.*, 45: 79–86.

Li, Y.N.B., Moore, D.E., and Tattam, B.N. (1999) Photodegradation of amiloride in aqueous solution, *Int. J. Pharm.*, 183: 109–116.

Ljunggren, B. and Möller, H. (1977) Phenothiazine phototoxicity: an experimental study on chlorpromazine and its metabolites, *J. Invest. Dermatol.*, 68: 313–317.

Lovell, W.W. (1993) A scheme for *in vitro* screening of substances for photoallergic potential, *Toxic. In Vitro*, 7: 95–102.

Magnus, I.A. (1976) *Dermatological Photobiology*, pp. 213–216. London: Blackwell.

Mason, R.P., and Chignell, C.F. (1982) Free radicals in pharmacology and toxicology — selected topics, *Pharmacol. Rev.*, 33: 189–211.

Moan, J., Petterson, E.O., and Christensen, T. (1979) The mechanism of photodynamic inactivation of human cells *in vitro* in the presence of haematoporphyrin, *Br. J. Cancer*, 39: 398–407.

Moore, D.E. (1976) Antioxidant efficiency of polyhydric phenols in photo-oxidation of benzaldehyde, *J. Pharm. Sci.*, 65: 1447–1451.

Moore, D.E. (1977) Photosensitization by drugs, *J. Pharm. Sci.*, 66: 1282–1284.

Moore, D.E. (2002) Drug-induced cutaneous photosensitivity, *Drug Safety*, 25: 345–372.

Moore, D.E. and Burt, C.D. (1981) Photosensitization by drugs in surfactant solutions, *Photochem. Photobiol.*, 34: 431–439.

Moore, D.E. and Chappuis, P.P. (1988) A comparative study of the photochemistry of the non-steroidal anti-inflammatory drugs, naproxen, benoxaprofen and indomethacin, *Photochem. Photobiol.*, 47: 173–181.

Moore, D.E. and Hemmens, V.J. (1982) Photosensitization by antimalarial drugs, *Photochem. Photobiol.*, 36: 71–77.

Moore, D.E. and Sithipitaks, V. (1983) Photolytic degradation of frusemide, *J. Pharm. Pharmacol.*, 35: 489–493.

Moore, D.E. and Zhou, W. (1994) Photodegradation of sulfamethoxazole: a chemical system capable of monitoring seasonal changes in UVB intensity, *Photochem. Photobiol.*, 59: 497–502.

Moore, D.E., Roberts–Thomson, S., Dong, Z., and Duke, C.C. (1990) Photochemical studies on the antiinflammatory drug diclofenac, *Photochem. Photobiol.*, 52: 685–690.

Navaratnam, S., Parsons, B.J., Phillips, G.O., and Davies, A.K. (1978) Flash photolysis study of the photoionization of chlorpromazine and promazine in solution, *J. Chem. Soc. Faraday Trans. I*, 74: 1811–1819.

Parrish, J.A., Anderson, R.R., Urbach, F., and Pitts, D. (1978) *UV-A: Biological Effects of Ultraviolet Radiation with Emphasis on Human Responses to Long-Wave Ultraviolet*, p. 121. New York: Plenum Press.

Patterson, M.S., Madsen, S.J., and Wilson, B.C. (1990) Experimental tests of the feasibility of singlet oxygen luminescence monitoring *in vivo* during photodynamic therapy, *J. Photochem. Photobiol. B: Biol.*, 5: 69–84.

Proctor, P.H. and Reynolds, E.S. (1984) Free radicals and disease in man. *Physiol. Chem. Phys. Med. NMR*, 16: 175–195.

Pryor, W.A. (1986) Oxy-radicals and related species: their formation, lifetimes, and reactions, *Annu. Rev. Physiol.*, 48: 657–667.

Rahn, R.O., Landry, L.C., and Carrier, W.L. (1974) Formation of chain breaks and thymine dimers in DNA upon photosensitization at 313 nm with acetophenone, acetone or benzophenone, *Photochem. Photobiol.*, 19: 75–78.

Reszka, K. and Chignell, C.F. (1983) Spectroscopic studies of cutaneous photosensitizing agents. IV. The photolysis of benoxaprofen, an anti-inflammatory drug with photo-toxic properties, *Photochem. Photobiol.*, 38: 281–291.

Rodgers, M.A.J. and Snowden, P.T. (1982) Lifetime of $O_2(^1D_g)$ in liquid water as determined by time-resolved infrared luminescence measurements, *J. Am. Chem. Soc.*, 104: 5541–5543.

Song, P. and Tapley, K.J., Jr., (1979) Photochemistry and photobiology of psoralens, *Photochem. Photobiol.*, 29: 1177–1197.

Spikes, J.D. (1989) Photosensitization, in Smith, K.C. (Ed.), *The Science of Photobiology*, pp. 79–110. New York: Plenum Press.

Tamat, S.R. and Moore, D.E. (1983) Photolytic decomposition of hydrochlorothiazide, *J. Pharm. Sci.*, 72: 180–184.

Thorington, L. (1985) Spectral, irradiance and temporal aspects of natural and artificial light, *Ann. N.Y. Acad. Sci.*, 453: 28–54.

Trush, M.A., Mimnaugh, E.S., and Gram, T.E. (1982) Activation of pharmacological agents to radical intermediates. Implications for the role of free radicals in drug action and toxicity, *Biochem. Pharmacol.*, 31: 3335–3346.

Zhou, W. and Moore, D.E. (1994) Photochemical decomposition of sulfamethoxazole, *Int. J. Pharm.*, 110: 55–63.

CHAPTER 3

Standardization of Kinetic Studies of Photodegradation Reactions

Douglas E. Moore

CONTENTS

0-415-30323-0/04/$0.00+$1.50
© 2004 by CRC Press LLC

3.1 NEED FOR UNIFORMITY IN PHOTODEGRADATION STUDIES

Thermal stability studies on pharmaceutical formulations have been formalized for many years. Specific protocols have been developed to provide data from which a shelf-life determination can be made (Carstensen, 1995). Thus, procedures followed in one laboratory are readily reproduced in another, and shelf-life is a consistent estimation for the product, independent of the laboratory in which the data were gathered. In thermal stability studies, the principal consideration is how long the drug substance or formulation is exposed to a particular temperature. The nature of the apparatus to be used is not important as long as the temperature of the sample is uniform. Thus, the sample may be contained in a flask, bottle, or tube, or held in a water thermostat or an air incubator — whatever is most convenient for the study. Also, the concentration of the drug studied is not crucial because a thermal degradation usually proceeds by first-order kinetics for which the half-life is independent of the starting concentration.

The same is not true with respect to the investigation of photodegradation. The problem is that the exposure of a pharmaceutical to a UV- or visible-light source is a more difficult situation to control and quantify. It is not simply a matter of how long a drug or its formulation is exposed to the source, but also the wavelength range involved and the radiation intensity at the surface of the tested sample. Also important is the concentration of the sample, because the percentage change does depend on the starting concentration, even though first-order kinetics may apply to the reaction.

As a consequence of these factors, the manner in which drug photostability has been determined in pharmaceutical laboratories (Anderson et al., 1991) has varied greatly. Only general statements such as "protect from light" are to be found in the monographs of drugs that are decomposed in some way following exposure to various types of light sources.

In order to bring uniformity to the studies, as well as to gain more useful information transferable from one environment to another, standardized protocols covering all variables that can enter into photochemical reactions must be introduced. To this end, the apparatus must be defined in detail and the procedures to be followed fully explained. The discussion of these issues can be approached under three headings, namely, (1) the light or photon source; (2) the measurement of the number of photons absorbed by the sample by physical and chemical methods; and (3) the treatment of experimental data.

3.2 PHOTON SOURCES FOR PHOTODEGRADATION STUDIES

The first question to be answered relates to the nature of the irradiation source to which the drug should be exposed. Some may argue that, when handled and stored correctly, a pharmaceutical should never be exposed to direct sunlight, and therefore sunlight should not be contemplated as an irradiation source. Although that may be possible in the best of practices, it is an erroneous view; studies have shown that numerous situations occur when various formulations are exposed to direct or filtered sunlight in hospital pharmacies while they are held in clear glass or plastic containers (Tønnesen and Karlsen, 1995). Inadvertent exposure to sunlight during manufacture and transport, or by the consumer in the course of a treatment can also occur. The bathroom windowsill is a well-known storage place for medicines — it is considered more important for medicines to be out of the reach of children than protected from sunlight exposure! Because of these possibilities, photostability studies should be undertaken to include exposure to sunlight-simulating conditions, whether unfiltered or filtered by glass and/or plastic.

In principle, photodegradation studies could be performed by exposing samples to natural sunlight and analyzing after varying times. However, the intensity of sunlight, particularly the UV component, varies according to weather; latitude; time of day; and season of the year. Nonetheless, it would be a realistic situation to set samples in a window where direct sunlight could fall on them to varying extents in the course of the day. If the study were continued for a period of at least a year, the conditions would average over all seasons, but would only apply to a particular region. Thus, the quantitative data from experiments on the same formulation performed in different laboratories are unlikely to be in agreement in most cases. The use of natural sunlight is not a viable option, given the variability of conditions involved.

For consistency one needs to use an artificial photon source that has an output with a spectral power distribution as near as possible to that of sunlight. The output from the source may then be filtered, for example, with the glass or plastic required to simulate that type of packaging. Chapter 2 gave the spectral power distribution of sunlight and transmission characteristics of certain glass and plastic filters.

The two main types of photon source are (1) arc lamps and (2) fluorescent tubes, which have specific applications in photochemistry and photobiology because of their resemblance to the sunlight spectrum. The third major type, incandescent (filament) lamps, has a spectral output with relatively high infrared and low ultraviolet components compared to sunlight (see Chapter 2) and is not used for this purpose. Arc lamps generate a high intensity from a relatively small size, so the source can be readily focused in optical systems for accelerated studies. On the other hand, fluorescent tubes are simpler and cheaper to use, with the contrasting performance of being able to irradiate a large area at a lower intensity. A brief description of these sources is given here.

3.2.1 Arc Lamps

Mercury arc lamps can be constructed in three ways, with the mercury vapor at low, medium, or high pressure; each variant has specific characteristics in emphasizing certain aspects of the mercury emission spectrum. The low-pressure arc emits 90% of its energy as a line at 254 nm and so is of no direct use in a sunlight-simulating experiment. The medium-pressure arc is also a line source, producing greater intensities at the other characteristic mercury emission wavelengths: 302, 313, 334, 366, and 405 nm. Because this arc lamp is moderate in cost, has a long life, and gives a good representation of emission in the UV region, it has been widely used in drug photostability studies with a glass filter to shield the sample from the 254-nm

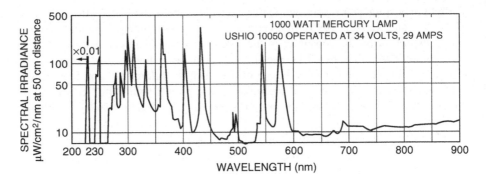

Figure 3.1 Spectral power distribution of a high pressure mercury arc lamp. (From Jagger, J., *Solar-UV Actions on Living Cells*, Praeger Scientific, New York, pp. 159–176, 1985.)

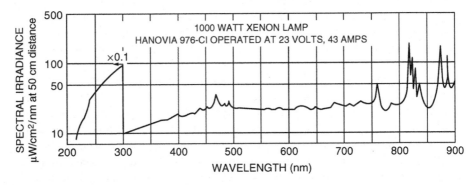

Figure 3.2 Spectral power distribution of a xenon arc lamp. (From Jagger, J., *Solar-UV Actions on Living Cells*, Praeger Scientific, New York, pp. 159–176, 1985.)

radiation (Moore, 1987). The principal application is as an irradiation source to determine degradation pathways. The high-pressure mercury arc emits the same lines, as well as a continuous background radiation right across the solar UV and visible regions. The emission spectrum of the high-pressure form is shown in Figure 3.1.

The arc lamp with the best resemblance to sunlight is the xenon arc lamp, although the development of new metal-halide lamps has led to competition for that claim. The spectral outputs of these two types are shown in Figure 3.2 and Figure 3.3. The xenon arc has a relatively smooth continuous output spectrum with some line emissions superimposed in the region of 450 to 500 nm, whereas the metal-halide lamp is more uniform across the 350 to 550 nm region.

One disadvantage of arc lamps is the high heat output, as seen by the continued output above 500 nm; however, this can be dissipated by the using a heat filter, which usually contains water. Because photochemical reactions are generally initiated by UV radiation, adjustment of the intensity above 500 nm is most unlikely to lead to erroneous photostability data. On the other hand, overheating of the sample by the lamp may lead to thermal decomposition processes that will complicate the issue. The other principal disadvantage of xenon and metal-halide sources is their

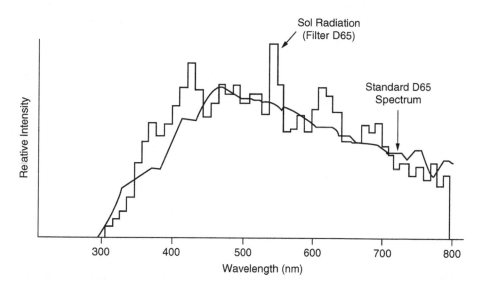

Figure 3.3   Spectral power distribution of a metal-halide arc lamp compared to sunlight.

comparatively short life span — 750 hours for the metal-halide and 1500 to 2000 hours for the xenon arc. Their initial cost is high, and the focus of their irradiation is such that only a relatively small number of samples can be irradiated at one time.

3.2.2 Fluorescent Lamps

Fluorescent lamps have been used in photostability testing by a number of labora-tories. The operating principle of fluorescent lamps is based on mercury vapor discharge at very low pressure, producing the 254-nm emission converted to higher wavelengths by the phosphor coating on the inside surface of the tube. Emission characteristics are determined by the particular phosphor used in the manufacture (some examples were given in Chapter 2). For possible application to drug photo-stability studies, the "daylight" and "cool-white" and near-UV fluorescent tubes have the advantage that they can be set up in large banks at relatively low cost to irradiate large numbers of samples at one time.

It is not possible to achieve a sunlight-simulating spectrum with only one type of fluorescent lamp; a combination must be used to get the appropriate amounts of UVA, UVB, and visible components. Another suggested combination involves using a "black-light" UVA source with a daylight fluorescent tube, for which the output is shown in Figure 3.4. This combination is considered unsatisfactory because the correct balance between the UV and visible region intensities is not achieved for all the irradiated samples without diffusers, which will reduce the overall intensity (Tønnesen and Moore, 1993). The fluorescent lamps have long lifetimes (in excess of 20,000 hours) and do not cause a problem with respect to heat output. However, the intensity developed is low compared to the xenon arc and exposure times of the order of several weeks may be needed to gain a change that occurs in a few hours with the xenon arc.

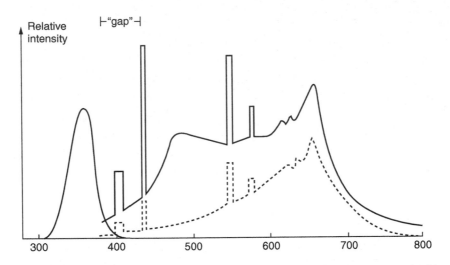

Figure 3.4   Relative spectral distribution of a UVA "black lamp" and two representative "daylight" fluorescent lamps. (Redrawn from the Osram Product Catalogue.)

At the present time, qualitative photostability testing of drugs in a number of laboratories involves a combination of a UVA fluorescent lamp with a daylight fluorescent lamp. The sample is irradiated for a fixed period of time and, if any changes in the physical or chemical properties of the sample are observed compared to a reference sample, it is recommended that the drug be protected from light. Such a test is inexpensive and easy to perform, but must be regarded as a minimum level of exposure; more stringent testing should be performed in appropriate cases when significant changes are observed. Reproducible results will not be obtained, nor practical conclusions reached, unless the output spectra of the recommended photon sources are standardized.

3.3 MEASUREMENT OF IRRADIATION INTENSITY

The intensity of the photon source is a major variable that depends very much on the nature of the lamp (factors such as its age and design). Its position with respect to the sample will determine the incident dose that the sample receives. Any attempt to standardize photostability testing must define the type of photon source to be used and include in the experimental arrangement a means of monitoring the quantity of photons falling on the sample so that the total exposure reaches a certain predetermined value. Photon source intensity measurement is termed radiometry and can be achieved by physical instruments such as a radiometer, a lux meter, or a thermopile, or by chemical actinometry using a reaction of known photochemical efficiency (quantum yield). First, it is necessary to define briefly the terms involved. Fuller accounts are given by Jagger (1985) and Thorington (1986).

3.3.1   Photon Intensity: Irradiance, Fluence, and Dose

Radiant intensity is the power emitted per unit of solid angle of the source, whereas radiance is the intensity per unit area of the source. Thus, a fluorescent lamp has intensity similar to a filament lamp, but a comparatively low radiance.

The measurement of the radiation incident upon a sample can be expressed in terms of the number of photons of a particular wavelength crossing a unit area in a unit time. The characteristic frequency ν or wavelength λ of the photon can be converted into its energy equivalent using the Planck Equation,

$$E = h\nu = hc/\lambda \qquad (3.1)$$

so that the energy fluence, or irradiance, is obtained expressed in units of joules m^{-2} sec^{-1}, or watts m^{-2}. This definition is complicated when the irradiating source covers a range of wavelengths, i.e., a range of energies. In that case, the response is integrated over the wavelength range, but it is difficult to obtain an absolute value for the irradiance without reference to a calibrated system of measurement.

In the visible region, the terminology is somewhat different; it is called photometry and is based on the human response to visible light. The unit of source intensity is the candle and the radiance of the source becomes the brightness, while the irradiance is called the illuminance (measured in lux) or the illumination (measured in footcandles). The lumen is the unit of power (in watts) while lux is the power per unit area, expressed as lumen m^{-2} or watts m^{-2}. Note that the units of photometry relate to standardized human perception. Thus, a monochromatic UV source may have high irradiance but zero illuminance because none of the energy can be perceived by the human eye. However, many sources of UVR are also powerful visible light sources and are rated by manufacturers in photometric terms.

The total dose is determined from the length of time the sample is irradiated, expressed in watthours m^{-2}. A fundamental difference between irradiance and dose is that the former describes a beam of photons, whereas the latter relates to the irradiated sample. In other words, the irradiance of a photon beam may be constant, but the dose will vary according to how an irregularly shaped sample is oriented with respect to the beam. In most experimental work, the irradiance is measured, although, strictly, the dose incident on the sample is more important. The radiation may be absorbed, transmitted, scattered, or reflected, but in terms of photochemical reaction, that which is absorbed is the critical quantity. Measuring the quantity of photons absorbed can only be achieved for liquid samples that transmit the unabsorbed photons to be measured by a detector behind the sample.

3.3.2   Physical Instrumentation for Intensity Measurement

Radiometers are devices based on various types of photocells that generate a current when photons fall upon them. The bimetallic photovoltaic cell, the basic component of solar energy converters, is the first and simplest example, but these have been replaced by photodiodes in recent times. The photomultiplier tube can also be used, but is regarded as too cumbersome for incorporation into a portable meter. These devices have a response that varies with the wavelength of the incident radiation, but their use can be designed for a particular spectral region by means of a specific filter. Each filter–photocell combination requires a separate calibration, as does each different source used. Thus, UV filter–radiometers are designed to measure incident radiant power in the UV region transmitted by the particular filter. By appropriate choice of filter, it is possible to get a reading of the UVB or the UVA intensity. Examples of these detectors are shown in Figure 3.5, and their relative spectral response in Figure 3.6.

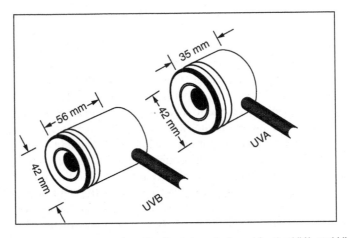

Figure 3.5   Schematic representation of radiometers designed for the UVA and UVB regions. (International Light, Inc.)

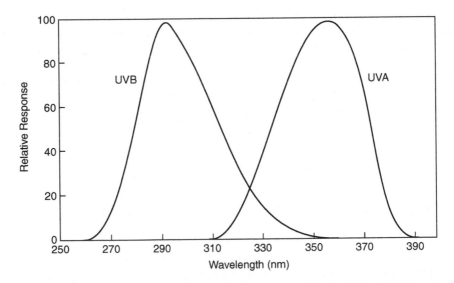

Figure 3.6   Relative spectral sensitivity of radiometers designed for the UVA and UVB regions. (International Light, Inc.)

Although this provides a convenient means of measuring the output of a photon source, response is dictated by the characteristics of the filter, which differ among manufacturers and also suffer from change with time. Due to the lack of standardized filters, different meters may measure different fractions of the radiant energy. They must be calibrated regularly to allow for changing responses due to alterations in performance of the filter and photodiode. A considerable amount of study of this type of instrument has been promoted by measurement of the changes in UVB intensity associated with the depletion of ozone in the stratosphere (De Luisi et al., 1992; Smith et al., 1993). The quantity measured is an instantaneous measurement of the irradiance in watts per square meter. From the viewpoint of drug photostability testing, the radiometers do not give the desired output of an integrated exposure reading over the testing period.

The lux meter is used in photometry and is simply a radiometer with a spectral responsiveness that closely matches the visual response of the human eye, thus measuring incident radiant power in the visible region of the electromagnetic spectrum. In this case, the unit of measurement is the illuminance in lux and is calibrated against a specific tungsten lamp. The lux meter should be provided with a set of correction factors to enable compensation for differences in spectral response for lamps with emission spectra different from the calibration lamp.

The spectroradiometer combines a monochromator with a photomultiplier detector, as in a spectrophotometer, to gain a detailed estimate of the photon intensity as a function of wavelength. It is used for checking radiometer calibration and monitoring sunlight intensity variations with season, particularly in the UVB region (De Luisi et al., 1992).

A thermopile is a thermal detector capable of measuring the total incident radiant flux (in W/m^2) in the UV, visible, and IR regions of the spectrum, with a response that is essentially independent of wavelength. It is the reference against which the other devices can be calibrated, but is not recommended for direct use because it is expensive and not a straightforward instrument to use correctly (Jagger, 1985).

3.4 CHEMICAL ACTINOMETRY

Although physical instrumentation is more convenient, lack of an integrated output and the need for regular calibration are hindrances to its widespread application to measurement of the number of photons absorbed by a sample. The alternative is to use a chemical actinometer system in which a photochemical reaction of known characteristics is monitored when subjected to the same irradiation conditions as the test sample. When used in the fashion described next, the chemical actinometer measures the number of photons actually absorbed by the sample, rather than the incident total.

The most important property of an actinometer system is that it gives a response over the full range of wavelength to which the various samples are tested. In the case of drug photostability studies, this can mean from the UVB through the visible regions, although in practice it would be sufficient if only the UVB and UVA regions were covered because very few drug substances absorb in the visible.

The various actinometer systems applicable to all types of photoprocesses have been summarized by Kuhn et al. (1989) for the International Union of Pure and Applied Chemistry Commission on Photochemistry. The fundamental problem is that actinometry is a technically demanding task, and it is then a complicated matter to relate the photon absorption rate to a reaction rate applicable to a particular experimental setting. In the following sections, the procedure will be briefly described for chemical actinometry and quantum yield determination; then it will be explained that the rate of a photochemical reaction can be predicted from knowledge of the quantum yield and the incident photon intensity. Finally, treatment of experimental rate data for a photostability study will be covered in detail.

3.4.1 Quantum Yield of a Photochemical Reaction

Chemical actinometer systems have been widely used in basic photochemical studies to enable the determination of the quantum yield of a photochemical reaction. The quantum yield or quantum efficiency φ is defined as:

$$\varphi = \frac{\text{number of molecules reacted/unit volume/unit time}}{\text{number of photons absorbed/unit volume/unit time}} \qquad (3.2)$$

Photochemists are interested in the quantum yield of a photochemical reaction because it is the measure of the amount of reaction corresponding to the photons actually absorbed by the sample; therefore, it is the true constant describing the rate or efficiency of that reaction independent of the experimental arrangement used. When correctly measured, the quantum yield of a photochemical reaction should be the same, irrespective of whether it is determined in Oslo, Sydney, or Kalamazoo. When the photochemical reactivity of a drug is reported in a quantitative sense, the quantum yield should be quoted for each reaction (Calvert and Pitts, 1966; Moore, 1987).

3.4.2 Actinometer Experimental Design

In order to determine the number of photons absorbed by a drug when irradiated in solution, an experiment is set up with a two-cell arrangement of the drug solution and the actinometer solution, as shown in Figure 3.7. Cell B is filled with the actinometer solution, and two sets of irradiations are performed. In set 1, cell A contains the solution of drug to be tested at a concentration such that its absorbance is about 0.4 at the irradiating wavelength; in set 2, cell A is filled with the solvent. The contents of the cells should be efficiently stirred throughout. The samples are analyzed after a series of different times of irradiation and the response of the actinometer determined for the two sets. A plot such as that given in Figure 3.8 should be obtained; the difference between the slopes for sets 1 and 2 represents what is absorbed by the drug sample. Thus, the number of photons absorbed per unit time by the drug (N_{drug}) is calculated by:

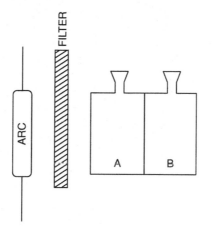

Figure 3.7   Arrangement of reaction vessels for actinometry experiments.

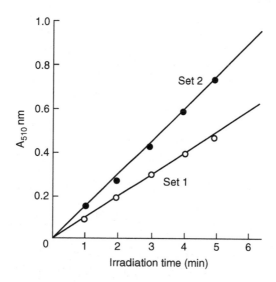

Figure 3.8   Typical plot of results of an actinometry experiment.

$$N_{drug} = R \, [\, slope(Set \, 2) - slope(Set \, 1) \,]/\varphi_{act} \qquad (3.3)$$

where φ_{act} is the quantum yield of the actinometer system, and R is the factor necessary to convert the analytical response of the actinometer to the number of molecules reacted. For example, if the actinometer is analyzed in terms of a product formation detected by a change in absorbance, then the factor R would include the molar absorptivity of the actinometer product, together with Avogadro's number and any dilution factors arising from the analysis procedures.

In parallel, the extent of reaction of the drug sample in cell A (set 1) is determined by the appropriate method of analysis (usually HPLC). The number of drug molecules reacted per unit time can be calculated from the reaction extent, and the quantum yield of the drug photodegradation is determined according to Equation 3.2. This experiment has been routinely performed by photochemists with chemical actinometers; however, in principle, there is no reason why a reliable integrating physical instrument could not be used in the same way.

The experimental arrangement described here applies to samples that transmit some of the incident radiation, implying a homogeneous solution. For many pharmaceutical formulations such as solids, suspensions, ointments, etc., the incident radiation is likely to be reflected or scattered, in which case the actinometer solution or physical instrument would need to be placed beside the sample. Although that placement will not give a measure of what is actually absorbed by the sample, it is at least a measure of the incident radiation from which an estimate of the relative reactivity can be gained.

3.4.3 Ferrioxalate Actinometer

The most widely used actinometer system is that based on potassium ferrioxalate because it fulfils the condition of applicability over a broad wavelength range. Potassium ferrioxalate is readily prepared by reaction of ferric chloride with potassium oxalate. The details of the photochemistry were worked out by Hatchard and Parker (1956). When acidic solutions (6 mM) of potassium ferrioxalate are irradiated by UV or visible light in the range of 250 to 570 nm, 99% of the incident radiation is absorbed; Fe(III) is reduced, while oxalate is oxidized. The progress of the reaction is monitored by detecting the Fe(II) as a complex with 1,10-phenanthroline at 510 nm. The product Fe(II) and its oxalate complex do not absorb the incident radiation to a measurable extent, so no back reaction occurs. The quantum yield of Fe(II) formation varies with wavelength of the irradiation, but for the 300- to 400-nm region, the value can be taken as 1.2; the stoichiometry of the reaction indicates that two atoms of Fe(III) are reduced per photon absorbed, i.e., a theoretical maximum quantum yield of 2.

Two disadvantages affect ferrioxalate use in drug photostability studies with sunlight-simulating irradiation. One is minor, in that the procedure to be followed to measure the extent of the photoreaction of ferrioxalate is technically cumbersome. The major concern is that, although the photoreaction occurs with a constant quantum yield, it is rapid in relation to most drug photodecomposition processes when used with a wide range of wavelengths. That is, the actinometer reacts sufficiently in 5 to 10 minutes to enable precise determination of the radiation falling on the sample, yet the exposure time required for the drug sample to show a suitable change for detection is often of the order of several hours.

If it can be assumed that the photon source is uniform in its output, this would not affect the result; unfortunately, that is not the case because xenon arc lamps may vary by as much as 20% over a period of 5 to 6 hours. Fundamentally, the ferrioxalate system remains the actinometer of choice when a narrow range of wavelength is used for the irradiation. A possible approach to extend the range of application is to use neutral density filters to reduce by a known factor the intensity of the photon source with respect to the actinometer, but not the sample being tested.

3.4.4   Other Actinometer Systems

Although the review of actinometers by Kuhn et al. (1989) covered a vast array of well-documented gaseous-, liquid-, and solid-phase systems, few might be considered useful for the drug photostability situation, principally because they are not applicable to the sunlight wavelength range. There is considerable interest in the development of solid-state "sunburn dosimeters" for the measurement of UVB, based on the photochemical change of a polysulfone film (Diffey, 1987; Herlihy et al., 1994). However, the wavelength range is too narrow for the actinometry application in stability studies. Uranyl oxalate is the only other chemical solution system that responds over a wide wavelength range, including the UVA and visible regions. Its reactivity is slower than ferrioxalate, but any advantage there is outweighed by its expense and toxicity.

Quinine hydrochloride solution has been investigated as an actinometer system with slower reactivity than ferrioxalate in sunlight-simulating conditions (Yoshioka et al., 1994). It has been suggested that the increase in absorbance at 400 nm of a 4% solution of quinine hydrochloride in water be used as a measure by which exposure times can be determined for formulations. The quantum yield for quinine photodecomposition has not been determined, but the proposal is for a unit of exposure time expressed as ΔA (400 nm). However, to measure incident radiation in terms of one species implies that its absorption spectrum covers the full range of incident wavelengths to be used. This is not the case for quinine, whose absorption maximum is at 335 nm, rapidly falling off as wavelength increases. The relatively high concentration of 4% has been used as an attempt to extend the absorption, as well as to ensure that all radiation in its absorption range is absorbed. The quinine system might be satisfactory for many drugs whose absorption spectrum does not extend much beyond the UVB region; however, many other examples can be given (e.g., tetracycline, nifedipine) for which the quinine spectrum does not overlap and is therefore unsatisfactory. At this time, no chemical actinometer appears ideal in all respects.

3.5 KINETIC TREATMENT OF PHOTOCHEMICAL REACTIONS

Testing the photostability of a drug substance at the preformulation stage invariably involves a study of the drug's rate of degradation in solution when exposed for a period of time to a source of irradiation. The experimenter wishes to know the kinetics of the process; however, he will find that the value of the rate constant depends very much on the design of the experiment, for the same reasons as those outlined earlier.

The factors that determine the rate of a photochemical reaction are simply the rate at which the radiation is absorbed by the test sample (i.e., the number N of photons absorbed per second) and the efficiency of the photochemical process (i.e., the quantum yield of the reaction, φ). If one is using a monochromatic photon source, the number of photons absorbed depends upon the intensity of the photon source and the absorbance at that wavelength of the absorbing species. As explained previously, the quantum yield is the characteristic constant for the process in question. Thus, the rate of a photochemical reaction is defined as:

$$\text{Rate} = \text{number of molecules transformed per second} = N\ \varphi \qquad (3.4)$$

In the first instance, the rate can be determined for a homogeneous liquid sample in which the only photon absorption is due to the drug molecule undergoing transformation, with the restriction that the concentration is low so that the drug does not absorb all of the available radiation in the wavelength range corresponding to its absorption spectrum.

The value of N can be derived at a particular wavelength λ and is given by:

$$N_\lambda = I_\lambda - I_t = I_\lambda\ (1 - 10^{-A}) \qquad (3.5)$$

where I_λ and I_t are the incident and transmitted radiation intensities, respectively, and A is the absorbance of the sample at the wavelength of irradiation. This expression can be expanded as a power series:

$$N_\lambda = 2.303\ I_\lambda\ (A + A^2/2 + A^3/6 + \ldots) \qquad (3.6)$$

When the absorbance is low ($A < 0.02$), the second- and higher-order terms are negligible and the expression simplifies to the first term in Equation 3.6. Given the Beer's law relation between absorbance and concentration, N can be seen to be directly proportional to concentration:

$$N_\lambda = 2.303\ I_\lambda\ A = 2.303\ I_\lambda\ \varepsilon_\lambda\ b\ C \qquad (3.7)$$

where ε_λ is the molar absorptivity at wavelength λ; C the molar concentration of the absorbing species; and b the optical path length of the reaction vessel. Now I_λ and ε_λ vary with wavelength, so the expression must be integrated over the relevant wavelength range where each has a nonzero value:

$$N = 2.303\ b\ C \int (I_\lambda\ \varepsilon_\lambda)\ d\lambda \text{ integrated from } \lambda_1 \text{ to } \lambda_2 \qquad (3.8)$$

Thus,

$$\text{Rate} = 2.303\ b\ C\ \varphi \int (I_\lambda\ \varepsilon_\lambda)\ d\lambda \qquad (3.9)$$

Now, the overlap integral ($\int I_\lambda\ \varepsilon_\lambda\ d\lambda$) is a constant for a particular combination of photon source and absorbing substance; b is determined by the reaction vessel chosen; and φ is a characteristic of the reaction. Thus, by grouping the constant terms into an overall constant k_1, the expression is simplified to a first-order kinetic equation:

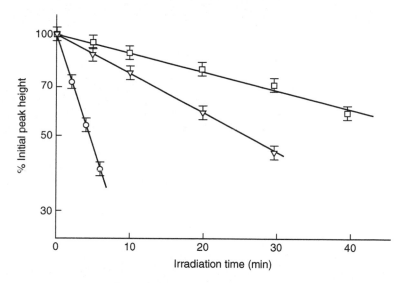

Figure 3.9   Photodegradation kinetics of sulfamethoxazole (5×10^{-5} M) in buffered aqueous solution analyzed by HPLC. (O): pH 3.0; (∇): pH 7.0; (□): pH 9.0. (From Moore, D.E. and Zhou, W., *Photochem. Photobiol.*, 59, 497–502, 1994.)

$$\text{Rate} = -d \, [\text{Drug}]/dt = k_1 \, C \qquad (3.10)$$

The integrated form of Equation 3.10 can be expressed in exponential form (Equation 3.11) or logarithmic form (Equation 3.12):

$$[\text{Drug}]_t = [\text{Drug}]_0 \, e^{-k_1 t} \qquad (3.11)$$

$$\ln \, [\text{Drug}]_t = \ln \, [\text{Drug}]_0 - k_1 \, t \qquad (3.12)$$

Verification of first-order kinetics is obtained when a plot of the logarithm of the concentration of drug remaining is found to be linear with slope equal to ($-k_1$). Data given in Figure 3.9 show the photodegradation of sulfamethoxazole in aqueous solutions of varying pH.

Strictly speaking, the overlap integral determines the rate, and therefore the rate constant, for the reaction. When the substance examined has only a relatively small absorption in the UVB and UVA regions, the overlap integral is small and first-order kinetics are observed, even though a large amount of the absorbing substance is present in the system. In Figure 3.10, the very small overlap of sulfamethoxazole absorption and sunlight is shown for Sydney (latitude 33.5°S) in the different seasons of the year. The different overlaps mean that sunlight-induced degradation of sulfamethoxazole will vary greatly, depending upon location and season. On the other hand, Figure 3.11 reveals that nifedipine has a more extensive overlap integral through the UVA region that varies to a lesser extent with location and season.

Although Equation 3.10 predicts that a photodegradation reaction studied at low concentrations in solution will follow first-order kinetics, the rate constant derived from

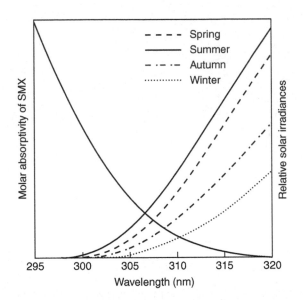

Figure 3.10   Molar absorption spectrum of sulfamethoxazole and sunlight intensity, as a function of season, at midday (latitude 33.5°). (Calculated by Moore, D.E. and Zhou, W., *Photochem. Photobiol.*, 59, 497–502, 1994, from data in Leifer, A., *The Kinetics of Environmental Aquatic Chemistry — Theory and Practice,* American Chemical Society, New York, pp. 255–264, and Zepp, R.G. and Cline, D.M., *Environ. Sci. Technol.*, 11, 359–366, 1977.)

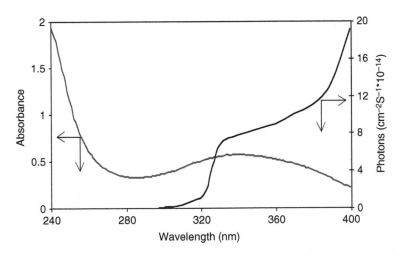

Figure 3.11   Absorption spectrum of nifedipine (5×10^{-5} *M*), and sunlight intensity at midday in midsummer as in Figure 3.10.

a study performed in one laboratory will not be the same as that recorded in another. This is because of the inherent difficulty in reproducing exactly the experimental arrangement of photon source and sample irradiation geometry. An obvious question that arises is whether any value exists in kinetic measurements in an uncalibrated apparatus. The answer is that relative values are useful in a given experimental arrangement for making comparisons of degradation of the absorbing substance in different formulations, e.g., those containing ingredients designed to inhibit the photoreaction.

The use of rate constants is helpful for comparative purposes when studying a number of different reaction mixtures under the same irradiation conditions, such as the effect of pH in Figure 3.9. However, the reaction order and numerical values of the rate constants are relative to the specific conditions used. Quantitative expressions of photochemical rate should be given in terms of the quantum yield φ, which will normally have a value between 0 and 1, except for the rare case when a chain reaction is involved, thus causing multiple degradation events for every photon absorbed. The quantum yield is a wavelength-independent quantity and thus can be determined at any wavelength that the substance absorbs, although clearly a greater precision can be gained near the absorption maximum.

3.5.1 Deviations from First-Order Kinetics

It should be recognized that the derivation in Equation 3.10 is an approximation for low values of the absorbance A and that, at higher values, second- and third-order terms of the power series would come into play, leading to a deviation from the first-order kinetic equation. Figure 3.12 is a theoretical plot that shows how the expression $(1 - 10^{-A})$ in Equation 3.5 changes with increasing values of the absorbance A. Clearly, at low values of A, the relationship is linear, showing a direct dependence on A. This gradually changes so that when the absorbance reaches 2, the drug absorbs 99% of the incident radiation. Here the rate-limiting factor is the intensity of the incident radiation, and the reaction tends toward pseudo zero-order kinetics.

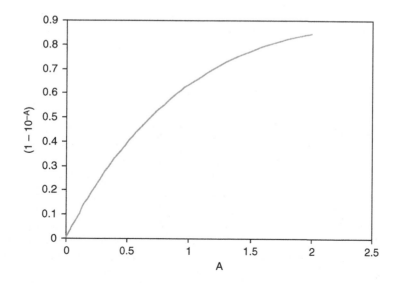

Figure 3.12 Calculated plot of the expression $(1 - 10^{-A})$ in Equation 3.5 for varying values of A.

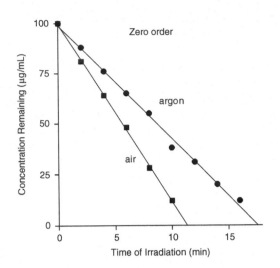

Figure 3.13   Zero order (linear) plot for the photodegradation of ketrolac tromethamine (initial concentration 100 µg/mL) in ethanol solution under argon and air atmospheres. (Replotted data from Gu et al., 1988.)

This type of kinetic order variation is referred to as saturation kinetics, commonly seen in reaction situations in which one of the important factors is limiting; an example is simple enzyme-catalyzed reactions. In between the two extremes, the reaction order is nonintegral, so neither first- nor zero-order kinetic plots will adequately describe the data. The zero-order kinetic relation is expressed by Equation 3.13.

$$\text{Rate of degradation} = -d\,[\text{Drug}]/dt = k_0 \tag{3.13}$$

The integrated form of Equation 3.13 is given in Equation 3.14, which shows that the concentration of the irradiated drug will fall linearly with time, as demonstrated for the photodegradation of ketorolac tromethamine in Figure 3.13 (Gu et al., 1988).

$$[\text{Drug}]_t = [\text{Drug}]_0 - k_0\,t \tag{3.14}$$

Distinguishing between zero- and first-order kinetics appears to be a simple task. However, a significant complication arises because a reaction must be taken to at least 50% conversion before an unequivocal delineation of the order can be made, even with good-quality analytical data for the photodegradation. This point is demonstrated in Figure 3.14, in which the early data of Figure 3.9 and Figure 3.13 are replotted according to the alternative kinetic equation (i.e., the sulfamethoxazole data are plotted on a linear scale in Figure 3.14A and the ketrolac data plotted on a logarithmic scale in Figure 3.14B). As can be seen, the linear plots show a reasonable agreement with the data in both cases, indicating that the sulfamethoxazole photodegradation could be zero order, while for ketrolac it could be considered a first-order process.

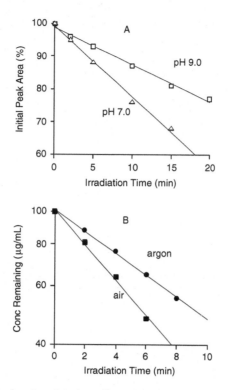

Figure 3.14 A. Replot of early points for sulfamethoxazole photodegradation (Figure 3.9) on a linear scale. B. Replot of early points for ketrolac tromethamine photodegradation (Figure 3.13) on a logarithmic scale.

Strictly speaking, any order between one and zero might be applied, unless the experimenter takes into account where the test solution fits in Figure 3.12. In an experiment using a higher concentration of drug, it would be rare to follow the reaction to the extent (>50% conversion) necessary to decide clearly which kinetic order is applicable. In most cases, the data have been treated according to the first-order equations, which may not be a correct interpretation, but is acceptable because the "relative" or "apparent" rate constants derived are not absolute and apply only to the apparatus used.

It is therefore recommended that preformulation studies of photodegradation be conducted with low solution concentrations (solution absorbance < 0.4 at the irradiation wavelength) so that first-order kinetics do apply and reaction rate is limited by the drug concentration rather than the light intensity. For the two examples given, sulfamethoxazole can be studied at concentrations up to 1 mM and produce a first-order plot with sunlight-simulating radiation because of its small overlap integral. In contrast, nifedipine can be used at no more than 5×10^{-5} M for first-order kinetics to apply. Experimentally, it is important that the irradiated solution is stirred effectively to ensure that a uniform concentration is maintained throughout the solution, and that products are dispersed to minimize secondary degradation.

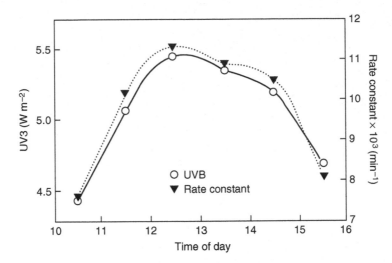

Figure 3.15   **Photodegradation** of midazolam at different concentrations in ethanolic solution. ● 20 m*M*; ■ 15 m*M*; ▲ 10 m*M*; ▼ 7.5 m*M*; ♦ 5 m*M*. (Replotted on a logarithmic scale from Andersin and Tammilehto, 1989.)

The preceding discussion implies that, from a kinetic viewpoint, no useful information is to be gained from a study of photodegradation of drug solutions of varying initial concentrations — the kinetic order is likely to be changing across the concentration range used. An additional point to observe in data treatment is the need to represent the change in concentration of the drug correctly. This is best illustrated in Figure 3.15 by an example showing the photodegradation of midazolam (Andersin and Tammilehto, 1989). Here, the extent of reaction has been incorrectly expressed as "residual drug as percent of initial concentration." As the initial drug concentration is increased, this presentation of the data shows that the slope of the plot (the apparent rate constant) diminishes, i.e., an inverse concentration relation appears to hold. The reason for this anomaly is that the concentrations studied are so high that most of the UVR in the absorption range of the drug is absorbed. If the data were to be replotted in terms of the "amount of drug reacted," it would be found that no dependence on drug concentration exists at all, i.e., zero-order kinetics apply. In other words, the intensity of the photon source has the dominant effect on rate of reaction.

3.5.2   Systems with More Than One Absorbing Species

The derivation of Equation 3.10 would need to be modified if other components of the system absorb light. Two possibilities arise, depending upon whether the other components participate in subsequent reactions. If the other absorbing components do not react, the only effect is reduction or filtering of the incident light. The fraction of the light absorbed by the drug can be expressed as:

$$F_C = \varepsilon_\lambda \, C/(\alpha_\lambda + \varepsilon_\lambda) \tag{3.15}$$

where α_λ is the attenuation coefficient of the medium at wavelength λ (sum of the absorbances of components other than the drug of interest).

When other absorbers react and influence the degradation of the drug (e.g., an impurity or additive that sensitizes the photoreaction), the overall kinetics will be a combination of both pathways. Thus, the reaction rate may depend on the drug and the sensitizer concentrations. Photochemical stability of a drug in a formulation may not necessarily be predicted only from the absorption spectrum and stability studies in a pure solvent; the final form of the drug in the formulation should also be considered.

3.5.3   Application of Derived Rate Constant Expression to Sunlight Exposure

In practice, the integration in Equation 3.9 is replaced by a summation of the values measured for finite wavelength bands ($\Delta\lambda$) across the region of interest so that the rate constant expression becomes:

$$k = 2.303 \, \varphi \, b \, \Sigma \, (I_\lambda \, \varepsilon_\lambda \, \Delta\lambda) \text{ summed from } \lambda_1 \text{ to } \lambda_2 \tag{3.16}$$

Using Equation 3.16, the reaction rate constant for the photodegradation of a particular compound can be predicted from the quantum yield and the molar absorptivities at specific wavelengths, together with the radiation source intensity data — for example, when the sample is exposed to direct sunlight. On the other hand, determination of the rate constant of a well-documented photodegradation occurring in a particular experimental arrangement can provide information about the radiant intensity of the source used, in the region of absorption by the solute. In other words, it can act as an actinometer system.

An example of the use of Equation 3.16 is that the rate constant can be calculated for outdoor exposure of a sample because the average intensity of sunlight at midday on a clear day has been recorded for a full range of latitudes (Leifer, 1988). This can be demonstrated by a study of the degradation kinetics of sulfamethoxazole exposed to sunlight (Moore and Zhou, 1994).

In this study, the first step was to determine quantum yield of the photodegradation of sulfamethoxazole in aqueous solution, using narrow wavelength range irradiation (xenon arc + monochromator). As indicated by the data in Figure 3.9, the reaction is strongly dependent on the pH of the solution, and the measured quantum yield was found to be 0.47 ± 0.05 at pH 3 (268 nm) and 0.084 ± 0.016 at pH 9 (257 nm) by ferrioxalate chemical actinometry. Relative to other drugs, this is a very high quantum yield in acid solution. The fact that sulfamethoxazole shows only slight absorptivity to UVR above 290 nm might lead one to conclude that it should not be very susceptible to sunlight. However, the spectrum is broad and extends into the UVB region, so direct photoreaction is observed when the drug is exposed to natural sunlight.

As discussed earlier, the rate constant of the direct photoreaction can be calculated according to Equation 3.16, using the quantum yield and the sunlight intensity. Values for the average solar irradiance for clear skies at midday Z_λ (photon cm^{-2} s^{-1} $(2.5 nm)^{-1}$) at 2.5-nm bandwidths over the relevant wavelength range of 297.5 to 320 nm as a function of season are available in the published literature only for 40°N (Leifer, 1988). On the other hand, day-average solar irradiance data are available for all latitudes (Zepp and Cline, 1977). The ratio of the 40°N values was used to estimate the midday irradiances for all other latitudes expressed for 2.5-nm bandwidths as U_λ in millieinstein cm^{-2} min^{-1} $(2.5 nm)^{-1}$ U_λ is therefore equivalent to $I_\lambda \Delta\lambda$ in Equation 3.16, which now becomes

$$k_1 = 2.303 \; \varphi \; b \; \Sigma \; (\varepsilon_\lambda \; U_\lambda) \tag{3.17}$$

by which the photodegradation rate constant can be calculated using the molar absorptivity of sulfamethoxazole measured at the midpoint of the specific wavelength intervals corresponding to those for the solar irradiance. The summation term represents the area under the overlap integral of the sulfamethoxazole absorption spectrum and the midday sunlight intensity, given in Figure 3.10 for Sydney (33.5°S) in various seasons.

Thus, the photochemical shelf-life t_{10} (time for 10% decomposition) of dilute acidic solutions of sulfamethoxazole in direct sunlight can be predicted as a function of latitude and season of year. These are shown for Sydney (33.5°S) in Table 3.1, together with the results from degradation experiments performed at midday on clear sunny days at the summer and winter solstices and the autumn and spring equinoxes. The agreement between the predicted and experimental values is within experimental error (except for the winter measurement, when the sunlight UVB was relatively low and the extent of reaction small). This is a significant result, considering the variation of weather condition; uncertainty of atmospheric pollution; and the possible difference in atmospheric aerosols and ozone levels between the Northern and Southern Hemispheres.

Table 3.1   Predicted and Experimental Photochemical Shelf-Life of Sulfamethoxazole in Acidic Aqueous Solution at Midday in Direct Sunlight as a Function of Season in Sydney[a]

Season 1992–1993	Measured Midday UVB[b] 290–320 nm (W m^{-2})	Shelf-life (mins)	
		Predicted[c]	Measured[d]
Summer	6.08 ± 0.31	9.02	8.87 ± 0.2
Autumn	3.81 ± 0.20	16.0	16.3 ± 0.3
Winter	2.09 ± 0.12	42.7	87.9 ± 8.1
Spring	4.39 ± 0.22	11.3	12.7 ± 0.3

[a] Latitude 33.5°S.
[b] Instantaneous UVB intensity recorded with an international light radiometer.
[c] Calculated using Equation 3.17.
[d] By HPLC analysis of sunlight-exposed solutions.

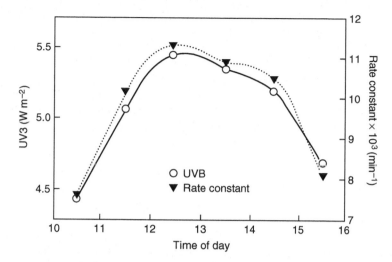

Figure 3.16   Rate constant for sulfamethoxazole photodegradation and UVB intensity as a function of time of day, measured at Sydney (latitude 33.5°S) on February 4, 1993. The weather conditions were sunny with generally clear sky, air temperature 34°C. (From Moore, D.E. and Zhou, W., *Photochem. Photobiol.*, 59, 497–502, 1994.)

Another component of this study is shown in Figure 3.16. The rate constant was measured at various times of the day, and a close correlation can be seen with the UVB intensity measured with the UVB radiometer. In other words, the rate constant for this particular reaction provides a means of determining the component of the incident radiation that sulfamethoxazole is able to absorb. Sulfamethoxazole can therefore serve as an actinometer for the UVB region.

This successful demonstration that laboratory testing can correlate with outdoor exposure points to the way in which determination of the quantum yield for the drug degradation can be used to predict the shelf-life of the product when exposed to sunlight. It is applicable to all solution preparations and all lighting conditions for which irradiance has been measured. An assumption implicit in the calculation is that all of the incident radiation within the overlap integral is absorbed by the substance. Therefore, the principle can be extended to other formulations for which the absorption characteristics are known and the quantum yield can be determined with sufficient precision.

3.5.4   Activation Energy of Photochemical Reactions

The Arrhenius equation expresses the variation with temperature of the rate constant of a chemical reaction in the form:

$$k = A \exp(-E_a/R\,T) \tag{3.18}$$

where E_a is the activation energy of the reaction, interpreted simplistically as the energy barrier over which the molecule must be raised to bring about the reaction.

For a thermal reaction such as ester hydrolysis, E_a has a value between 50 and 150 kJ mol^{-1}; the higher values indicate a slower reaction rate. When more complex mechanisms apply, the E_a value applies to the rate-determining step. On the other hand, in a chain reaction such as a peroxide-initiated oxidation, the high energy initiation step is compensated by the much lower energy propagation step. Once the reaction reaches steady state, kinetic analysis shows the overall activation energy is composed of the E_p for the propagation step plus half the E_i for the initiation step. Values of the order of 20 to 50 kJ mol^{-1} are typical for chain reactions (Dainton, 1966).

In a photochemical reaction, the photon absorbed by a molecule provides the energy to raise the molecule to the excited or reactive state from which the initial products are formed. The energy corresponding to a 300-nm photon is 400 kJ mol^{-1}, a value well in excess of thermal activation energy. Subsequent processes from the excited state represent the dissipation of energy. Thus, a true photochemical process will be independent of temperature. However, the initial products, which may be peroxides in the case of photo-oxidation, are likely to undergo secondary reactions of a thermal nature, so temperature dependence would then be observed. In a chain reaction initiated photochemically, the overall energy of activation would be that corresponding to the propagation step.

3.5.5   Use of Accelerated Tests

In the study of thermal stability, accelerated testing in the form of elevated temperatures has been used by many pharmaceutical companies to minimize time involved in the testing process. This procedure is only valid for simple formulations in which the single major ingredient is broken down by a thermal reaction. In practice, regulatory authorities demand that a shelf-life determined by extrapolation of accelerated test data should be supported by actual stability data obtained by normal temperature storage (Carstensen, 1995). This is because degradation of a product by microbial contamination may well be inhibited at elevated temperatures.

For photostability testing, there appears to be no similar reason why accelerated testing in the form of exposure of samples to high intensity sources should not be effective in providing data for direct consideration in shelf-life determination. If the study is correctly set up, the extent of degradation should be directly proportional to the number of photons absorbed. The principal experimental factors would be to ensure (1) adequate mixing of liquid formulations and turning of solids; and (2) the samples are maintained at a uniform temperature. Certainly, irradiation with a high-intensity photon source is a necessary form of stress testing in the preformulation stage; it is designed to evaluate the overall sensitivity of the material for purposes of analytical method development aimed at determining the presence of potential degradation products.

REFERENCES

Andersin, R. and Tammilehto, S. (1989) Photochemical decomposition of midazolam. II. Kinetics in ethanol, *Int. J. Pharm.,* 56, 175–179.

Anderson, N.H., Johnston, D., McLelland, M.A., and Munden, P. (1991) Photostability testing of drug substances and drug products in U.K. pharmaceutical laboratories, *J. Pharm. Biomed. Anal.,* 9, 443–449.

Calvert, J.G. and Pitts, J.N. (1966) Experimental methods in photochemistry, in *Photochemistry,* John Wiley & Sons, New York, pp. 783–804.

Carstensen, J.T. (1995) *Drug Stability,* 2nd ed., Marcel Dekker, New York, pp. 438–537.

Dainton, F.S. (1966) *Chain Reactions,* Metheun, London, pp. 52–53.

De Luisi, J., Wendell, J., and Kreiner F. (1992) An examination of the spectral response characteristics of seven Robertson–Berger meters after long-term field use, *Photochem. Photobiol.,* 56, 115–122.

Diffey, B.L. (1987) A comparison of dosimeters used for solar ultraviolet radiometry, *Photochem. Photobiol.,* 46, 55–60.

Gu, L., Chiang, H.-S., and Johnson, D. (1988) Light degradation of ketrolac tromethamine, *Int. J. Pharm.,* 41, 105–113.

Hatchard, C.G. and Parker, C.A. (1956) The photochemistry of the ferrioxalate actinometer, *Proc. R. Soc. London,* A235, 518–530.

Herlihy, E., Gies, P.H., Roy, C.R., and Jones, M. (1994) Personal dosimetry of solar UV radiation for different outdoor activities, *Photochem. Photobiol.,* 60, 288–294.

Jagger, J. (1985) *Solar–UV Actions on Living Cells,* Praeger Scientific, New York, pp. 159–176.

Kuhn, H.J., Braslavsky, S.E., and Schmidt, R. (1989) Chemical actinometry, *Pure Appl. Chem.,* 61, 187–210.

Leifer, A. (1988) *The Kinetics of Environmental Aquatic Chemistry — Theory and Practice,* American Chemical Society, New York, pp. 255–264.

Moore, D.E. (1987) Principles and practice of drug photodegradation studies, *J. Pharm. Biomed. Anal.,* 5, 441–453.

Moore, D.E. and Zhou, W. (1994) Photodegradation of sulfamethoxazole: a chemical system capable of monitoring seasonal changes in UVB intensity, *Photochem. Photobiol.,* 59, 497–502.

Smith, G.J., White, M.G., and Ryan, K.G. (1993) Seasonal trends in erythemal and carcinogenic ultraviolet radiation at mid-southern latitudes 1989–1991, *Photochem. Photobiol.,* 57, 513–517.

Thorington, L. (1986) Spectral, irradiance and temporal aspects of natural and artificial light, *Ann. N.Y. Acad. Sci.,* 453, 28–54.

Tønnesen, H.H. and Karlsen, J. (1995) Photochemical degradation of components in drug formulations, *Pharmeuropa,* 7, 137–141.

Tønnesen, H.H. and Moore, D.E. (1993) Photochemical degradation of components in drug formulations, *Pharm. Technol. Int.,* 5, 27–33.

Yoshioka, S., Ishihara, Y., Terazono, T., Tsunakawa, N., Murai, M., Yasuda, T., Kitamura, M., Kunihiro, Y., Sakai, K., Hirose, Y., Tonooka, K., Takayama, K., Imai, F., Godo, M., Matsuo, M., Nakamura, K., Aso, Y., Kojima, S., Takeda, Y., and Terao, T. (1994) Quinine actinometry as a method for calibrating ultraviolet radiation intensity in light stability testing of pharmaceuticals, *Drug Dev. Ind. Pharm.,* 20, 2049–2062.

Zepp, R.G. and Cline, D.M. (1977) Rates of direct photolysis in aquatic environment, *Environ. Sci. Technol.,* 11, 359–366.

CHAPTER 4

Rationalizing the Photochemistry of Drugs

Angelo Albini and Elisa Fasani

CONTENTS

4.1 INTRODUCTION

Some families of drugs, such as glucocorticosteroids and vitamin D derivatives, were the subject of detailed studies in the 1960s and 1970s; indeed, they have been important issues in the development of modern organic photochemistry. For drugs in general, the interest has been only occasional until recently. In the last decade, work in this field has expanded for reasons discussed elsewhere in this book, and the basic photoreaction paths have been recognized for most classes. In the first edition of this book, Greenhill reviewed and classified the most important photoreactions of drugs, with the aim of answering the question of whether the photochemistry of drugs is predictable and thus establishing chemical features that induce photolability (Greenhill, 1996). Additional discussions are also available (Greenhill, 1995; Albini and Fasani, 1998).

Not unexpectedly, the general patterns of photoreactions that are well established for each chromophore of organic molecules have been found nowadays also among drugs, in which most classes of organic compounds are represented. Nevertheless, photochemistry of drugs has autonomic characteristics and has also revealed new facets of interest for the advancement of photochemical science. In particular, the effort to characterize the photochemistry of drugs under the actual conditions of use has led to many studies of the drugs in aqueous solution and in the solid state, rather than in organic solvents as usually studied. Thus some unexpected reactions have been identified. For example, heterolysis of aromatic halides in polar solvents and hydrogen transfer reactions in crystals have a more important role in the photochemistry of drug molecules than elsewhere in organic photochemistry.

On the basis of the literature available, it is certainly possible to predict photolability of a given molecule and to guess possible reaction paths. The prediction of photolability should therefore be considered at an early stage when planning new drugs, along with other characteristics leading to insufficient stability. In this chapter, the main photoreaction patterns are presented, though with no pretense of completeness. Because the target is to recognize the possible reactions, little attention is given to conditions under which the reactions have been studied. Thus, not all of the reactions presented occur under conditions of actual use and not all of them need to worry medicinal chemists.

Several factors play a role in this respect. First, the extent of photolability depends on the absorption of a significant portion of the light (natural or artificial) to which the drug is exposed. Thus, the fact that a drug is highly absorbing in the UV-C and reacts under such conditions has little practical significance because this wavelength range is barely present in the environment. Second, the reactivity may depend on the wavelength used (e.g., when two independent chromophores are present); the physical state (reactivity is usually lower in the solid than in dilute solution and may follow a completely different path); the presence of oxygen (which quenches some reactions via long-lived triplet states, but may accelerate some photoinitiated radical decompositions); and other experimental parameters. Third, taking care of the light stability of a drug preparation may be relatively easy (typically, by using an opaque container, although this is not simply applied to some forms of administration). However, avoiding a photoreaction in the skin after administering the drug is obviously a different matter (and depends *inter alia* on the localization of the drug in the tissues). In the latter case, some form of phototoxicity may result, and it may well be that this is the actual stumbling block because it prevents safe clinical use of the drug.

A prediction of the preceding effects is possible, but will not be considered in the present chapter, which is merely devoted to presentation of the most typical photoreactions. As will be seen, photoreactivity is rather ubiquitous and a large fraction of the drugs used in therapy are known to be photolabile to some degree (as indeed indicated by the mention "protect from light" in the lemmas of various pharmacopoeias and on the labels of bottles or boxes of commercial preparations).

Obviously, in order to photoreact, a molecule must absorb light (the first law of photochemistry). Thus, the first move for predicting photoreactivity is identifying the chromophores present. Inspection of the UV spectrum will also show whether ambient light is absorbed. In practice, compounds containing conjugated C=C double bonds, or a C-heteroatom double bond and aromatic and heteroaromatic compounds, absorb strongly and usually photoreact. In the following, the reactions are classed in the preceding categories. Molecules possessing two (nonconjugated) chromophores are generally expected to present both photoreactivities, possibly under different conditions. Furthermore, reactions in the solid state and in solution are often different, whether due to the constraint to the molecular movement imposed by the lattice or because the reaction occurs only at the crystal surface.

4.2 CARBON–CARBON DOUBLE BONDS

Simple alkenes do not absorb ambient light, but may react upon photosensitization by an added sensitizer (which may be an impurity or a component of the drug formulation). As an example, photosensitized oxidation may be quite efficient. Conjugated alkenes and polyenes absorb strongly in the UV (in some cases also in the visible) and photoreact.

4.2.1 Isomerization and Rearrangement

Conjugated olefins undergo *E/Z* isomerization (Scheme 4.1a), as shown in the case of vitamin A, retinal, and its derivatives, e.g., retinoic acid used as a keratolytic (Curley and Fowble, 1988). The same holds for aryl- or heteroaryl-conjugated alkenes. Examples are the topical antimycotic naftifin (**1**, Scheme 4.2), (Thoma et al., 1997); alkylidenethioxanthines employed as antipsychotic agents such as flupenthixol (**2**); clopenthixol (**3**); thioxitene (**4**); chlorprothixene (**5**, Scheme 4.3) (Po and Irwin, 1980); and the tricyclic antidepressant drugs doxepin and dothiepin (Tammilehto and Torniainen, 1989). Geometrical isomerization is often accompanied by sigmatropic (Scheme 4.1b) and electrocyclic (Scheme 4.1c) rearrangements. As an example, with stilbenes, isomerization to the Z form is followed by reversible cyclization to dihydrophenanthrenes that, in turn, usually undergo oxidation by dissolved oxygen to give oxidized phenathrenes. This is observed with clomiphene (Frith and Phillipou, 1986) and tomoxifen (Mendenhall et al., 1978). It is also observed

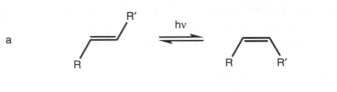

a

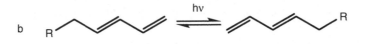

b

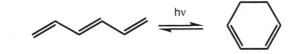

c

Scheme 4.1

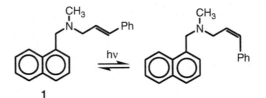

1

Scheme 4.2

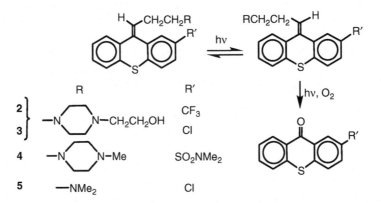

	R	R′
2	—N⌒N—CH₂CH₂OH	CF₃
3		Cl
4	—N⌒N—Me	SO₂NMe₂
5	—NMe₂	Cl

Scheme 4.3

with stilboestrol (**6**), from which a rearomatized dihydroxystilbene or a polyunsaturated diketone (resulting from hydrogen shift) is obtained, depending on conditions (Scheme 4.4), and with dienoestrol, where a sigmatropic shift also occurs (Doyle et al., 1978).

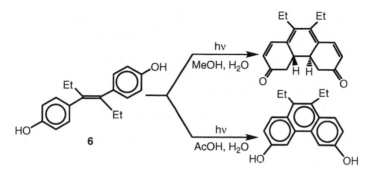

Scheme 4.4

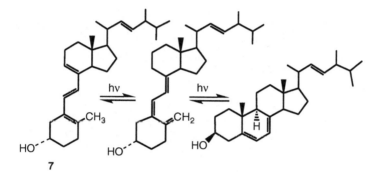

7

Scheme 4.5

Geometrical isomerization and sigmatropic and electrocyclic rearrangements are all observed in the family of vitamin D, as shown for the case of calciferol (vitamin D_2, **7**) in Scheme 4.5 (Jacobs and Havinga, 1979; Mermet–Bouvier, 1973).

4.2.2 Addition Reactions

On the C=C double bond, 2+2 cycloaddition occurs with conjugated olefins (Scheme 4.6), e.g., with the antidepressant protriptyline (Gasparro and Kochevar, 1982) and in particular with α,β-unsaturated ketones (e.g., with menadione (**8**), which yields the *syn*-cyclobutane dimer in the solid state and the *anti*-isomer by irradiation in acetone solution (Scheme 4.7; (Marciniec and Witkowska, 1988). The epoxide is obtained in the presence of oxygen (Jones and Sharples, 1984) (see later in this chapter for other oxidations). Polar addition to arylalkenes has also been observed,

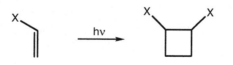

Scheme 4.6

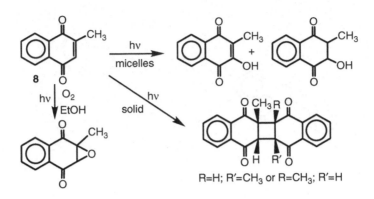

Scheme 4.7

as in the hydration taking place by irradiation of the vasoconstrictor ergotamine in water to yield an alcohol (Karliceck and Klimesova, 1977).

4.2.3   Reactions with Oxygen

Irradiation of alkenes in the presence of oxygen often leads to oxidation reactions (Scheme 4.8), such as the oxidative cleavage of the double bond in $\Delta^{9(11)}$-dehydro-estrone methyl ether (**9**, Scheme 4.9) (Lupon et al., 1984) or *ene* addition as with

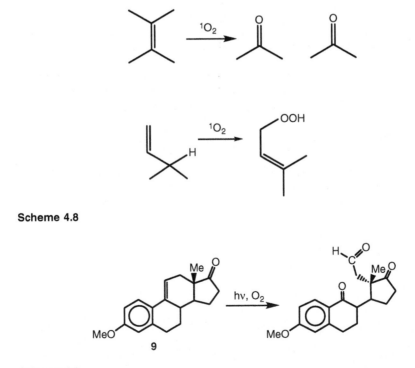

Scheme 4.8

Scheme 4.9

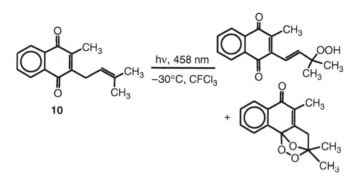

Scheme 4.10

allylnaphthoquinones of the group of vitamin K, e.g., menaquinone-1 (vitamin $K_{2(1)}$, **10**, Scheme 4.10) (Ohmae and Katsui, 1969; Teraoka and Matsuda, 1993; Wilson et al., 1980).

4.3 AROMATICS AND HETEROAROMATICS

Aromatic and heteroaromatic rings are often a feature of drug molecules and lead to a strong absorption in the UV. With carbocyclic aromatics, the reactions observed involve a substituent, not the ring, and are classified accordingly. Heterocycles give the same types of reactions and, furthermore, undergo ring cleavage and fragmentation.

4.3.1 Dehalogenation

Ring halogenated aromatics and heterocycles often undergo substitution of the halogen or reduction upon irradiation (Scheme 4.11). Perhaps unexpectedly, in view of the strength of the aryl–F bond, defluorination is the main process with antibacterial fluoroquinolones. Three types of defluorination reactions have been identified in water. Substitution of a hydroxyl for a fluoro group (Scheme 4.12) results in monofluoro (in position 6) derivatives such as enoxacin (**11**), norfloxacin (**12**), and ciprofloxacin (**13**) (Monti et al., 2001; Fasani et al., 1999; Sortino et al., 1998a; Bilski et al., 1998; Mella et al., 2001). This occurs via the addition elimination mechanism (S_NAr2^*).

$$Ar\text{-}X \xrightarrow{h\nu} Ar^{\cdot} \xrightarrow{[H]} Ar\text{-}H$$

$$Ar\text{-}X \xrightarrow{h\nu} Ar^{+} \xrightarrow{NuH} Ar\text{-}Nu$$

Scheme 4.11

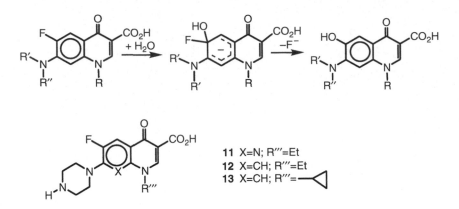

11 X=N; R'''=Et
12 X=CH; R'''=Et
13 X=CH; R'''=◁

Scheme 4.12

The process does not take place with quinolones bearing an electron-donating group, as observed for 8-alkoxy substituted ofloxacin (**14**) and 8-thioalkoxy substituted rufloxacin (**15**). In these cases, other reactions, such as decarboxylation or degradation of the aminoalkyl side chain in position 7, are observed with a very low quantum yield (Scheme 4.13) (Condorelli et al., 1999; Navaratnam and Claridge, 2000a). The same drugs in the presence of various salts undergo a different reaction, namely, reductive defluorination in sulfite buffer, or reductive defluorination and degradation of the alkylamino side chain (Scheme 4.14) in phosphate buffer (Fasani et al., 2001). A third type of defluorination occurs with 6,8-difluoroquinolones and involves unimolecular heterolytic fragmentation preferentially from position 8 forming an aryl cation. This intermediate is trapped by anions (e.g., chloride) when present (Morimura et al., 1997a) or inserts into a neighboring group, leading to functionalization or degradation of the alkyl chains in positions 1 or 7. Such reactions occur with lomefloxacin (**16**) (Fasani et al., 1997a); fleroxacin (**17**) (Martinez et al., 1997); orbifloxacin (**18**) (Morimura et al., 1997a, b); sparfloxacin (**19**) (Engler et al., 1998) (see Scheme 4.15); and related derivatives (Robertson et al., 1991).

Dechlorination is observed with the fungicide and bactericide chlorothen (2,4,5,6-tetrachloro-1,3-benzenedicarbonitrile); the topical antibacterial hexachlorophene (Schaffer et al., 1971); the bacteriostatic agent 3,3',4,5-tetrachlorosalicylanilide (Davies et al., 1975); the topical antimycotic chlorphenesin (Thoma et al., 1997); and the phenothiazine derivative chlorpromazine (**20**, Scheme 4.16) (Li and Chignell, 1987b; Moore and Tamat, 1980; Davies et al., 1976). The most important group in this class is that of diuretic drugs, for which again photohydrolysis and reductive photodehalogenation (followed in some cases by radical coupling) are observed. Examples are chlorpropamide (**23**, Scheme 4.17) (Vargas et al., 1995); frusemide (Moore and Sithipitaks, 1983a); amiloride (Li et al., 1999); chlorothiazide; hydrochlorothiazide (Revelle et al., 1997); and trichlormethiazide (Ulvi et al., 1996). The anti-inflammatory agent 2-(2,6-dichlorophenylamino)phenylacetic acid (**24**) undergoes sequential loss of the two chlorine atoms and ring closure (Poiger et al., 2001) (Scheme 4.18). Meclofenamic acid (2-(2,6-dichloro-3-methylamino)benzoic acid)

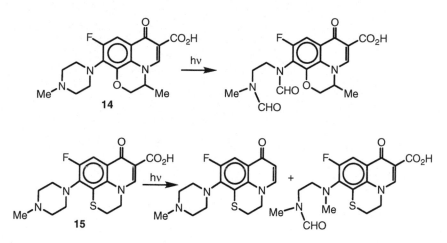

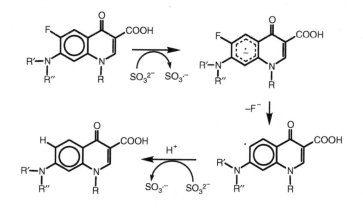

Scheme 4.13

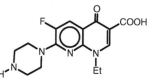

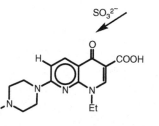

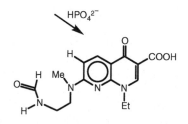

Scheme 4.14

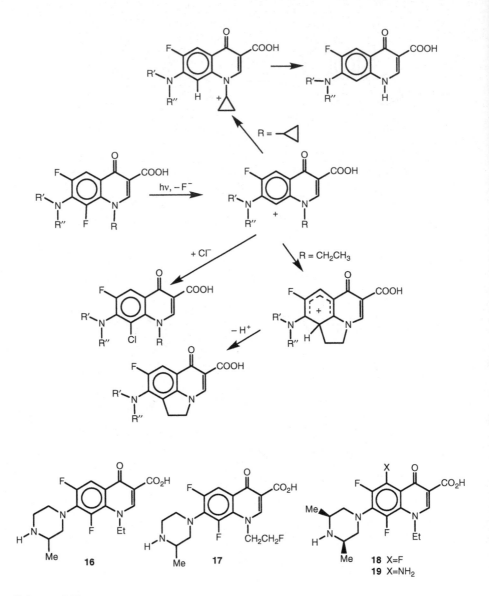

Scheme 4.15

similarly undergoes dechlorination and cyclization to give an indole derivative (Philip and Szulczewski, 1973).

Deiodination occurs by irradiation of the antiarrhythmical and antianginal agent amiodarone in water (Paillous and Verrier, 1988). Contrary to the case of the C–F bond, C–I, and at least a part of C–Cl fragmentations are homolytic processes; the products result from aryl radicals (Li and Chignell, 1987a).

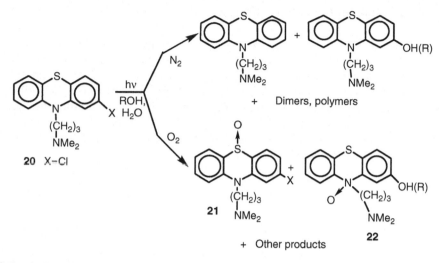

Scheme 4.16

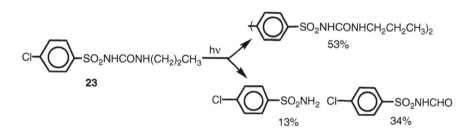

Scheme 4.17

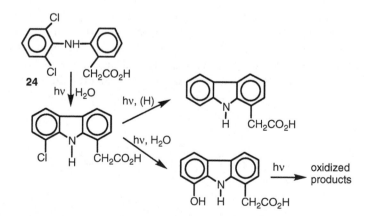

Scheme 4.18

4.3.2 Detachment of Other Groups

Sulfa drugs of general structure (**25**) undergo homolytic cleavage finally resulting in SO_2 production with formation of aniline and the corresponding amine (Schemes 4.19a, 4.20), though the efficiency of the reaction and the yield of the products isolated depend on the structure (Golpashin et al., 1984; Motten and Chignell, 1983). Decarboxylation (Scheme 4.19b) takes place with nonfluorinated quinolones such as nalidixic acid (Detzer and Huber, 1975), as well as for some electron-donating substituted fluoroquinolones such as ofloxacin (**14**) and rufloxacin (**15**) (see Scheme 4.13), for which the reaction is quite slow.

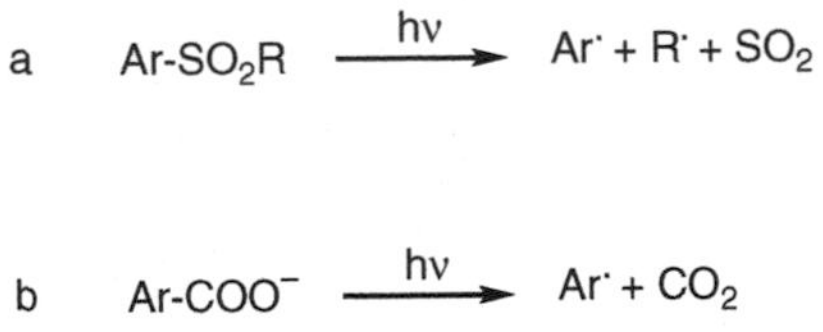

$$a \quad Ar\text{-}SO_2R \xrightarrow{\ h\nu\ } Ar^{\cdot} + R^{\cdot} + SO_2$$

$$b \quad Ar\text{-}COO^- \xrightarrow{\ h\nu\ } Ar^{\cdot} + CO_2$$

Scheme 4.19

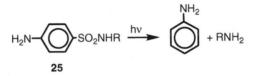

25

Scheme 4.20

4.3.3 Reactions of Nitro(hetero)aromatics

Nitro-substituted (hetero)aromatics are generally photoreactive. Two reactions occur from the $n\pi^*$ states. The first is rearrangement to nitrito group, which in turn fragments, liberating NO (Scheme 4.21). The second is hydrogen abstraction resulting in stepwise reduction to nitroso, azoxy, azo, hydrazo, and amino derivatives. The competition between the two paths may depend on conditions, as shown for flutamide (**26**) (Scheme 4.22), a non-steroidal antiandrogen drug. In water- or hydrogen-donating organic solvents, this yields phenol (**27**) via rearrangement to the nitrite

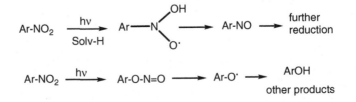

Scheme 4.21

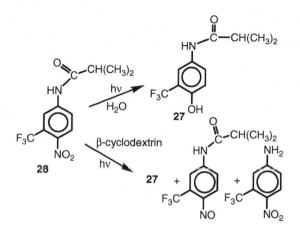

Scheme 4.22

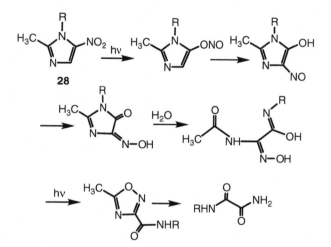

Scheme 4.23

and homolytic fragmentation. In β–cyclodextrin it reacts faster and leads to reduction of the nitro to nitroso group and to amide group hydrolysis in addition to the previous process (Vargas et al., 1998; Sortino et al., 2001). In the case of metronidazole (**28**), rearrangement to nitrito initiates a complex sequence involving shift of the NO group and ring rearrangement to yield a 1,2,4-oxadiazole-3-carboxyamide (Scheme 4.23). In turn, this product is photoreactive via ring cleavage (Wilkins et al., 1987; Pfoertner and Daly, 1987).

Reduction of the nitro group via hydrogen abstraction from the medium occurs, for example, with nitrated benzodiazepines such as the anticonvulsant nitrazepam and the hypnotic flunitrazepan. Stepwise reduction down to the amino level takes place under anaerobic conditions; different processes occur in the presence of oxygen (Busker et al., 1987; Givens et al., 1986). The more common case, however, involves

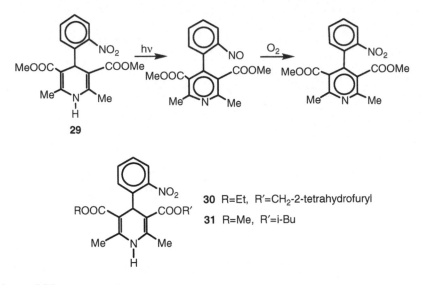

30 R=Et, R'=CH$_2$-2-tetrahydrofuryl

31 R=Me, R'=i-Bu

Scheme 4.24

intramolecular hydrogen abstraction, in particular with 4-nitrophenyl-1,4-dihydro-pyridines used as vasodilators. In solution (Pietta et al., 1981; Al-Turk et al., 1988; Sadana and Goghare, 1991) and in the solid state (Marciniec and Rychcik, 1994; Thoma and Kerker, 1992; Teraoka et al., 1999; Hayase et al., 1994), 4-(2-nitrophenyl) derivatives, e.g., nifedipine (**29**), react efficiently to yield the corresponding 2-nitrosophenylpyridine (Scheme 4.24).

The nitroso derivative may in turn be oxidized to the nitropyridine upon pro-longed irradiation or during the isolation procedure (Vargas et al., 1992). Related furnidipine (**30**) and nisoldipine (**31**) react similarly (Alvarez–Lueje et al., 1998). When the nitro group is in position 3, as in nitrendipine (**32**) (Squella et al., 1990); nimodipine (**33**) (Zanocco et al., 1992); nilvadipine (**34**) (Mielcarek et al., 2000); and nicardipine (**35**) (Bonferoni et al., 1992), the reaction is less efficient. The main products identified have undergone aromatization of the pyridine ring, but the nitro group is not reduced. Alkyl substituents may be involved in the hydrogen abstraction, as in the case of nilvadipine (see Scheme 4.25). The benzofurazanyldihydropyridine isradipine (**36**) is likewise photoreactive (Mielkarek and Daczkowska, 1999).

4.3.4   Reactions of Phenols and Anilines

Phenols are susceptible to photoinduced oxidation, as is the case for ethinyl estradiol (**37**) (Sedee et al., 1985) and estrone (**38**) (Sedee et al., 1984), which form hydro-peroxycylohexadienones (**39**) (Scheme 4.26), as well as in the analogous reaction α–tocopherol (vitamin E) (Clough et al., 1979). These reactions involve singlet oxygen but, because phenols are liable also to radicalic photo-oxidation (Segmuller et al., 2000), the distinction is not always clear. The same holds for the oxidation of polyphenols to quinones, as with adrenergics adrenaline (**40**) and isoprenaline (**41**, Scheme 4.27) (De Mol and Beijersbergen van Henegouwen, 1979; Kruk, 1985).

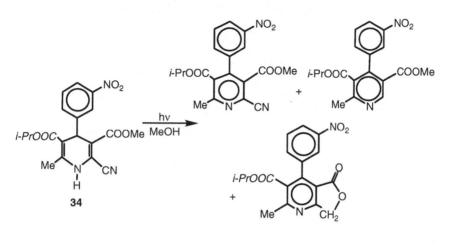

Scheme 4.25

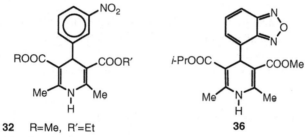

32 R=Me, R′=Et
33 R=i-Pr, R′=CH₂CH₂OCH₃
35 R=Me, R′=CH₂CH₂N(Me)CH₂Ph

36

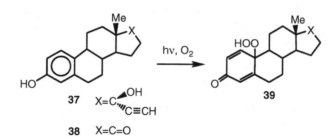

Scheme 4.26

37 X=C$\overset{OH}{\underset{C\equiv CH}{}}$
38 X=C=O

39

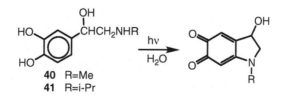

40 R=Me
41 R=i-Pr

Scheme 4.27

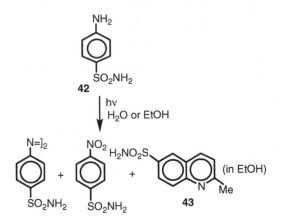

Scheme 4.28

Anilines are also liable to photo-oxidation, as shown for 5-aminosalicilic acid (Jensen et al., 1992) and sulfanilamide (**42**). The latter gives the azo and nitro derivatives in water (Pawlaczyck and Turowska, 1972); in ethanol, quinoline (**43**) is also formed through trapping of photoformed acetaldehyde (Scheme 4.28) (Reisch and Niemayer, 1972).

4.3.5 Side-Chain Fragmentation and Oxidation

The stabilization of benzylic radical and cation makes homolytic or heterolytic fragmentation at the benzylic position a common occurrence in the photochemistry of these derivatives; C–H and C–C fragmentation take place (Scheme 4.29). An example of the first case is the oxidation of the alcohol function to ketone in the anthistaminic drug terfenadine (**44**; dehydration to form a conjugated alkene also takes place, Scheme 4.30) (Chen et al., 1986). Another is the oxidation of a CH_2 to

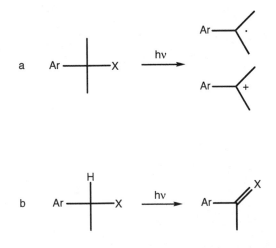

Scheme 4.29

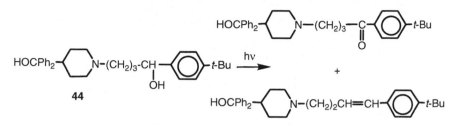

Scheme 4.30

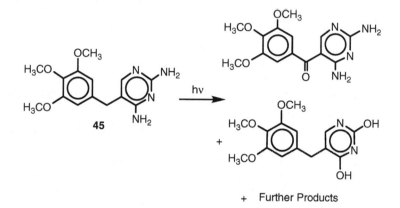

+ Further Products

Scheme 4.31

a CO group in the antibacterial trimethoprim (**45**, Scheme 4.31) (Dedola et al., 1999). Oxidation and fragmentation occur also with the antineoplastic agent methotrexate, a diaminopteridine derivative (Chatterji and Gallelli, 1978).

Benzylic C–C fragmentation typically occurs with arylacetic acids and with some phenethylamines. Arylacetic and β-arylpropionic acids are a large family of anti-inflammatory and analgesic drugs and easily decarboxylate. The benzyl radicals formed are reduced or trapped by oxygen. Typical examples are 2-phenylpropionic acids such as ibuprofen (Castell et al., 1987a); butibufen (Castell et al., 1992); and flurbiprofen, as well as the related naphthylpropionic acid naproxen (**46**, Scheme 4.32) (Moore and Chappuis, 1988).

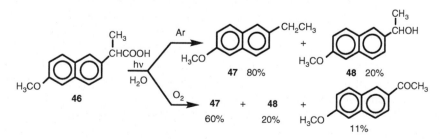

Scheme 4.32

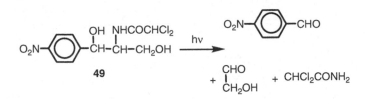

Scheme 4.33

The same process occurs when the aromatic moiety contains a (hetero)arylketone chromophore (Monti et al., 1998), as in the case of ketoprofen (Monti et al., 1997; Boscà and Miranda, 1998); suprofen (Sortino et al., 1998b); thiaprofenic acid (Encinas et al., 1998); tolmetin (Sortino et al., 1999a); and ketorolac tromethamine (Gu et al., 1988). The same is true with heterocyclic derivatives such as benoxaprofen (Castell et al., 1992; Moore et al., 1988; Navaratnam et al., 1985) and indomethacin (Vargas et al., 1991), although the latter drug is photostable in the solid state (Ekiz–Guecer et al., 1991).

As for phenethylamines, fragmentation occurs when an α–hydroxy group is present, as in the cases of adrenergic ephedrine (Usmanghani et al., 1975); adrenergic blocker labetalol (Andrisano et al., 2001a); and antibacterial chloramphenicol (**49**, Scheme 4.33) (Shih, 1971; Mubarak et al., 1983), which give the corresponding benzaldehydes via the hydroxybenzyl radicals. Related fragmentations occur in heterocycles bearing a β–aminoalkyl group, such as quinine (**50**), which undergoes cleavage of the hydroxybenzyl side chain by irradiation in an organic solvent (Epling and Yoon, 1977) and loss of the benzylic hydroxyl group to form product (**51**) in acidic aqueous solution (Scheme 4.34) (Sternberg and Travecedo, 1970); mefluoquine (Tønnesen and Grislingaas, 1990); and the ipecac alkaloid emetine (Schuijt et al., 1979). C–F bond cleavage occurs in the case of 4-trifluoromethyl-2-hydroxylbenzoic acid (the active form of the corresponding acetyl derivative, the platelet antiaggregant triflusal), with hydrolysis of the trifluoromethyl to methoxycarbonyl group (Boscà et al., 2001).

Some heteroatom-bonded substituents are liable to photoreaction. Fragmentation of the C–O σ bond in a phenyl ether has been reported for atenolol (Andrisano et al., 1999). Fenofibric acid (**52**) exhibits a medium-dependent photochemistry. With the salt, decarboxylation is accompanied by side-chain isomerization and cleavage to give products (**53** to **55**), while with the nondissociated acid reduction of the ketone function occurs (product **56**, Scheme 4.35) (Boscà and Miranda, 1999). Phenyl esters undergo the photo-Fries rearrangement, as found in the case of analgesic benorylate (**57**). This yields the hydroxyphenyl ketone (**58**), which then undergoes thermal shift of the acetyl group to the other ketone (**59**, Scheme 4.36) (Castell et al., 1987b). Analogous to the photo-Fries rearrangement is the shift of the alkyl chain from the oxygen to the nitrogen atom in position 2 occurring in the aminoalkoxybenzopyrazole benzydamine, an analgesic (Vargas et al., 1993). Cleavage of a heteroaryl-sulfur bond takes place in the immunosuppressant drug azathioprine (**60**), which undergoes formal hydrolysis of the C–S bond to give 6-mercaptopurine (of which it is considered a prodrug) and a pyrazolone. Loss of the nitro group also occurs, leading to a cyclized product (**61**, Scheme 4.37) (Hemmens and Moore, 1986).

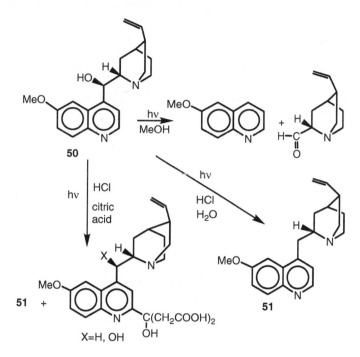

Scheme 4.34

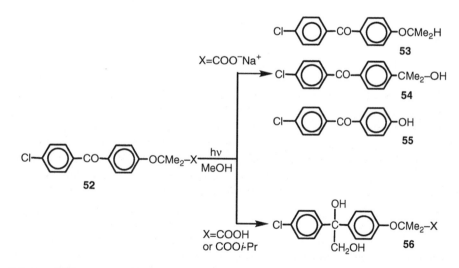

Scheme 4.35

4.3.6 Ring Rearrangement and Ring Cleavage in Heterocycles

The azetidinone ring is cleaved in penem (**62**) to give thiazole (**63**) and 3-hydroxy-butyric acid in methanolic solution; in the crystals, intramolecular hydrogen abstraction occurs and the biradical is trapped by oxygen to yield hydroperoxide (**64**) (see

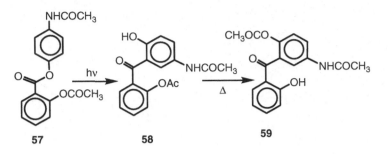

Scheme 4.36

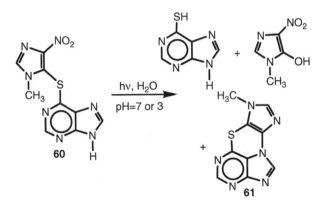

Scheme 4.37

Scheme 4.38) (Albini et al., 1995). Isoxazole–oxazole rearrangement has been observed for antibacterial sulfamethoxazole (**65**, Scheme 4.39) (Zhou and Moore, 1994). With other five-membered heterocycles, cleavage occurs. Pyrazolone analgesics, e.g., 4-aminopyrazolones, undergo cleavage of the N–N bond (Marciniec, 1983; Reisch et al., 1986a), accompanied by photo-oxidation in the presence of oxygen (Marciniec, 1985). In the solid state, intramolecular hydrogen abstraction and oxygen addition occur, as evidenced for antipyrine (Marciniec, 1983). Cleavage in solution occurs also with pyrazolo(1,2-*a*)benzo-1,2,4-triazine azapropazone (**66**, Scheme 4.40), while rearrangement via 1,3-shift of one of the C–N bonds takes place in the solid state (Reisch et al., 1989a).

    Further five-membered ring fragmentations are known for imidazolines such as vasoconstrictor naphazoline (Sortino et al., 1999b) and anticonvulsant phenyltoin (Parman et al., 1998); thiazoles as vitamin thiamine (van Dort et al., 1984); isoindolines as diuretic chlorthalidone (Vargas and Mendez, 1999); and sydnones as vasodilator molsidomide (Carney, 1987). The indole derivative melatonin is oxidized to a 2-acylformanilide (Andrisano et al., 2000).

    As for six-membered heterocycles, barbituric acid derivatives (**67**) undergo cleavage of the C-4–C-5 bond, giving an isocyanate (**68**) (spectroscopically identified in the solid state) (Jochym et al., 1988) and products resulting from it, such as ureides or urethanes via addition of external (Barton et al., 1982) or internal (as with

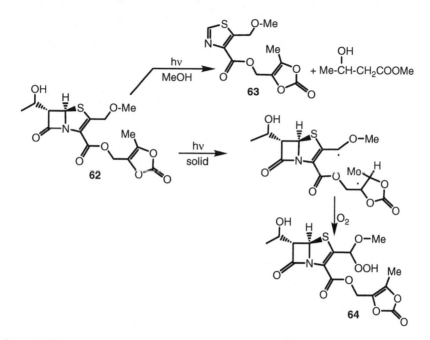

Scheme 4.38

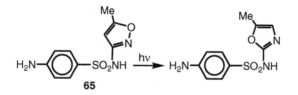

Scheme 4.39

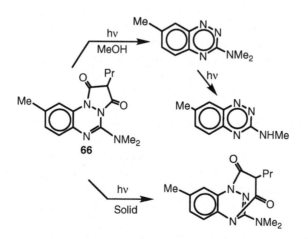

Scheme 4.40

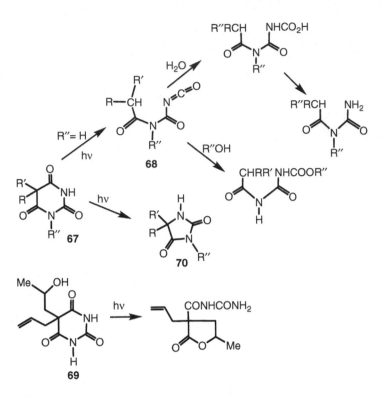

Scheme 4.41

proxibarbital, (**69**) (Mokrosz et al., 1982) nucleophile, or a hydantoin (**70**) through CO elimination (Barton et al., 1986) (Scheme 4.41).

Among seven-membered heterocycles, benzodiazepines are known to rearrange. Thus, diazepam (**71**) undergoes ring contraction (irradiation at 254 nm) to give a 1,4-dihydroquinazoline (**72**), in turn photoreactive, or ring cleavage to (**73**) (irradiation at 320 nm) (Moore, 1977) (Scheme 4.42). Contraction of the diazepine ring and cleavage of the imidazole moiety are observed with the imidazolobenzodiazepine midazolam (Andersin and Mesilaakso, 1995). Homolytic cleavage of the C–S bond in the seven-membered ring leads to ring contraction with dothiepin, a tricyclic antidepressant (Tammilehto and Torniainen, 1989).

4.3.7   Reactions of Heterocycles *N*-Oxides

Heterocycles *N*-oxides undergo photochemical rearrangement or deoxygenation. Thus, the nitrone-oxaziridine photochemical rearrangement takes place in sedative chlordiazepoxide (**74**) and the resulting fused oxaziridine (**75**) undergoes ring contraction and ring expansion to give benzoquinazoline (**76**) or benzo-1,5-diazocine (**77**), unless it is deoxygenated to the benzodiazepine (**78**) (Bakri et al., 1985) (Scheme 4.43). A different rearrangement to 4-phenylbenzodiazepin-5-one (**79**) and

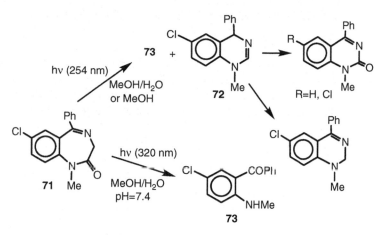

Scheme 4.42

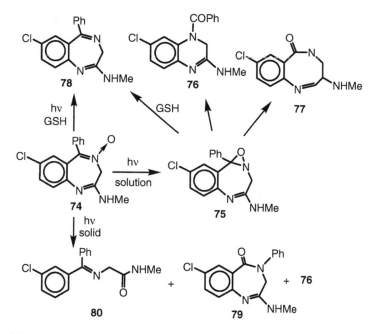

Scheme 4.43

ring cleavage (to compound **80**) occur in the solid state (Reisch et al., 1992a). Further photoreactive *N*-oxides include triaminopyrimidine 1-oxide minoxidil (Ekiz–Guecer and Reisch, 1990); 1-hydroxy-2-pyridinethione (or pyrithione) (Evans et al., 1975); and an antitumoral cyanoquinoxaline 1,4-dioxide (Fuchs et al., 1999).

4.4 CARBON–HETEROATOM DOUBLE BOND

Aldehydes and ketones are probably the most investigated class of compounds in photochemistry. Even nonconjugated derivatives absorb in the UV-B region, and conjugated compouds absorb well into UV-A. The reactive excited state is often an $n\pi^*$ state of radical character, the singlet, or more frequently, the triplet. Compounds containing a carbon–nitrogen (or sulfur) double bond often react similarly.

4.4.1 Fragmentation α to a C=O Double Bond

A typical reaction of ketones is α-cleavage (Norrish type I reaction, Scheme 4.44). This dominates when the departing alkyl radical is stabilized, e.g., by a hydroxyl group. A family of drugs containing such a feature is that of glucocorticosteroids, which bear an acyl side chain and a hydroxy or alkoxy group at C-17. These include cortisone (**81**); hydrocortisone (**82**); their acetates (**83, 84**, Scheme 4.45) (Reisch et al., 1992b); halogenated derivatives such as halomethasone (Reisch et al., 1992c); triamcinolone acetonide (**85**); fluocinolone acetonide (**86**); flumethasone; and further related drugs (Ekiz–Guecer et al., 1991). Elimination of the acyl side chain takes place in the solid and in solution (see products **87** and **88** in Scheme 4.46), at least when this chromophore absorbs significantly (irradiation at 310 nm; for different processes involving the conjugated ketone also present, see following) (Ricci et al., 2001).

Various drugs of different families contain the same feature and react accordingly, e.g., antineoplastic daunorubicin (**89**, Scheme 4.47) (Li and Chignell, 1987b); and, presumably, strictly related doxorubicin (Wood et al., 1990). Other radical-stabilizing substituents in the α-position likewise favor α-cleavage as it is the case for the

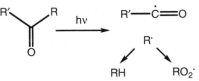

Scheme 4.44

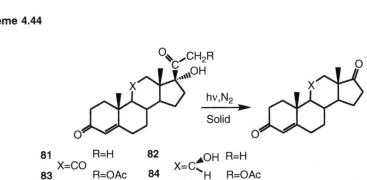

Scheme 4.45

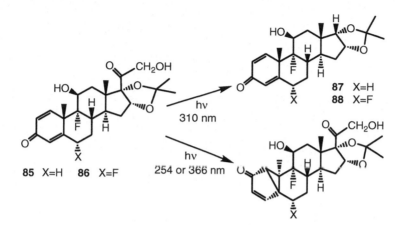

Scheme 4.46

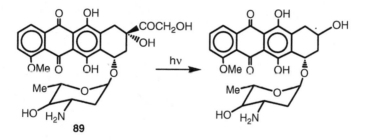

Scheme 4.47

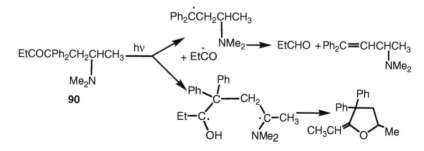

Scheme 4.48

narcotic analgesic drug methadone (**90**, Scheme 4.48) (Reisch et al., 1972; Denson et al., 1991) or for metyrapone (**91**, Scheme 4.49), where the radical pair that were formed recombine, giving in part an unsaturated compound that rapidly polymerizes (Fasani et al., 1997b). Alpha-cleavage further occurs in cyclic (particularly five-membered) ketones, e.g., androstan-17-one derivatives, which undergo epimerization at ring D (Scheme 4.50) via homolytic C–C bond cleavage and radical recombination

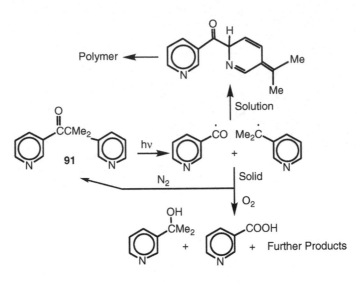

Scheme 4.49

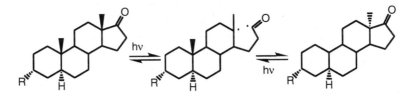

Scheme 4.50

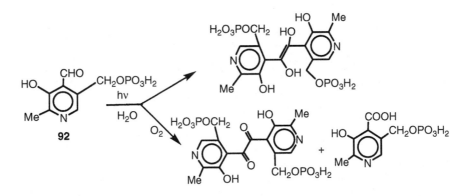

Scheme 4.51

(Wu et al., 1992). A particular case of α-cleavage is that concerning the C–H bond of aldehydes, which causes oxidation in the presence of oxygen and other reactions, e.g., coupling, in deoxygenated solution. This is shown in the case of pyridoxal-5-phosphate (**92**, Scheme 4.51) (Bazkulina et al., 1974).

4.4.2 Hydrogen Abstraction by Carbonyls

The other general reaction is hydrogen abstraction. This process occurs intermolecularly, thus forming a ketyl radical (Scheme 4.52a) and finally leading to reduction
of the carbonyl function, or intramolecularly, giving a biradical. In turn, the latter
fragments (Norrish type II reaction) or cyclizes (Yang cyclization, Scheme 4.52b).
Intermolecular hydrogen abstraction is frequent among diaryl ketones. An example
is the photo-oxidation of the antipsoriatic anthrone dithranol (**93**) to the corresponding
anthraquinone and dianthrone derivatives, probably occurring via hydrogen abstraction from the benzylic position (Scheme 4.53) (Thoma and Holzmann, 1998). Hydrogen abstraction has been observed in the solid state for various steroids, where it is
guided by the crystal conformation. The radicals give dimers, as reported for testosterone propionate (Reisch et al., 1989b); progestinic levonorgestrel (Reisch et al.,
1994); ethisterone; and norethisterone (Reisch et al., 1993); or disproportionate as
with testosterone or fragment as with 17α-methyltestosterone (Reisch et al., 1989c).

In many molecules, intramolecular hydrogen abstraction is favored by the conformation and occurs smoothly. The diradical formed may disproportionate, add

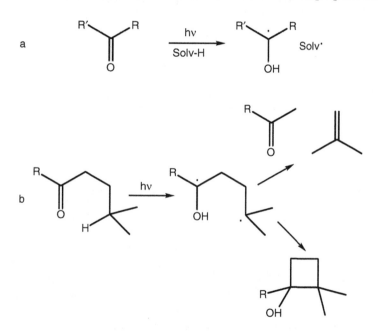

Scheme 4.52

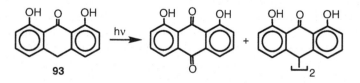

Scheme 4.53

oxygen, or further fragment, as in the case of the antitumor antibiotics hedamycin and kidamycin (Fredenhagen et al., 1985) or for methadone (see Scheme 4.48). A similar mechanism probably operates in the side chain oxidation taking place in the cyclohexenyl barbituric cyclobarbital (**94**, Scheme 4.54) (Bouche et al., 1978) and in the side chain elimination in the other barbiturics pentobarbital and secobarbital (Barton and Bojarski, 1983).

Alternatively, radical coupling ensues with formation of a new ring through a C–C bond. This has been reported for 10-(dimethoxymethyl)testosterone acetate (**95**), which yields epimeric bridged acetals (Karvas, 1974), and for some pregnan-20-ones (**96**, Scheme 4.55), which give hydroxycyclobutanes (Wehrli et al., 1980). Again, an intramolecular bond is formed by irradiation of taxicin diacetate in which the allylic hydrogen lies close to the enone chromophore (Kobayashi et al., 1972).

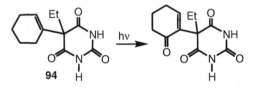

94

Scheme 4.54

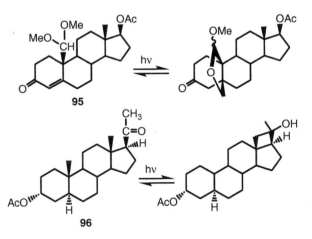

95

96

Scheme 4.55

4.4.3   Rearrangement of Conjugated Carbonyls

Cross-conjugated cyclohexenones and cyclohexadienones undergo the stereospecific "lumiketone" photorearrangement to bicyclo[3.1.0]hexanones (or hexenones, Scheme 4.56). These moieties are present in ring A in many steroids, which rearrange accordingly. An example is testosterone acetate (Bellus et al., 1969), for which rearrangement is accompanied by other processes such as cycloaddition and hydrogen abstraction. The product distribution depends on conditions because of the role of secondary photoprocesses and the easy occurrence of (acid-catalyzed) nucleophilic

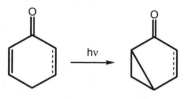

Scheme 4.56

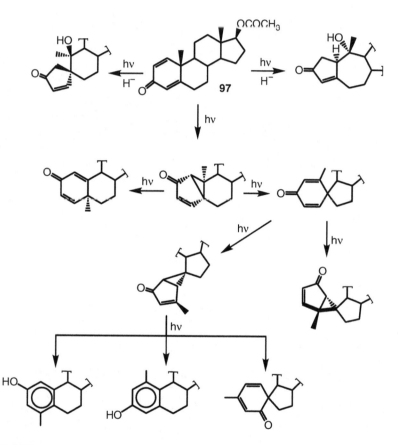

Scheme 4.57

trapping. An example of the complexity of path followed is shown for the case of 1-dehydrotestosterone acetate (**97**, Scheme 4.57) (Frei et al., 1966).

Many further steroidal dienones have been found to undergo similar photoreactions (Arakawa et al., 1985; Hidaka et al., 1980), including prednisone (Williams et al., 1979); prednisolone (Williams et al., 1980); dexamethasone (Reisch et al., 1986b); bethamethasone and some of its acetates (Thoma et al., 1987); and methyl prednisolone suleptanate (**98**, Scheme 4.58) (Ogata et al., 1998). In other families, fragmentation occurs with some anti-inflammatory drugs (**85, 86**, Scheme 4.46). Note

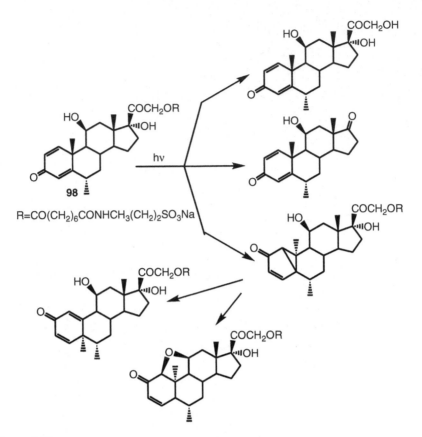

Scheme 4.58

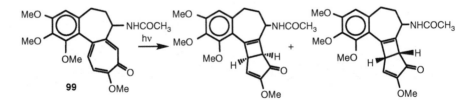

Scheme 4.59

that a wavelength-dependent photochemistry occurs: as established for the last derivatives, the cyclohexadienone in ring A absorbs exclusively and undergoes the lumiketone rearrangement at 254 and 360 nm, while at 310 nm the side-chain ketone is involved (Ricci et al., 2001). Cycloheptatrienones undergo a different photorearrangement, i.e., cyclization to cyclobutencyclopentenones, as observed in the case of alkaloid colchicine (**99**, Scheme 4.59), used as an antineoplastic agent (Nery et al., 2001).

4.4.4 Reactions of Compounds Containing a C=N or C=S Double Bond

Oximes, hydrazones, hydrazides, and related derivatives are subject to geometric isomerization or cleavage by irradiation (Scheme 4.60). Photoinduced *syn*-anti-isomerization occurs with oximes aztreonam (Fabre et al., 1992) and cefotaxime (Lerner et al., 1988); hydrazone (Z) 2-amino-5-chlorobenzophenoneamidinohydra-zone acetate (Schleuder et al., 1993); and semicarbazone nitrofurazone (**100**) (Quil-liam et al., 1987). Isomerization is a fast process and is accompanied by cleavage upon prolonged irradiation (see product **101**). Fragmentation has been observed with acylhydrazones nitrofurantoin (**102**) (Narbutt–Mering et al., 1980) and furazolidone (**103**, Scheme 4.61); triazene dacarbazine (Horton and Stevens, 1981); and hydrazide isoniazide (Ninomiya and Yamamoto, 1976). As for sulfurated π–functions, the reduction of the thioureido group has been observed in the irradiation in alcohols of 2-thiophenobarbital (**104**, Scheme 4.62) (Paluchowska and Bojarski, 1988).

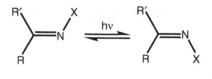

Scheme 4.60

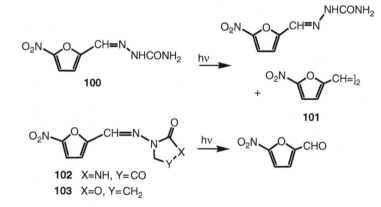

Scheme 4.61

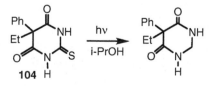

Scheme 4.62

4.5 OTHER PHOTOREACTIVE MOIETIES

The inducement of a photoreaction by the amino group, e.g., benzylic cleavage, has been mentioned earlier. Some reactions more directly concerning the amino function are discussed next, along with those (in part analogous) of sulfides and thioethers. A few relevant photoreactions of inorganic complexes are added.

4.5.1 Amines

Amino groups often undergo photochemical oxidation at the nitrogen or α-carbon or cleavage of the C–N bond. Thus, chlorpromazine and thioridazine give the *N*-oxides (**22**, see Scheme 4.16). Oxidation to amide is exemplified by the oxidation of one piperidino group in the anticoagulant dipyridamole (**105**, Scheme 4.63) (Kigasawa et al., 1984) and in the oxidative cleavage of the side chain in ofloxacin (**14**, see Scheme 4.13). Oxidative degradation of an aminoalkyl chain is observed with several aminoquinoline antimalarials, such as chloroquine (Nord et al., 1991); hydroxychloroquine (Tønnesen et al., 1988); amodiaquine (Owoyale, 1989); and primaquine (**106**) (Kristensen et al., 1998) (Scheme 4.64).

Different mechanisms are involved via singlet oxygen, superoxide anion, or hydroxyl radical, and product distribution depends on conditions (Motten et al., 1999; Navaratnam et al., 2000b). Related reactions take place with tetracycline (**107**) under oxygen (Scheme 4.65) (Davies et al., 1979; Moore et al., 1983b), with the antiarrhythmic drug mexiletine (the same products are also obtained by chemical oxidation) (Takacs et al., 2000) and with folic acid (Akhtar et al., 1999). *N*-Dealkylation has been observed with several heterocycles, such as the *N*-alkylphenothiazine

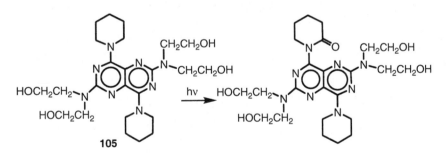

105

Scheme 4.63

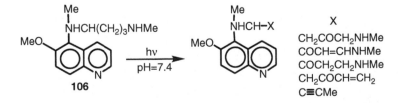

106

Scheme 4.64

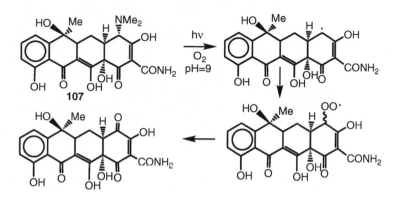

Scheme 4.65

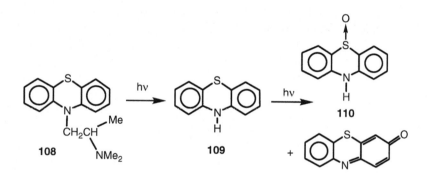

Scheme 4.66

promethazine (**108**, Scheme 4.66) (Hashiba et al., 1979) and vitamin riboflavin (Cairns and Metzler, 1971; Ahmad and Rapson, 1990; Moore et al., 1994).

4.5.2 Sulfur-Containing Compounds

Thiols are often oxidized to disulfides via a photoinitiated radical path, as with thiorfan (**111**, Scheme 4.67) (Gimenez et al., 1988), or up to sulfinate and sulfonate as with the antineoplastic 6-mercaptopurine (Hemmens and Moore, 1984). Both alkyl and aryl sulfides are photooxidized to sulfoxides, as in the formation of S-oxides (**110**) from dibenzothiazepin (**109**, Scheme 4.66), as well as (**21**) from chlorpromazine (**20**, Scheme 4.16) and similarly from thioridazine (Elisei et al., 2002) and diltiazem (Andrisano et al., 2001b).

$$PhCH_2CHCONHCH_2CO_2H \xrightarrow{\ h\nu\ } PhCH_2-CH-CONHCH_2CO_2H$$
$$\underset{SH}{|} \qquad\qquad\qquad \underset{S]_2}{|}$$
$$\mathbf{111}$$

Scheme 4.67

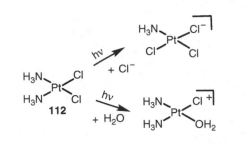

$$[Fe(NO)(CN)_5]^{2-} + 2H_2O \xrightarrow{h\nu} [Fe(H_2O)(CN)_5]^{3-} + 2H^+ + NO_2^-$$

113

Scheme 4.68

4.5.3 Inorganic Complexes

Solvation is often observed upon irradiation of inorganic complexes such as the antineoplastic cisplatin (**112**) (Zieske et al., 1991) and sodium nitroprusside (**113**, Scheme 4.68) (Zhelyaskov and Godwin, 1998).

4.6 CONCLUSIONS

The examples presented in this chapter cover a large variety of drug molecules and identify the main photoreaction paths characteristic of the main chromophores. Nonabsorbing molecules, e.g., aliphatic derivatives, do not react unless a sensitizer (which may be an impurity) promotes the reaction (typical examples are the oxidation of alkyl sulfides or of simple alkenes). Thus, predicting the photoreactivity pattern of new molecules is possible. Predicting how efficient the photoreaction will be requires further evaluation because the excited state partitions between the chemical paths and decay to the ground state. The quantum yield (number of molecules transformed per photon absorbed) depends on structure and on conditions that may further affect the competition between different photoreactions. A few examples of such effects have been given and many more are available in the literature. Such literature should be examined as early as possible when considering a new drug, or at least whenever a photoinduced process shows up during preparation, storage, or usage time of a drug, or when phototoxic effects are reported.

AKNOWLEDGMENT

This review has been prepared in the frame of a PRIN project sponsored by MIUR, Rome.

REFERENCES

Ahmad, I. and Rapson, H.D.C., 1990, Multicomponent spectrophotometric assay of riboflavine and photoproducts, *J. Pharm. Biomed. Anal.* 8, 217–223.

Akhtar, M.J., Khan, M.A., and Ahmad, I., 1999, Photodegradation of folic acid in aqueous solution, *J. Pharm. Biomed. Anal.* 19, 269–275.

Albini, A., Alpegiani, M., Borghi, D., Del Nero, S., Fasani, E., and Perrone, E., 1995, Solid state photoreactivity of a dioxolenonemethyl ester, *Tetrahedron Lett.* 36, 4633–4636.

Albini, A. and Fasani, E. (Eds.), 1998, *Drugs. Photochemistry and Photostability*. Cambridge: The Royal Society of Chemistry.

Al-Turk, W.A., Mjeed, I.A., Murray, W.J., Newton, D.W., and Othman, S., 1988, Some factors affecting the photodecomposition of nifedipine, *Int. J. Pharm.* 41, 227–230.

Alvarez–Lueje, A., Naranjo, L., Nunez–Vergara, L.J., and Squella, J.A., 1998, Electrochemical study of nisoldipine: analytical application in pharmaceutical forms and photodegradation, *J. Pharm. Biomed. Anal.* 16, 853–862.

Andersin, R. and Mesilaakso, M., 1985, Structure elucidation of 6-chloro-2-methyl-4(1H)-quinazolinone, a photodecomposition product of midazolam, *J. Pharm. Biomed. Anal.*, 13, 667–670.

Andrisano, V., Gotti, R., Leoni, A., and Cavrini, V., 1999, Photodegradation studies of atenolol by liquid chromatography, *J. Pharm. Biomed. Anal.* 21, 851–857.

Andrisano, V., Bertucci, V., Battaglia, A., and Cavrini, V., 2000, Photodegradation of melatonin and its determination in commercial formulations, *J. Pharm. Biomed. Anal.* 23, 15–23.

Andrisano, V., Ballardini, R., Hrelia, P., Cameli, N., Tosti, A., Gotti, R., and Cavrini, V., 2001a, Studies on the photostability and *in vitro* phototoxicity of labetalol, *Eur. J. Pharm. Sci.* 12, 495–504.

Andrisano, V., Hrelia, P., Gotti, R., Leoni, A., and Cavrini, V., 2001b, Photostability and phototoxicity studies on diltiazem, *J. Pharm. Biomed. Anal.*, 25, 589–597.

Arakawa, Y., Fukaya, C., Yamaouchi, K., and Yokoama, K., 1985, Photoreaction of desamethasone 21-acetate, *Yakugaku Zasshi* 105, 1029–1033.

Bakri, A., Beijersbergen van Henegouwen, G.M.J., and Chanal, J.L., 1985, Involvement of the N4-oxide group in the phototoxicity of chlordiazepoxide in the rat *Photodermatology* 2, 205–212.

Barton, H., Bojarski, J., and Mokrowsz, J., 1982, Photochemical ring opening of barbital, *Tetrahedron Lett.* 23, 2133–2134.

Barton, H. and Bojarski, J., 1983, Products of photolysis of 5-allyl-5-(1-methylbutyl)barbituric acid — secobarbital, *Pharmazie* 38, 630–631.

Barton, H., Bojarski, J., and Zurowska, A., 1986, Stereospecificity of the photoinduced conversion of methylphenobarbital to meptienytoin, *Arch. Pharm. (Weinheim)* 319, 457–461.

Bazkulina, N.P., Kirpichnikov, M.P., Morozov, Y.V., Savin, F.A., Sinyavina, L.B., and Florentiev, V.L., 1974, Photochemistry of the aldehyde form of pyridoxal, pyridoxal-5′-phosphate and their derivatives, *Mol. Photochem.* 6, 367.

Bellus, D., Kearns, D.R., and Schaffner, K., 1969, Photochemistry of α,β-unsaturated cyclic ketones: specific reaction of $n\pi^*$ and $\pi\pi^*$ triplet states of O-acetyltestosterone and 10-methyl-$\Delta^{1,9}$-octalone(2), *Helv. Chim. Acta* 52, 971–1009.

Bilski, P., Martinez, L.J., Koker, E.B., and Chignell, C.F., 1998, Effect of solvent polarity and proticity on the photochemical properties of norfloxacin, *Photochem. Photobiol.* 68, 20–24.

Bonferoni, M.C., Mellerio, G., Giunchedi, P., Caramella, C., and Conte, U., 1992, Photostability evalutation of nicardipine HCl solutions, *Int. J. Pharm.* 80, 109–117.

Boscà, F. and Miranda, M.A., 1998, Photosensitizing drugs containing the benzophenone chromophore, *J. Photochem. Photobiol.* 43, 1–26.

Boscà, F. and Miranda, M.A., 1999, A laser flash photolysis study on phenofibric acid, *Photochem. Photobiol.* 70, 853–857.

Boscà, F., Cuquerella, M.C., Marin, M.L., and Miranda, M.A., 2001, Photochemistry of 2-hydroxy-4-trifluoromethylbenzoic acid, major metabolite of the photosensible platelet antiaggregant drug triflunisal, *Photochem. Photobiol.* 73, 463–468.

Bouche, R., Draguet–Broughmans, M., Flandre, J.P., Moreaux, C., and Van Meersche, M., 1978, Ketonic oxidation products of cyclobarbital, *J. Pharm. Sci.* 67, 1019–1022.

Busker, R.W., Beiersbergen van Henegouwen, G.M.J., Kwee, B.M.C., and Winkens, J.H.M., 1987, Photobinding of flunitrazepam and its major photodecomposition product N-desmethylflunitrazepam, *Int. J. Pharm.* 36, 113.

Cairns, W. and Metzler, D.E., 1971, Photodegradation of flavin. IV. A new photoproduct and its use in the study of the photochemical mechanism, *J. Am. Chem. Soc.* 93, 2772–2777.

Carney, C.F., 1987, Solution stability of cyclosidomine, *J. Pharm. Sci.* 76, 393–397.

Castell, J.V., Gomez–Lechon, M.J., Miranda, M.A., and Morea, I.M., 1987a, Photolytic degradation of ibuprofen, *Photochem. Photobiol.* 46, 991–996.

Castell, J.V., Gomez–Lechon, M.J., Mirabet, V., Miranda, M.A., and Morea, I.M., 1987b, Photolytic degradation of benorylate: effect of the photoproducts on cultured hepato-cytes, *J. Pharm. Sci.* 76, 374–378.

Castell, J.V., Gomez–Lechon, M.J., Miranda, M.A., and Morea, I.M., 1992, Phototoxicity of non-steroidal anti-inflammatory drugs: *in vitro* testing of the photoproducts of buti-bufen and flurbiprofen, *J. Photochem. Photobiol. B-Biol.* 13, 71–81.

Chatterji, D.J. and Gallelli, J.F., 1978, Thermal and photolytic decomposition of methotrexate in aqueous solutions, *J. Pharm. Sci.*, 67, 526–531.

Chen, T.M., Sill, A.D., and Housmyer, C.L., 1986, Solution stability studies of terfenadine by high-performance liquid chromatography, *J. Pharm. Biomed. Anal.* 4, 533–539.

Clough, R.L., Yee, B.G., and Foote, C.S., 1979, Chemistry of singlet oxygen. 30. Unstable primary products of tocopherol photo-oxidation, *J. Am. Chem. Soc.* 101, 683–686.

Condorelli, G., De Guidi, G., Giuffrida, S., Sortino, S., Chillemi, R., and Sciuto, S., 1999, Photochemistry and photophysics of rufloxacin: an unusual photodegradation path for an antibacterial containing a fluoroquinolone-like chromophore, *Photochem. Photobiol.* 70, 280–286.

Curley, R.W. and Fowble, J.W., 1988, Photoisomerization of retinoic acid and its photopro-tection in physiologic-like solutions, *Photochem. Photobiol.* 47, 831–835.

Davies, A.K., Hilal, N.S., McKellar, J.F., and Phillips, G.O., 1975, Photochemistry of tetra-chlorosalicylanilide and its relevance to the persistent light reaction, *Brit. J. Dermatol.* 92, 143–147.

Davies, A.K., Navaratnam, S., and Phillips, G.O., 1976, Photolysis of chlorpromazine in propan-2-ol solution, *J. Chem. Soc. Perkin Trans.* 2, 25.

Davies, A.K., McKellar, J.F., Phillips, G.O., and Reid, A.G., 1979, Photochemical oxidation of tetracycline in aqueous solution, *J. Chem. Soc. Perkin Trans.* 2, 369–375.

Dedola, G., Fasani, E., and Albini, A., 1999, Photochemistry of trimethoprim in solution, *J. Photochem. Photobiol. A-Chem.* 123, 47–51.

De Mol, N.J. and Beijersbergen van Henegouwen, G.M.J., 1979, The extent of aminochrome formation from adrenaline, isoprenaline and noradrenaline induced by ultraviolet light, *Photochem. Photobiol.* 29, 479–482.

Denson, D.D., Crews, J.C., Grummich, K.W., Stirm, E.J., and Sue, C.A., Stability of methadone hydrochloride in 0.9% sodium chloride injection in single-dose plastic containers, 1991, *Am. J. Hosp. Pharm.* 48, 515–517.

Detzer, N. and Huber, B., 1975, Photolysis and thermolysis of nalidixic acid, *Tetrahedron* 31, 1937–1941.

Doyle, T.D., Benson, W.R., and Filipescu, N., 1978, Spectrophotometrical studies of dienestrol photoisomerization, *Photochem. Photobiol.* 27, 3–8.

Ekiz–Guecer, N. and Reisch, J., 1990, Photostability of minoxidil in the liquid and solid state, *Acta Pharm. Turc.* 32, 103–106.

Ekiz–Guecer, N., Reisch, J., and Nolte, G., 1991, Photostability of cross-conjugated gluco-corticosteroids in the crystal state, *Eur. J. Pharm. Biopharm.* 37, 234–237.

Ekiz–Guecer, N. and Reisch, J., 1991, Photostability of indomethacin in the crystal state, *Pharm. Acta Helv.* 66, 66–67.

Elisei, F., Latterini, L., Aloisi, G.G., Mazzucato, U., Viola, G., Miolo, G., Vedaldi, D., and Dall'Acqua, F., 2002, Excited state properties and *in vitro* phototoxicity studies of three phenothiazine derivatives, *Photochem. Photobiol.*, 75, 11–21.

Encinas, S.M., Miranda, M.A., Marconi, G., and Monti, S., 1998, Transient species in the photochemistry of tiaprofenic acid and its decarboxylation photoproducts, *Photochem. Photobiol.* 68, 633–639.

Engler, M., Ruensing, G., Soergel, F., and Holzgrabe, U., 1998, Defluorinated sparfloxacin as a new photoproduct identified by liquid chromatography coupled with UV detection and tandem mass spectrometry, *Antimicrob. Agents Chemother.* 42, 1151–1159.

Epling, G.A. and Yoon, U.C., 1977, Photolysis of cinchona alkaloids. Photochemical degradation to 5-vinylquinuclidine-2-carboxyaldehyde, a precursor to synthetic antimalarials, *Tetrahedron Lett.*, 2471–2474.

Evans, P.G., Sugden, J.K., and Van Abbe, N.J., 1975, Aspects of the photochemical behavior of 1-hydroxypyridine-1-thione, *Pharm. Acta Helv.* 50, 94–99.

Fabre, H., Ibork, H., and Lerner, D.A., 1992, Photodegradation kinetics under UV light of aztreonam solutions, *J. Pharm. Biomed. Anal.* 10, 645–650.

Fasani, E., Mella, M., Caccia, D., Tassi, S., Fagnoni, M., and Albini, A., 1997a, The photochemistry of lomefloxacin. An aromatic carbene as the key intermediate in the photodecomposition, *J. Chem. Soc., Chem. Commun.* 1329–1330.

Fasani, E., Mella, M., and Albini, A., 1997b, Photochemistry of metyrapone in the solid state, *Chem. Pharm. Bull.* 45, 394–396.

Fasani, E., Barberis Negra, F.F., Mella, M., Monti, S., and Albini, A., 1999, Photoinduced C–F bond cleavage in some fluorinated 7-amino-4-quinolone-3-carboxylic acids, *J. Org. Chem.* 64, 5388–5395.

Fasani, E., Mella, M., Monti, S., and Albini, A., 2001, Unexpected photoreactions of some 6-fluoro-7-aminoquinolones in phosphate buffer, *Eur. J. Org. Chem.* 391–397.

Fredenhagen, A. and Sequin, U., 1985, The structure of some products from the photodegradation of pluramycin antibiotics hedamycin and kidamycin, *Helv. Chim. Acta* 68, 391–402.

Frei, F., Ganter, C., Kaegi, K., Kocsis, K., Miljkovic, M., Siewinski, A., Wenger, R., Schaffner, K., and Jeger, O., 1966, Photoisomerization of 3-oxo-steroids in dioxane solution. Structure identification of the photoisomers and establishing the rearrangement sequence, *Helv. Chim. Acta* 45, 1049–1105.

Frith, R.G. and Phillipou, G., 1986, Applications of clomiphene photolysis to assay based on the derived phenanthrenes, *J. Chromatogr.* 367, 260–266.

Fuchs, T., Gates, K.S., Kwang, J.T., and Greenberg, M.M., 1999, Photosensitization of guanine specific DNA damage by a cyano-substituted quinoxaline di-*N*-oxide, *Chem. Res. Toxicol.* 12, 1190–1194.

Gasparro, F.P. and Kochevar, I.E., 1982, Investigation of protriptyline photoproducts which cause cell membrane disruption, *Photochem. Photobiol.* 35, 351–358.

Gimenez, F., Postaire, E., Prognon, P., LeHoang, M.D., Lecomte, J.M., Pradeau, D., and Hazebroucq, G., 1988, Study of thiorfan degradation, *Int. J. Pharm.* 43, 23–30.

Givens, R.S., Gingrich, J., and Meklemburg, S., 1986. Photochemistry of flunitrazepam: a product and model study, *Int. J. Pharm.* 29, 67–72.

Golpashin, F., Weiss, B., and Durr, H., 1984, Photochemical model studies on skin photosensitizing drugs: sulfonamides and sulfonylureas, *Arch. Pharm. (Weinheim)* 317, 906–913.

Greenhill, J.V., in Swarbrick, J. and Boylan, J.C. (Eds.), 1995, *Encyclopedia of Pharmaceutical Technology.* New York: Marcel Dekker, Vol. 12, pp. 105–135.

Greenhill, J.V., in Tønnesen, H.H. (Ed.), 1996, *Photostability of Drugs and Drug Formulations.* London: Taylor & Francis, pp. 83–110.

Gu, L., Chiang, H.S., and Johnson, D., 1988, Light degradation of ketorolac tromethamine, *Int. J. Pharm.* 41, 105–113.

Hashiba, S., Tatsuzawa, M., and Eijma, A., 1979, Photodegradation of chlorpromazine hydrochloride and promethazine hydrochloride, *Eisei Shikensko Hokoku* 97, 73–78, through *Chem. Abstr.* 93, 16854n (1980).

Hayase, N., Itagaki, Y., Ogawawa, S., Akutsu, S., Inagaki, S., and Abiko, Y., 1994, Newly discovered photodegradation products of nifedipine in hospital prescriptions, *J. Pharm. Sci.* 83, 532–538.

Hemmens, V.J. and Moore, D.E., 1984, Photo-oxidation of 6-mercaptopurine, *J. Chem. Soc. Perkin Trans.* 2, 209–211.

Hemmens, V.G. and Moore, D.E., 1986, Photochemical sensitization by azathioprine and its metabolites. II. Azathioprine and nitroimidazole metabolites, *Photochem. Photobiol.* 43, 257–262.

Hidaka, T., Huruumi, S., Tamaki, S., Shiraishi, M., and Minato, H., 1980, Behavior of betamethasone in acidic and alkaline medium, photolysis and oxidation, *Yakugaku Zasshi* 100, 72–80.

Horton, J.K. and Stevens, M.F.G., 1981, A new light on the photodecomposition of the antitumor drug DTIC, *J. Pharm. Pharmacol.* 33, 808–811.

Jacobs, H.C.J. and Havinga, E., 1979, Photochemistry of vitamin D and its isomers and of simple trienes, *Adv. Photochem.* 11, 305–373.

Jensen, J., Cornett, C., Olsen, C.E., Tjoernelund, J., and Hansen, S.H., 1992, Identification of major degradation products of 5-aminosalicylic acid formed in aqueous solution, *Int. J. Pharm.* 88, 177–187.

Jochym, K., Barton, H., and Bojarski, J., 1988, Photolysis of sodium salts of barbiturates in the solid state, *Pharmazie* 43, 623–624.

Jones, G.E. and Sharples, D., 1984, Structure elucidation of protriptyline photoirradiation products, *J. Pharm. Pharmacol.* 36, 46–48.

Karliceck, R. and Klimesova, V., 1977, Study of the effect of radiation on the stability of ergot alkaloids, *Cesk. Farm.* 26, 413–416 through *Chem. Abst.*, 89, 12047 (1978).

Karvas, M., Marti, F., Wehrli, H., Schaffner, K., and Jeger, O., 1974, Application of $\pi\pi^*$ induced cyclization of α,β-unsaturated-γ-dimethoxymethylketones to 19-methoxy and 19-dimethoxy steroid enones and dienones, *Helv. Chim. Acta* 57, 1851–1859.

Kigasawa, K., Shimizu, H., Hayashida, S., and Ohkubo, K., 1984, Photodecomposition and stabilization of dipyridamol, *Yakagaku Zasshi* 104, 1191–1197.

Kobayashi, T., Kurono, M., Sato, H., and Nakanishi, K., 1972, Nature of photoinduced transannular H abstraction of taxinines, *J. Am. Chem. Soc.* 94, 2863–2865.

Kristensen, S., Nord, K., Orsteen, A.L., and Tønnesen, H.H., 1998, Influence of oxygen on light induced reaction of primaquine, *Pharmazie* 53, 98–103.

Kruk, I., 1985, The identification by electron spin resonance spectroscopy of singlet oxygen formed in the photo-oxidation of catecholamines, *Z. Phys. Chem. (Leipzig)* 226, 1239–1242.

Lerner, D.A., Bannefond, G., Fabre, H., Mandou, B., and Simeon de Buochberg, M., 1988, Photodegradation paths of cefotaxime, *J. Pharm. Sci.* 77, 699–703.

Li, A.S.W. and Chignell, C.F., 1987a, A spin trapping study of the photolysis of amiodarone and desethylamiodarone, *Photochem. Photobiol.* 45, 191–197.

Li, A.S.W. and Chignell, C.F., 1987b, Photolysis of chlorpromazine metabolites: a spin trapping study, *Photochem. Photobiol.* 45, 695–701.

Li, Y.N., Moore, D.E., and Tattam, B.N., 1999, Photodegradation of amiloride in aqueous solution, *Int. J. Pharm.* 183, 109–116.

Lupon, P., Grau, F., and Bonet, J.J., 1984, Photooxygenation of $\Delta^{9(11)}$ dehydroestrone and its methyl ether, *Helv. Chim. Acta* 67, 332.

Marciniec, B., 1983, Photochemical decomposition of phenazone derivatives. Kinetics of decomposition in the solid state, *Pharmazie* 38, 848–850.

Marciniec, B., 1985, Photochemical decomposition of phenazone derivatives. Kinetics of photolysis and photo-oxidation in solution, *Acta Pol. Pharm.* 42, 448–454.

Marciniec, B. and Witkowska, D., 1988, Photodecomposition of vitamin K4 in solid state, *Acta Pol. Pharm.* 45, 528–534.

Marciniec, B. and Rychcik, W., 1994, Kinetic analysis of nifedipine photodegradation in the solid state, *Pharmazie* 49, 894–897.

Martinez, L.J., Li, G., and Chignell, C.F., 1997, Photogeneration of fluoride by the fluoro-quinolone antimicrobial agents lomefloxacin and fleroxacin, *Photochem. Photobiol.* 65, 599–561.

Mella, M., Fasani, E., and Albini, A., 2001, Photochemistry of ciprofloxacin in aqueous solution, *Helv. Chim. Acta* 84, 2508–2519.

Mendenhall, D.W., Kobayashi, H., Shih, F.M., Sternson, L.A.T., Higuchi, L.A., and Fabian, C., 1978, Clinical analysis of tamoxifen, an antineoplastic agent, in plasma, *Clin. Chem.* 1518–1524.

Mermet–Bouvier, R., 1973, Photochemistry of vitamin D2, *Bull. Soc. Chim. Fr.* 3023.

Mielcarek, J. and Daczkowska, S., 1999, Photodegradation of inclusion complexes of israd-ipine with methyl-β-cyclodextrin, *J. Pharm. Biomed. Anal.* 21, 393–398.

Mielcarek, J., Stobiecki, M., and Franski, R., 2000, Identification of photodegradation prod-ucts of nivaldipine using GC-MS, *J. Pharm. Biomed. Anal.* 24, 71–79.

Mokrosz, J., Zurowska, A., and Bojarski, J., 1982, Kinetics and TLC investigations of pho-tolysis of proxibarbital, *Pharmazie* 37, 832–835.

Monti, S., Sortino, S., De Guidi, G., and Marconi, G., 1997, Photochemistry of 2-(3-benzoyl-phenyl)propionic acid (ketoprofen). I. A picosecond and nanosecond time resolved study in aqueous solution, *J. Chem. Soc. Faraday Trans.* 93, 2269–2275.

Monti, S., Sortino, S., Fasani, E., and Albini, A., 2001, Multifaceted reactivity of 6-fluoro-7-aminoquinolones from the lowest excited states in aqueous media: a study by nano-second and picosecond spectroscopy, *Chem. Eur. J.* 7, 2185–2196.

Monti, S., Sortino, S., Encinas, S., Marconi, G., De Guidi, G., and Miranda, M.A. in Albini, A. and Fasani, E. (Eds.), 1998, *Drugs. Photochemistry and Photostability*. Cambridge: The Royal Society of Chemistry, p. 150–161.

Moore, D.E., 1977, Photosensitization by drugs, *J. Pharm. Sci.* 66, 1282–1284.

Moore, D.E. and Tamat, S.R., 1980, Photosensitization by drugs: photolysis of some chlorine-containing drugs, *J. Pharm. Pharmacol.* 32, 172–177.

Moore, D.E. and Sithipitaks, V., 1983a, Photolytic degradation of frusemide, *J. Pharm. Pharmacol.* 35, 489–493.

Moore, D.E., Fallon, M.P., and Burt, C.D., 1983, Photo-oxidation of tetracycline: a differential pulse polarographic study, *Int. J. Pharm.* 14, 133–142.

Moore, D.E. and Chappuis, P.P., 1988, A comparative study of the photochemistry of the non-steroidal anti-inflammatory drugs naproxen, benoxaprofen and indomethacin, *Photochem. Photobiol.* 47, 173–180.

Moore, D.E., Sik, R.H., Bilski, P., Chignell, C.F., and Reszka, K.J., 1994, A direct epr and spin trapping study of light-induced free radicals from 6-mercaptopurine and its oxidation products, *Photochem. Photobiol.* 60, 574–581.

Morimura, T., Kohno, K., Nobuhara, Y., and Matsukura, H., 1997a, Photoreaction and active oxygen generation by photosensitization of a new antibacterial fluoroquinolone, orbifloxacin, in the presence of chloride ion, *Chem. Pharm. Bull.* 45, 1828–1832.

Morimura, T., Nobuhara, Y., and Matsukura, H., 1997b, Photodegradation products of a new fluoroquinolone antibacterial, orbifloxacin, in aqueous solution, *Chem. Pharm. Bull.* 45, 373–377.

Motten, A.G. and Chignell, C.F., 1983, Spectroscopic studies of cutaneous photosensitizing agents. III. Spin trapping of photolysis products from sulfanilamide analogs, *Photochem. Photobiol.* 37, 17–26.

Motten, A.G., Martinez, L.J., Holt, N., Sik, R.H., Reska, K., Chignell, C.F., Tønnesen, H.H., and Roberts, J.E., 1999, Photophysical studies of anitmalarial drugs, *Photochem. Photobiol.* 69, 282–287.

Mubarak, S.I.M., Standford, J.B., and Sugden, J.K., 1983, Photochemical reactions of chloramphenicol with diols, *Pharm. Acta Helv.* 58, 343–347.

Narbutt–Mering, A.B. and Weglowska, W., 1980, Decomposition of 5-nitrofuran pharmaceuticals, *Acta Pol. Pharm.* 37, 301–308.

Navaratnam, S., Hughes, J.L., Parsons, B.J., and Phillips, G.O., 1985, Laser flash photolysis and steady state photolysis of benzoxaprofen in aqueous solution, *Photochem. Photobiol.* 41, 375–380.

Navaratnam, S. and Claridge, J., 2000a, Primary photochemical properties of ofloxacin, *Photochem. Photobiol.* 72, 283.

Navaratnam, S., Hamblett, I., and Tønnesen, H.H., 2000b, Formation and reactivity of free radicals in mefloquine, *J. Photochem. Photobiol. B-Biol.* 56, 25–38.

Nery, A.L.P., Quina, F.H., Moreira, P.F., Medeiros, C.E.R., Baader, W.J., Shimizu, K., Catalani, L.H., and Bechara, E.J.H., 2001, Does the photochemical conversion of colchicine into lumicolchicine involve triplet transients? *Photochem. Photobiol.* 73, 213–218.

Ninomiya, I. and Yamamoto, O., 1976, Photochemistry of isonicotinohydrazide and its analogs in alcohol, *Heterocycles* 4, 475–481.

Nord, K., Karlsen, J., and Tønnesen, H.H., 1991, Photochemical degradation of chloroquine, *Int. J. Pharm.* 72, 11–18.

Ogata, M., Noro, Y., Yamada, M., Tahara, T., and Nishimura, T., 1998, Photodegradation products of methylprednisolone suleptate in aqueous solution — evidence of a bicyclo[3.1.0]hex-3-en-2-one intermediate, *J. Pharm. Sci.* 87, 91–95.

Ohmae, M. and Katsui, G., 1969, Photolysis of vitamin K1. IV. Degradation of the oil and the alcoholic solution, *Vitamin* 39, 190–194.

Owoyale, J.A., 1989, Amodiaquine is less sensitive than chloroquine to photochemical reactions, *Int. J. Pharm.* 56, 213–215.

Paillous, N. and Verrier, M., 1988, Photolysis of amiodarone, an antiarrhythmic drug, *Photochem. Photobiol.* 47, 337–343.

Paluchowska, M.H. and Bojarski, H.J., 1988, Photochemical formation of primidone from 2-thiophenobarbital, *Arch. Pharm. (Weinheim)* 321, 343–344.

Parman, T., Chen, G., and Wells, P.G., 1998, Prostaglandin H-synthase catalyzed bioactivation, electron spin resonance spectrometry and photochemical product analysis, *J. Biol. Chem.* 273, 25079–25088.

Pawlaczyck, J. and Turowska, W., 1976, Identification of certain photoproducts of sulfanilamide photolysis in aqueous solutions, *Arch. Pol. Pharm.* 33, 505–509.

Pfoertner, K.H. and Daly, J.J., 1987, Photochemical rearrangement of N-substituted 2-ethyl-5-nitro-1H-imidazoles in the presence of water, *Helv. Chim. Acta* 70, 171–174.

Philip, J. and Szulczewski, D.H., 1973, Photolytic decomposition of mefloquine in water, *J. Pharm. Sci.* 62, 1479–1482.

Pietta, P., Rava, A., and Biondi, P.A., 1981, High-performance liquid chromatography of nifedipine, its metabolites and photochemical degradation products, *J. Chromatogr.* 210, 516–521.

Po, A.L.W. and Irwin, W.J., 1980, The photochemical stability of *cis*- and *trans*-isomers of tricyclic neuroleptic drugs, *J. Pharm. Pharmacol.* 32, 25–29.

Poiger, T., Buser, H.R., and Muller, M.D., 2001, Photodegradation of the pharmaceutical drug diclofenac in a lake: pathway, field measurements and mathematical modeling, *Environ. Toxicol. Chem.* 20, 256–263.

Quilliam, M.A., McCarry, B.E., Hoo, K.H., McCalla, D.R., and Vaitekunas, S., 1987, Identification of the photolysis products of nitrofurazone irradiated with laboratory illumination, *Can. J. Chem.* 65, 1128–1132.

Reisch, J. and Schildgen, R., 1972, Light induced fragmentation of D-(-)methadone hydrochloride; radiolysis of aqueous solutions of D-(-)-methadone hydrochloride, *Arch. Pharm. (Weinheim)* 305, 49–53.

Reisch, J. and Niemayer, D.H., 1972, Photochemical studies of sulfanilamide, sulfacetamide and ethyl 4-aminobenzoate, *Arch. Pharm. (Weinheim)* 305, 135–140.

Reisch, J., Ekiz-Guecer, N., and Guneri, T., 1986a, Photodegradation of propylphenazone. Photo and radiochemical studies, *Pharmazie* 41, 287–288.

Reisch, J., Topaloglu, Y., and Henkel, G., 1986b, Photostability of pharmacopeial drugs and excipients. II. Cross-conjugated corticoids of the European Pharmacopeia, *Acta Pharm. Technol.* 32, 115–121.

Reisch, J., Ekiz–Guecer, N., Takacs, M., Gunaherath, G.M., and Kamal, B., 1989a, The photoisomerization of azapropazone, *Arch. Pharm. (Weinhemi)* 322, 295–296.

Reisch, J., Ekiz–Guecer, N., Takacs, M., and Henkel, G., 1989b, Photodimerization of testosterone propionate in the crystal state and crystal structure of testosterone propionate, *Liebigs Ann. Chem.* 595–597.

Reisch, J., Ekiz–Guecer, N., and Takacs, M., 1989c, Studies on the photostability of testosterone and methyltestosterone in the crystal state, *Arch. Pharm. (Weinhemi)* 322, 173–175.

Reisch, J., Ekiz–Guecer, N., and Tewes, G., 1992a, Photostability of some 1,4-benzodiazepines in the solid state, *Liebigs Ann. Chem.* 69–70.

Reisch, J., Iranshani, L., and Ekiz–Guecer, N., 1992b, Photostability of glucocorticosteroids in the crystal state, *Liebigs Ann. Chem.* 1199–1200.

Reisch, J., Henkel, G., Ekiz–Guecer, N., and Nolte, G., 1992c, Studies on the photostability of halomethasone and prednicarbamate in the crystal state and crystal structure of halomethason, *Liebigs Ann. Chem.* 63–67.

Reisch, J., Zappel, J., Enkel, G., and Ekiz–Guecer, N., 1993, Studies on the photostability of ethisterone and norethisterone and their crystal structures, *Monatsh. Chem.* 124, 1169–1175.

Reisch, J., Zappel, J., Rao, A.R., and Henkel, G., 1994, Dimerization of levonorgestrel in solid state ultraviolet light irradiation, *Pharm. Acta Helv.* 69, 97–100.

Revelle, L.K., Musser, S.M., Rowe, B.J., and Feldman, I.C., 1997, Identification of chlorothiazide and hydrochlorothiazide UV-A photolytic products, *J. Pharm. Sci.* 86, 63–64.

Ricci, A., Fasani, E., Mella, M., and Albini, A., 2001, Noncommunicating photoreaction paths
in some pregna-1,4-dien-3,20-diones, *J. Org. Chem.* 66, 8086–8093.

Robertson, D.G., Epling, G.A., Kiely, J., Bailey, D.L., and Song, B., 1991, Mechanistic studies
of the phototoxic potential of PD117596, a quinolone antibacterial compound, *Toxicol. Appl. Pharmacol.* 111, 221–232.

Sadana, G.S. and Goghare, A.B., 1991, Mechanistic studies on photolytic degradation of
nifedipine by use of ^{1}H NMR and ^{13}C NMR sectroscopy, *Int. J. Pharm.* 70, 195–199.

Schaffer, G.W., Nikawitz, E., Manowitz, M., and Daeniker, H.U., 1971, Photodegradation of
hexachlorophene and related polychlorinated phenols, *Photochem. Photobiol.* 13,
347–355.

Schleuder, M., Richter, P.H., Keckeis, A., and Jira, T., 1993, The stability of (Z) 2-amino-5-
chlorobenzophenone amidinohydrazone acetate in solution, *Pharmazie* 48, 33–37.

Schuijt, C., Beijersbergen van Henegouwen, G.M.J., and Gerritsma, K.W., 1979, Isolation
and identification of decomposition products of emetine, *Pharm. Weekblad Sci. Ed.*
1, 186–195.

Sedee, A.G.J., Beijersbergen van Henegouwen, G.M.J., De Mol, N.J., and Lodder, G., 1984,
Interaction of the photosensitized products of estrone with DNA: comparison with
the horseradisch peroxidase catalyzed reactions, *Chem. Biol. Interact.* 51, 357–363.

Sedee, A.G.J. and Beijersbergen van Henegouwen, G.M.J., 1985, Photosensitized decomposition of contraceptive steroids: a possible explanation of the observed (photo)allergy
of the oral contraceptive pill, *Arch. Pharm. (Weinheim)* 318, 111–119.

Segmuller, B.E., Armstrong, B.L., Dunphy, R., and Oyler, A.R., 2000, Identification of
autoxidation and photodegradation products of ethynylestradiol by on-line HPLC-
NMR and HPLC-MS, *J. Pharm. Biomed. Anal.* 23, 927–937.

Shih, I.K., 1971, Photochemical products of chloramphenicol in aqueous medium, *J. Pharm.
Sci.* 60, 1889–1890.

Sortino, S., De Guidi, G., Giuffrida, S., Monti, S., and Velardita, A., 1998a, pH effect on the
spectroscopy and photochemistry of enoxacin: a steady state and time-resolved study,
Photochem. Photobiol. 67, 167–173.

Sortino, S., De Guidi, S., Marconi, G., and Monti, S., 1998b, Triplet photochemistry of
suprofen in aqueous environment and in β-cyclodextrin inclusion complexes, *Photochem. Photobiol.* 67, 603–611.

Sortino, S., Scaiano, J.C., De Guidi, G., and Monti, S., 1999a, Effect of β-cyclodextrin
complexation on the photochemical and photosensitizing properties of tolmetin: a
steady state and time-resolved study, *Photochem. Photobiol.* 70, 549–556.

Sortino, S., Giuffrida, S., and Scaiano, J.C., 1999b, Phototoxicity of naphazoline: evidence
that hydrated electrons, nitrogen-centered radicals, and OH radicals trigger DNA
damage: a combined photocleavage and laser flash photolysis study, *Chem. Res.
Toxicol.* 12, 971–978.

Sortino, S., Giuffrida, S., De Guidi, G., Chillemi, R., Petralia, S., Marconi, G., Condorelli,
G., and Sciuto, S., 2001, The photochemistry of flutamide and its inclusion complexes
with β-cyclodextrin, *Photochem. Photobiol.* 73, 6–11.

Squella, J.A., Zanocco, A., Perna, S., and Nunez–Vergara, L.J., 1990, A polarographic study
of the photodegradation of nitrendipine, *J. Pharm. Biomed. Anal.* 8, 43–47.

Sternberg, V.E. and Travecedo, E.F., 1970, A new photochemical reaction of the cinchona alkaloids, quinine, quinidine, cinchonidine and cinchonine, *J. Org. Chem.* 35, 4131–4136.

Takacs, M., Vamos, J., Toth, G., and Miko–Hideg, Z., 2000, Photochemical and chemical oxidation of mexiletine and tocainide. Structure elucidation of the major products, *Arch. Pharm. Med. Chem.* 333, 48–52.

Tammilehto, S. and Torniainen, K., 1989, Photochemical stability of dothiepin in aqeuous solutions, *Int. J. Pharm.* 52, 123–128.

Teraoka, R. and Matsuda, Y., 1993, Stabilization oriented preformulation study of photolabile menatetrenone (vitamin K2), *Int. J. Pharm.* 93, 85–90.

Teraoka, R., Otsuka, M., and Matsuda, Y., 1999, Evaluation of photostability of solid state dimethyl 1,4-dihydro-2,6-dimethyl-4-(2-nitrophenyl)-3,5-dicarboxylate by using Fourier-transform reflection-absorption IR spectroscopy, *Int. J. Pharm.* 184, 35–43.

Thoma, K., Kerker, R., and Weissbach, C., 1987, Effect of irradiation methods on the stability of betamethasone, *Pharm. Ind.* 49, 961–965.

Thoma, K. and Kerker, R., 1992, Studies of the degradation products of nifedipine, *Pharm. Ind.* 54, 465–468.

Thoma, K., Kuebler, N., and Reimann, E., 1997, Photodegradation of topical antimycotics, *Pharmazie* 52, 362–373.

Thoma, K. and Holzmann, C., 1998, Photostability of dithranol, *Eur. J. Pharm. Biopharm.* 46, 201–208.

Tønnesen, H.H., Grislingaas, A.L., Woo, S.O., and Karlsen, J., Photochemical stability of antimalarials. I. Hydroxychloroquine, 1988, *Int. J. Pharm.* 43, 215–219.

Tønnesen, H.H. and Grislingaas, A.L., 1990, Photochemical decomposition of mefloquine in water, *Int. J. Pharm.* 60, 157–162.

Ulvi, V., Mesilaakso, M., and Matikainen, J., 1996, Isolation and structure elucidation of the main photodecomposition product of trichlorthiazide, *Pharmazie* 51, 774–775.

Usmanghani, K., Ahmad, I., and Zoha, S.M.S., 1975, Studies on the quantitative determination and photodegradation of ephedrine, *Pakistan J. Sci. Ind. Res.* 18, 229–230.

van Dort, H.M., van der Linde, L.M., and de Rijke, D., 1984, Identification and synthesis of new odor compounds from photolysis of thiamin, *J. Agric. Food. Chem.* 32, 454–457.

Vargas, F., Rivas, C., Miranda, M.A., and Boscà, F., 1991, Photochemistry of non steroidal anti-inflammatory drugs, propionic acid derivatives, *Pharmazie* 46, 767–771.

Vargas, F., Rivas, C., and Machado, R., 1992, Photodegradation of nifedipine under aerobic conditions: evidence of formation of singlet oxygen and radical intermediate, *J. Pharm. Sci.* 81, 399–400.

Vargas, F., Rivas, C., Machado, R., and Sarabia, Z., 1993, Photodegradation of benzydamine: phototoxicity of an isolated photoproduct on erythrocytes, *J. Pharm. Sci.* 82, 371–372.

Vargas, F., Matskevitch, V., and Sarabia, Z., 1995, Photodegradation and *in vitro* phototoxicity of the antidiabetic drug dichlorpropamide, *Arzneimittel.* 45, 1079–1081.

Vargas, F., Martinez Volkmar, I., Sequera, J., Mendez, H., Rojas, J., Fraile, G., Velasquez, M., and Medina, R., 1998, Photodegradation and phototoxicity studies of frusemide. Involvement of singlet oxygen in the hemolysis and lipid peroxidation, *J. Photochem. Photobiol. B-Biol.* 42, 219–225.

Vargas, F. and Mendez, H., 1999, Study of the photochemical and *in vitro* phototoxicity of chlortalidone, *Pharmazie* 54, 920–922.

Wehrli, H., Cereghetti, M., Schaffner, K., and Jeger, O., 1960, Influence on the photochemical behavior of 20-ketopregnane compounds by acyloxy substituent on C21, *Helv. Chim. Acta* 43, 367–371.

Wilkins, B.J., Gainsford, G.J., and Moore, D.E., 1987, Photolytic rearrangement of metronidazole to N-(2-hydroxyethyl)-5-methyl-1,2,4-oxadiazole-3-carboxamide. Crystal structure of its 4-nitrobenzoate derivative, *J. Chem. Soc. Perkin Trans. 1*, 1817–1820.

Williams, J.R., Moore, R.H., and Blount, J.F., 1979, Structure and photochemistry of lumi-prednisone and lumiprednisone acetate, *J. Am. Chem. Soc.* 95, 5019–5025.

Williams, J.R., Moore, R.H., Li, R., and Weeks, C.M., 1980, Photochemistry of 11α- and 11β-hydroxy steroidal 1,4-dien-3-ones and 11α- and 11β-hydroxy steroidal bicyclo[3.1.0]hex-3-en-2-ones in neutral and acidic media, *J. Org. Chem.* 45, 2324–2331.

Wilson, R.M., Walsh, T.F., and Gee, S.K., 1980, Laser photochemistry: the wavelength dependent oxidative photodegradation of vitamin K analogs, *Tetrahedron Lett.* 21, 3459–3463.

Wood, M.J., Irwin, W.J., and Scott, D.K., 1990, Photodegradation of doxorubicin, daunorubicin and epirubicin measured by high-performance liquid chromatography, *J. Clin. Pharm. Ther.* 15, 291–300.

Wu, Z.Z., Nash, J., and Morrison, H., 1992, Photoepimerization of 3-α(dimethylphenylsilyloxy)-5α-androstane-6,17-dione and its 3β isomer through bond exchange energy transfer, *J. Am. Chem. Soc.* 114, 6640–6648.

Zanocco, A.L., Diaz, L., Lopez, M., Nunez–Vergara, L.J., and Squella, J.A., 1992, Polarographic study of the photodegradation of nimodipine, *J. Pharm. Sci.* 81, 920–924.

Zhelyaskov, V.R. and Godwin, D.W., 1998, Photolytic generation of nitric oxide through a porous glass partitioning membrane, *Nitric Oxide* 2, 454–459.

Zhou, W. and Moore, D.E., 1994, Photochemical decomposition of sulfamethoxazole, *Int. J. Pharm.* 110, 55–63.

Zieske, P.A., Koberda, M., Hines, J.L., Knight, C.C., Sriram, R., Raghavan, N.V., and Rabinow, B.E., 1991, Characterization of cisplatin degradation as affected by pH and light, *Am. J. Hosp. Pharm.* 48, 1500–1506.

CHAPTER 5

Technical Requirements and Equipment for Photostability Testing

Jörg Boxhammer and Artur Schönlein

CONTENTS

5.1 INTRODUCTION

In the past, various publications on photostability testing of drugs under simulated conditions were based on a wide range of quite different artificial light sources. Much of this work remains questionable because the real stress conditions were partially unknown. In 1996, publication of ICH guideline Q1b — *Photostability Testing of New Drug Substances and Products* — improved this situation. This ICH guideline formed an annex to the main ICH stability guideline and gave guidance on the basic testing protocol required to evaluate light sensitivity and stability of new drugs and products. However, the current contents of the guideline (still four lamp options and only rather general requirements concerning real testing conditions) has raised questions about the correct interpretation of the guideline.

This situation clearly reflects years of research and continuous permanent revision and redefinition of standardized test methods in the general field of photostability testing on materials and products in nonpharmaceutical industries. A review of the present state of experiences in this testing field may help users and manufacturers of testing equipment to define concrete stress factors and conditions necessary for carrying out reliable tests based on the existing ICH guideline. It may also be helpful as a basis for a possible revision of the guideline. The main objective of tests with laboratory light sources is to conduct them under accelerated, more controlled conditions *vs.* natural exposure. The accelerated conditions should result in good correlation as well as good repeatability and reproducibility of test results. The most important stress conditions must be defined precisely in standardized test methods to meet this objective.

Specification of the spectral distribution of radiation; level of irradiance; surface temperature; and description of necessary measuring devices are focal points of the newest international standards. Various improvements in equipment technology are necessary to meet these requirements. Comparison of the laboratory light sources used in general material testing with those for photostability testing of drugs, as well as contents of international standards to the present state of discussion in the pharmaceutical industry, provides a basic concept for a well-defined guideline for performance tests of drugs concerning technical requirements for the equipment.

5.2 SPECTRAL ENERGY DISTRIBUTION OF SOLAR RADIATION

Most drug compounds and drug formulations will be subjected to light during production, storage, distribution, and use by the patient. It is obvious that various kinds of artificial light sources as well as sunlight present in rooms behind window glass may be major stress factors and therefore may initiate the degradation processes. Real room conditions will include large variances; sunlight through the window may represent the worst case of stress on drugs. Generally, as a basis for interpretation of test results under daylight, as well as for establishing test methods for simulated and accelerating conditions with artificial light sources, a thorough knowledge about sunlight is necessary.

It is well known that the spectral distribution and irradiance of the solar radiation at the Earth's surface depend on the location and is subjected to seasonal and diurnal variations. Therefore, a "reference spectrum" is needed as a basis for comparison with the spectral energy distribution of artificial light sources. Data from CIE No. 15: 1971 (colorimetry; official recommendations of the International Commission on Illumination) that recommend a standard illuminant D65 with a scheduled color temperature of approximately 6500 K have been used as a basis over the years.

The first recommendation for integrated irradiance and spectral distribution of simulated solar radiation — especially for testing purposes — was published in 1972 as the publication of Commission Internationale de L'Eclairage (CIE) No. 20. The listed spectral energy distribution was given in 40-nm steps up to 800 nm and included the infrared part of radiation up to 3000 nm. The maximum value of total irradiance has been specified for testing resistance of technical objects to deterioration. As an international criterion for comparison of artificial sources with natural daylight, specific spectral data more precise compared with CIE No. 20 have been published in CIE No. 85: 1989 and are widely used today as a basis for simulation.

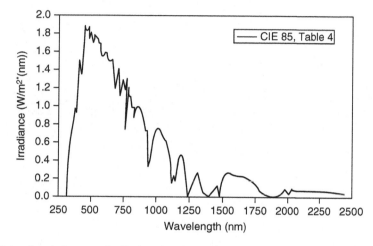

Figure 5.1   Spectral energy distribution of global solar radiation according to CIE No. 85, Table 4 (1989).

The sum of direct and diffuse solar radiation on a horizontal plane at the Earth's surface, called global solar radiation, in the emission range between 300 and 2450 nm is given as 1090 W/m² (relative air mass = 1). The spectral energy distribution is shown in Figure 5.1. Compared with outdoor measurements, the curve is smoothed because the spectral data (5-nm steps in the UV range) are centered spectral irradiance values averaged over 20 nm. The increase of the spectral data in UV is somewhat equivalent to mean values of the UV spectral irradiance as measured in Arizona (Figure 5.2). For specific testing purposes, spectral energy data may be required that are still more precise compared to CIE No. 85 data, especially in the UV region respectively in the cut-off region of radiation (Kockott, 1999).

The total radiation on a horizontal plane is the sum of direct radiation and the part scattered by the Earth's atmosphere. Depending on the altitude of the sun in the sky, radiation may be between 30 and 90% of total radiation. At very low altitudes of the sun, the total energy level decreases and spectral distribution is shifted toward the red. In sky radiation, the shorter wavelengths are scattered more than the longer ones; therefore, sky radiation shows a spectral distribution shifted toward the blue. The irradiance of the total radiation may be reduced to 20% on a totally cloudy day compared with cloudless days.

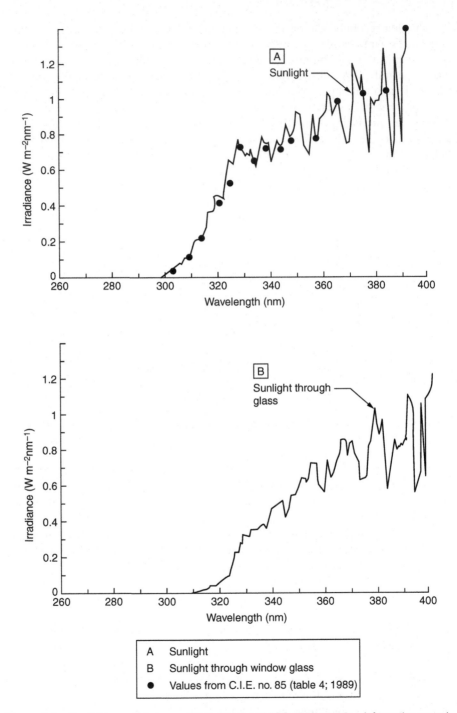

Figure 5.2   Sunlight spectral energy distribution in UV (ISO 4892, part 3 — informative annex).

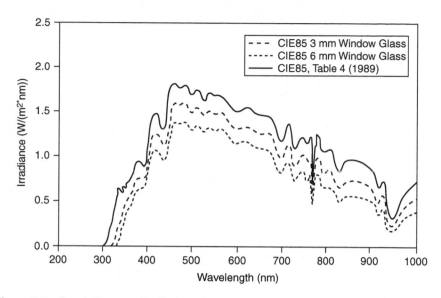

Figure 5.3 Spectral energy distribution of global solar radiation behind window glass.

Based on the spectral energy distribution of global solar radiation, the spectral energy data behind a window glass depend on the type and thickness of the window glass used (see examples in Figure 5.3). With increasing thickness of the window glass, the total irradiance is reduced and the cut off in the UV range shifts toward longer wavelengths.

5.3 ABSORPTION OF RADIATION IN MATERIALS

Based on the incident spectral irradiance of radiation, e.g., sunlight, only the radiation absorbed by a material according to its spectral absorption characteristics, $\varepsilon(\lambda)$, may produce an effect in the material. A very small part of the absorbed radiation will result in primary photochemical processes. The UV range and, partially, the visible wavelength range, of solar radiation have a major influence. Determination of the "spectral sensitivity" (Trubiroha, 1987) and the "activation spectrum" (Searle, 1985) may give valuable information on the wavelength bands of incident radiation that cause photodegradation. Thus, it may not always be necessary to simulate the reference spectrum over the whole wavelength range of light emission. On the other hand, the most accurate simulation minimizes the risk of anomalous reactions that may not occur in an actual use environment.

The major part of incident radiation absorbed by the material leads to heating of the surface (especially of solid samples above ambient temperature), depending on the spectral distribution of radiation; the irradiance level; other climatic environmental conditions (wind velocity, humidity); and thickness as well as optical and thermal

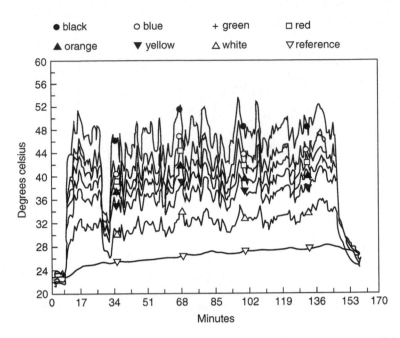

Figure 5.4   Temperature profile for colored samples as collected from the multichannel recorder (Minnesota 30° backed, July 23, 1991). (According to Fischer, R.M. and Ketola, W.D., *ASTM, STP* 1202, 1994.)

properties of the exposed material. A specific situation is given for translucent samples in which very complex relationships between heat absorption by radiation and heat emission can occur (Wendisch, 1966).

Nearly all photochemical aging processes are influenced by temperature. Considering that a temperature increase of $\Delta T = 10$ K may double the reaction rate, the increased temperature of the exposed surface is a more important parameter than the temperature of the ambient air. Measurements on differently colored PVC films in outdoor exposure have shown temperature differences between black and white samples of 15 to 25 K, colored samples ranging in-between (Fischer and Ketola, 1994). An example is shown in Figure 5.4 (one example of temperature measurement sequence) and Figure 5.5 ("typical" temperature profile).

5.4 ACCELERATED PHOTOSTABILITY TESTS WITH ARTIFICIAL LIGHT SOURCES

Various publications on photostability testing of drugs under simulated and accelerating conditions based on a wide range of quite different artificial light sources have been published (e.g., Sandeep Nema et al., 1995). The current ICH guideline has improved the situation but questions can still be raised about the correct interpretation of the guideline contents.

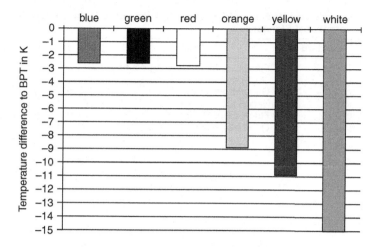

Figure 5.5   "Typical" temperature profile for colored samples exposed 45° open (BPT = 53°C) exposure (from Minnesota 45° open temperature model). (According to Fischer, R.M. and Ketola, W.D., *ASTM, STP* 1202, 1994.)

With respect to stress conditions leading to reactions in materials caused by daylight, only a few, but very complex, factors need to be defined as precisely as possible in a standardized test method. It is of highest importance that these parameters are specified for the location or volume in which the substances or products are really exposed. In the exposed samples, still other factors concerning the entire test procedure may influence the reactions and/or the reaction rate and must be described and/or possibly specified in detail. In the following discussion, only technical requirements for the equipment for photostability tests will be discussed. This may also provide impetus for defining detailed performance criteria based on the contents of the ICH guideline. Table 5.1 shows some of the most important, thoroughly revised international standards in the field of testing under simulated conditions on plastics, paints, and textiles.

The *General Guidance — Part 1* of ISO 4892: 1999 provides general information on testing procedures under simulated natural conditions that may be useful in other industries, especially for plastics. The introduction of this standard states that "accelerated exposure in artificial light devices is conducted under more controlled conditions that are designed to accelerate degradation and product failures." The introduction also states that generally valid correlations between the aging processes that occur during artificial and natural exposure cannot be expected because of the large number of factors involved. The prerequisites for running tests, considering correlation to tests under natural conditions as well as repeatability and reproducibility of tests under accelerated conditions, are described. The focal points for accelerated tests are:

- Use of CIE Publication No. 85 as a basis for comparison of artificial light sources with natural daylight
- Statements concerning factors tending to decrease the degree of correlation because of using:

Table 5.1   Materials and Methods of Exposure to Laboratory Light Sources — International Standards

ISO/TC	Standard	Title	Valid ed.	Note
61	ISO 4892	Plastics — Methods of exposure to laboratory light sources		
	Part 1	General guidance	1999	EN ISO 2000
	Part 2	Xenon-arc sources	1994	EN ISO 1999
	Part 3	Fluorescent UV lamps	1994	ISO, parts 2 to
	Part 4	Open flame carbon-arc lamps	1994	4 in revision since 1999
35	ISO 11 341	Paints and varnishes — Artificial weathering and exposure to artificial radiation — Exposure to filtered xenon arc radiation	1994	EN ISO 2001 ISO in revision since 1999
	ISO 11 507	Paints and varnishes — Exposure of coatings to artificial weathering in apparatus — Exposure to fluorescent UV and water	1997	EN ISO 2001
38	ISO 105-B06[a]	Textiles — Tests for color fastness Color fastness and aging to artificial light at high temperatures: xenon-arc fading lamp test	1998	Amendment (AMD 1) included 2002
61	ISO 9370	Plastics — Instrumental determination of radiant exposure in weathering tests — General guidance and basic test methods	1997	In revision since 2002
42	ISO 10 977	Photography — Processed photographic color films and paper prints — Methods for measuring image stability	1993	
	ISO/DIS 18 909	Imaging materials — Processed photographic color films and paper prints — Methods for measuring image stability	2002	Revision of ISO 10 977

Note: ISO — International Organization for Standardization; TC — technical committee; DIS — draft international standard.
[a]Automotive applications.

– Ultraviolet radiation of wavelengths shorter than those occurring in natural exposure
– Spectral distribution that differs widely from that of daylight
– Very high light flux
– High specimen temperatures, particularly with materials that readily undergo changes from thermal effects alone

- Statement that all testing conditions need to be ensured on specimen area where the materials are really exposed because of:
  - Spectral distribution of radiation
  - Level of irradiance
  - Surface temperature level
- Description and recommendation of methods and devices for determining irradiance and radiant exposure as well as for measuring temperature

From the various artificial light sources that were used in the past in the general field of material light stability testing, specially filtered xenon arc sources and specific fluorescent UV lamps were found to be useful and are therefore specified in the basic standards.

Still other international standards have photostability tests as part of the contents. The ISO 10 977: 1993 (Table 5.1) is an example. In this standard, relative spectral distributions for simulated indoor indirect daylight (Figure 5.6 compared with CIE No. 85 and 6-mm window glass), as well as different artificial light sources, have also been given, especially under the aspect of radiation behind window glass. This international standard will be replaced in the future by the ISO 18 909 (*Imaging Materials — Processed Photographic Color Films and Paper Prints — Methods for Measuring Image Stability*) containing still more test details. The document has the status of a draft international standard (DIS) at the present time. Only in some national standards are specific metal halide lamps currently described as the basis for simulation of solar radiation (DIN 75220: 1992, *Ageing of Automotive Components in Solar Simulation Units* serves as an example).

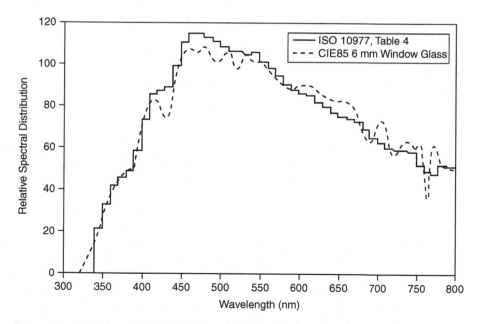

Figure 5.6   Relative spectral distribution of CIE No. 85 with 6-mm window glass compared with the data of ISO 10 977 simulated indoor indirect daylight.

Table 5.2   Radiation Functions (Basis for Simulation of Global Radiation with Filtered Xenon Arc Lamps) as Specified in International Standards

Standard	ISO 4892, Part 2 : 1994[a] ISO 11 341 : 1994[a]		ISO 105-B06:1998
Method resp. set of exposure conditions	A (artificial weathering)	B (daylight behind window glass)	1 to 3
Wavelength (nm)	**Relative irradiance (%)[b]**		
<290	0[c]		0
<300		0	<0.05
280–320	0.6 ± 0.2	<0.1	<0.1
320–360	4.2 ± 0.5	3.0 ± 0.5	3.0 ± 0.85
360–400	6.2 ± 1.0	6.0 ± 1.0	5.7 + 2.0/–1.3
400–520			32.2 + 3.0/–5.0
520–640			30.0 ± 3.0
640–800			29.1 ± 6.0
<800	100	100	100

[a]Standards are under revision (status DIS in 2003); spectral functions have changed.
[b]As a percentage of the total irradiance in the wavelength range up to 800 nm.
[c]Xenon arcs operating as specified in method A emit a small amount of radiation below 290 nm. In some cases, this can cause degradation reactions that do not occur in outdoor exposures.

Some basic aspects concerning requirements for simulation of solar radiation are best discussed on the basis of technical contents of the ISO Standards. Independent of the type of radiation system used, the spectral energy distribution of global solar radiation according to Table 4 in CIE No. 85 represents the basis for an evaluation of the quality of simulation.

For daylight behind window glass, the CIE data are integrated with the spectral transmittance of a window glass. The spectral distribution of radiation depends on the type and thickness of the window glass, especially in the lowest UV region (see Table 5.2 and Figure 5.3). Different from ISO 10 977: 1993 (window glass with a thickness of 6 mm), the reference spectrum for daylight behind window glass in ISO 4892-2 and ISO 11 341 is based on a window glass with a thickness of 3 and 4 mm in ISO 105-B06, respectively. Consequences of aging processes occurring that are caused by radiation in the lowest UV region must be considered.

Relative spectral functions have been specified using the values of CIE No. 85 in 40-nm steps in the UV range and a wavelength range up to 800 nm set to 100% (Figure 5.7) — especially for filtered xenon arc sources as artificial radiation systems for simulating natural conditions outdoors as well as behind window glass. In an additional step (ISO 105-B06), the spectral function is extended to separated wavelength steps in the visible wavelength range. The data specified in the ISO standards are shown in Table 5.2. These data are still valid but standards 4892, part 2, and ISO 11 341 are now under revision (status DIS at present time), and the spectral functions will be replaced by new ones. These will be limited to the UV range of radiation containing minimum and maximum values that are based on numerous spectral irradiance measurements on concrete lamp systems (Table 5.4; only data

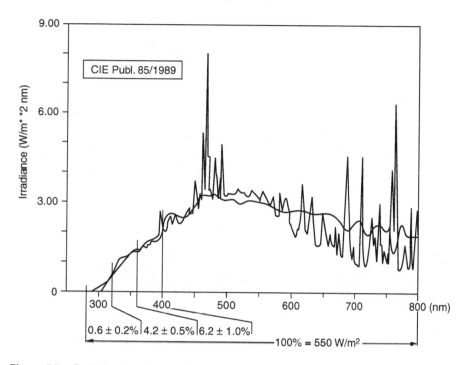

Figure 5.7   Radiation function — ISO 4892, part 2 (1994); filtered xenon arc radiation for comparison.

for daylight behind window glass). However, the CIE data will always serve as target values.

The most important requirement is that the spectral functions be realized within the given tolerances (minimum/maximum limits) on the sample area of the test cabinets independent of occurring effects such as changing of laboratory voltage or lamp/filter age of the used xenon light systems. For fluorescent UV lamps, the situation is not as clear as for filtered xenon arc sources. In the still valid version of the international standard, only different lamp types are described and preliminary spectral data are listed and compared with global radiation in an annex for information (Table 5.3). However, the standard is under revision and will also contain spectral UV functions with minimum/maximum limits for specified UV fluorescent lamps (Table 5.4; only data for daylight behind window glass). For filtered xenon radiation, the irradiance level has been set to 550 W/m^2 in the wavelength range up to 800 nm for the purpose of reference. In the currently revised version, specific UV broadband (300 to 400 nm) and adequate narrowband levels (340/420 nm) have been recommended. Other values may be used but must be reported. It is most important that the irradiance used not vary by more than ±10% when any two points in the sample plane parallel to the lamp axis are compared.

When using fluorescent lamps, especially those with emission of radiation only in the UV part of the spectrum, the differences in temperature between specimen surfaces and the ambient air are very small as experienced under natural exposure.

Table 5.3   Spectral Data for Fluorescent UV Lamps as Described in ISO 4892[a]

Spectral Band (nm)	Type I (340-nm peak)	Type I (351-nm peak)	Type II (313-nm peak)	Lamp Combination
<270	0	0	0	0
270–300	0.1	0	5.2	0.3
301–320	3.0	0.8	13.1	3.0
321–360	25.1	22.6	12.1	22.0
361–400	11.0	12.7	1.1	18.0

Note: Data shown in this table are preliminary only. Additional work is under way in ASTM Committee G03, Durability of Nonmetallic Materials, to develop more complete and technically valid data on irradiance specifications for fluorescent UV lamps.

[a] Part 3: 1997, informative annex.

Table 5.4   Radiation Functions as Described in ISO/DIS 4892[a]

ISO/DIS 4892	Part 2: Xenon Arc with Window Glass Filters (Method B)			Part 3: Fluorescent UV Lamps (Method B), Type I UVA-351 Lamps		
Spectral Bandpass, Wavelength λ in nm	Minimum %	CIE No. 85, Table 4 plus wg %	Maximum %	Minimum %	CIE No. 85, Table 4 plus wg %	Maximum %
$\lambda < 300$			0.29			0.2
$301 \leq \lambda \leq 320$	0.1	≤1	2.8	1.1	≤1	3.3
$321 \leq \lambda \leq 360$	23.8	33.1	35.5	60.5	33.1	66.8
$361 \leq \lambda \leq 400$	62.4	66.0	76.2	30.0	66.0	38.0

Notes: Data are the irradiance in the given bandpass expressed as a percentage of the total irradiance from 290 to 400 nm; wg — window glass.

[a] Parts 2 and 3: 2003, only for "daylight behind window glass."

For laboratory light sources with an amount of radiation at longer wavelengths, especially in the infrared region (for example, the emission of xenon arcs reduced by specific filter elements), the temperature level must be thoroughly considered and is specified as the surface temperature of a described specific measuring element to be discussed later. Specific requirements for artificial tests are necessary in the general field of material testing to obtain an improved correlation to natural exposure as well for improved repeatability and reproducibility of accelerated tests.

The characteristics of artificial light sources and most of the applied filter elements are subject to change during use due to aging and other influencing factors lending to a change in spectral distribution, irradiance level, and surface temperature on the sample area. To meet the discussed requirements, improved equipment technology was necessary in already existing devices and is now realized in a range of today's commercially available testing equipment, especially with xenon arc sources. Some focal points that can be solved by different approaches are:

- Improvements in lamp and filter technology
- Variable irradiances on specimen area

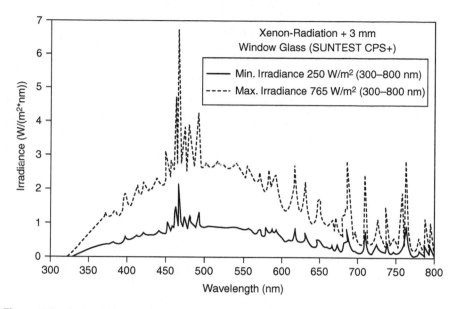

Figure 5.8 Spectral distribution (min/max) on sample level in a xenon equipment.

- Measuring and controlling of preset values for irradiance and surface temperature at the sample level
- Development of specific measuring devices for irradiance/radiant exposure and surface temperature

These essentials should always be considered when designing test systems for photostability tests on the mere basis of any artificial light sources and measuring devices available in the market.

Design and validation characteristics of environmental chambers for photostability testing have been described in the literature (e.g., Boxhammer, 1998). Bench top light exposure cabinets (Atlas Company), which are widely used by many pharmaceutical and cosmetic companies (Thoma and Kübler, 1994), may serve as an example for the high equipment technology available today. These units complete a range of equipment with filtered xenon arc radiation and contain most of the features previously mentioned. The focal points are:

- The irradiance level on the sample area can be varied in a wide range without changing the spectral energy distribution (Figure 5.8).
- Specific UV cut-off filters are used for realizing specific requirements on the simulation of solar radiation (Figure 5.9).
- The preset irradiance is continuously measured and controlled by a filter radiometer calibrated in the wavelength range 300 to 800 nm.
- Tests may be run to preset values of radiant exposure (300 to 800 nm).
- The installed filter radiometer is calibrated/recalibrated with instruments that are traceable to national standards (see Table 5.5).

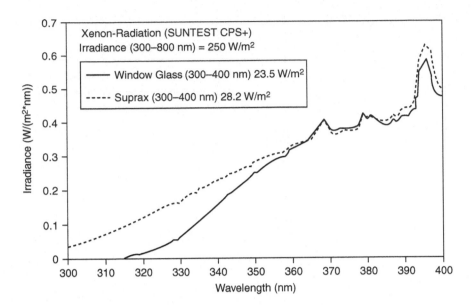

Figure 5.9   Spectral energy distribution on sample level with different filter elements in xenon equipment.

Table 5.5   Relative Irradiance Values and Conversion of Illuminance into Irradiance

	Relative Irradiance for Different Wavelength Bands (nm) in %			Conversion Factor = Illuminance in lux/Irradiance (300–800 nm) in W/m²
	300–400	400–800	300–800	
CIE No. 85, Table 4	10.0	90.0	100	182
CIE No. 85, Table 4 + 3-mm window glass	8.1	91.9	100	189
CIE No. 85, Table 4 + 6-mm window glass	7.3	92.7	100	194
ISO 10 977, Table 4	7.2	92.8	100	199
Fluorescent lamp artificial daylight	10.9	89.1	100	220
Metal halide lamp (Ultratest[a]) + 6-mm window glass	20.9	79.1	100	200
Xenon arc lamp (Suntest[b]) + 6-mm window glass	8.4	91.6	100	225

[a] Heraeus Company.
[b] Atlas MTT GmbH.

- The temperature on the sample area can be varied in a wide range and is measured and controlled constant by a black standard thermometer.
- The measured values are continuously monitored and can be documented on a chart recorder or data logger.

5.5 MEASUREMENT OF RADIATION AND TEMPERATURE AT SAMPLE LEVEL

5.5.1 Irradiance/Radiant Exposure

The available technical data for various artificial radiation sources in use for photo-stability testing show some confusion concerning the use of photometric and radio-metric units. Both have their proper place in science and engineering; however, it is necessary to be aware of the differences in terminologies and uses and determine which is most appropriate for each type of study. Based on a given relative spectral distribution of light sources, radiometric units for a specific wavelength interval may be calculated from photometric units by the use of determined conversion factors. Based on the spectral energy distribution of an artificial daylight lamp as shown in Figure 5.10, an irradiance of 8.4 W/m^2 in the wavelength 300- to 700-nm range will be measured; 2 klx will be measured by a photometer. Even if a filter will cut off the UV range, the photometer still measures 2 klx. Because radiometric units are not limited to visual response and the UV part of solar radiation is most important for photoreactions, radiometric units are recommended and widely used in photo-stability tests.

Two approaches to the measurement of radiation (with emphasis on the ultra-violet wavelength range) are commonly used. The first is to use a physical standard (e.g., chemical actinometer), which is a reference material (substance) that shows a change in property proportional to the dose of incident radiation (UV or other specific

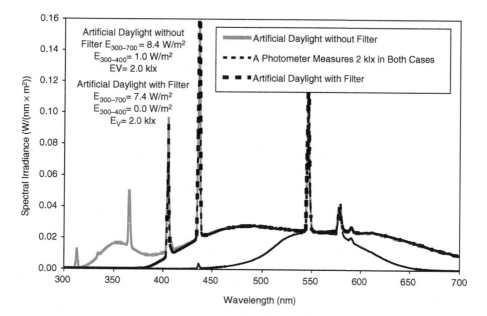

Figure 5.10 Examples of photometric and radiometric quantities.

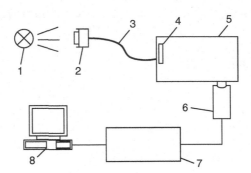

1: Lamp-filter system; 2: Entrance system with cosine adaption; 3: Optical fiber;
4: Filter set; 5: Polychromator; 6: Line camera; 7: Controller; 8: PC

Figure 5.11   Spectroradiometric measurements on sample area (principle).

wavelength bands). The second (preferred) approach is to use a radiometer that responds to the defined wavelength band (ISO 9370). Precise information on the spectral irradiance of incident radiation on a sample surface is only available using spectroradiometers. A schematic view of a measuring system that can be used for measurements on sample area in any equipment is shown in Figure 5.11. These systems are very expensive and not suitable for the daily practice in stability testing. Therefore, radiation detectors for measuring only in the UV or UV plus visible wavelength range (filter radiometers, broadband, narrowband, or wideband; see ISO 9370) described in the basic ISO standards are used in practice. A schematic view of a measuring system is shown in Figure 5.12.

The spectral sensitivity of filter radiometers is not constant over the sensitive wavelength region. Two typical spectral sensitivities of UV radiometers are shown in Figure 5.13. The response of such radiometers depends on the measured spectral power distribution. If the calibration distribution differs from the measured distribution, a spectral mismatch must be taken into account. If the measured spectral distribution is known, the mismatch can be calculated.

ISO 9370 (Table 5.1) contains valuable information and recommendations on important characteristics for the instruments used and provides a guide for selection and use as well as calibration procedures of these radiometers. The guide includes natural and simulated exposure testing. Instrumental techniques include the continuous measurement of irradiance in specific wavelength bands and the accumulation (or integration) of instantaneous data to provide a total radiant exposure (dosage).

Modern commercially available equipment for photostability testing in which the level of irradiance can be varied over a wide range, especially devices with filtered xenon arc radiation, is equipped with filter radiometers to control the preset level of irradiance as well as to accumulate current data. Depending on the type of apparatus, the preferred wavelength bands are 300 to 400 nm, 300 to 800 nm, or measurements at a specific wavelength (e.g., 340 or 420 nm) (Boxhammer, 1994). Still other filter radiometers are sold for different wavelength bands. When measured results are compared, these differences must be considered thoroughly. Filter

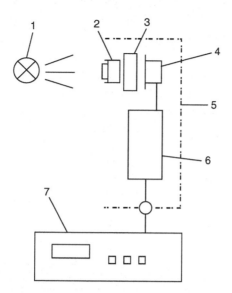

1: Lamp-filter system; 2: entrance system with cosine adaption; 3: band filter; 4: photo-diode; 5: measuring head case; 6: measuring amplifier with memory unit; 7: controller

Figure 5.12 Measurement of irradiance and radiant exposure (bandpass) on sample area (principle).

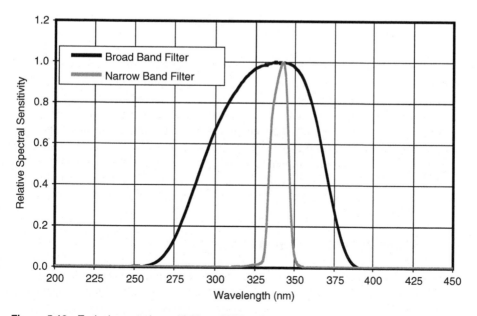

Figure 5.13 Typical spectral sensitivities of UV radiometers.

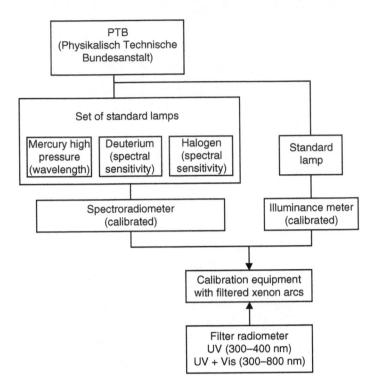

Figure 5.14 Calibration scheme — filter radiometer.

radiometers should be calibrated to a given source. When used with sources that have different spectral distribution of radiation, the radiometer must be recalibrated to that source.

Calibration of spectrally selective radiometers should be by substitution, using a spectroradiometer calibrated to standard lamps in the appropriate wavelength range traceable to a national primary standard lamp, as well as a radiation source identical to the source that will be measured. One example for a calibration procedure is given in Figure 5.14. Calibration standard lamps traceable to national primary standard lamps may also be used (Figure 5.15). Filter radiometers must be equal to operating requirements in weathering equipment. Therefore, many improvements have been carried out in the last couple of years (nonsensitive diffuser against contamination; filter protection against humidity; temperature compensation; miniaturization of electronic components).

5.5.2 Material Surface Temperature

Because it is not practical to monitor the surface temperature of individual exposed samples, a specified black-coated flat plate sensor — so-called black panel thermometer (BPT) — has been used for decades to measure and control temperature on the sample surface in material testing. Currently, a so-called black standard thermometer (BST; sometimes called "insulated black panel thermometer" in the

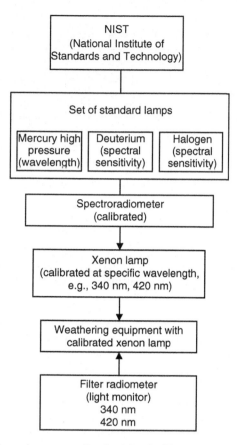

Figure 5.15 Calibration scheme — calibration standard lamp.

literature) is described and specified in the basic ISO standards for characterization of the highest possible surface temperature of a dark solid sample with poor thermal conductivity under a given incident radiation and other surrounding conditions (air temperature, air velocity, and humidity) (Boxhammer, 1994). Alternatively, BPTs may still be used, but differences in the indicated temperatures between the two types of thermometers must be taken into account.

Samples with different colored or even white surfaces show different deviations to the black standard temperature. A so-called white standard thermometer (WST), indicating the lowest surface temperature of a sample whose surface is directly absorbing and reflecting (also described in the basic ISO standard), can be used. Alternatively, a white panel thermometer (WPT) may be used. The temperature of samples with colored surfaces ranges between the black and white standard temperatures. Constant black and white standard temperatures under given exposure conditions during a test indicate constant surface temperature of any exposed sample — even if, in the field of pharmaceutical substances, the real surface temperature of specifically prepared samples may be above or below the characterizing standard temperatures.

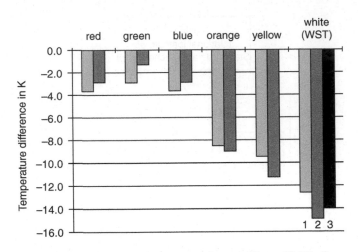

Figure 5.16 Temperature differences in xenon equipment (E_{UV} = 60 W/m^2) compared with outdoor exposure (45° open): 1 — xenon equipment; 2 — outdoor exposure; 3 — measured white standard temperature. (According to Fischer, R.M. and Ketola, W.D., *ASTM, STP* 1202, 1994.)

Various measurements at varied conditions in equipment with filtered xenon arc radiation have shown a linear function between standard temperatures and level of irradiance (Boxhammer et. al., 1993), and also that the surface temperatures of colored samples measured in equipment can be correlated to those under natural conditions (Figure 5.16) (Boxhammer, 1995).

At present, no calibration procedure for black and white standard thermometer is available that includes all stress factors (air temperature, air velocity, and humidity). Today, calibration traceability is guaranteed by a contact thermometric procedure. It would be preferable to measure the temperature at the surface of the coated sensor because this is the temperature of interest. A contactless surface temperature measurement requires knowing the emission ratio of the material and a minimization of the reflected and scattered radiation. For a minor error contact surface temperature measurement, a known method is the multiprobe measurement with extrapolation to the surface temperature.

5.6 SPECTRAL DISTRIBUTIONS OF ARTIFICIAL LIGHT SOURCES PROPOSED FOR DRUG PHOTOSTABILITY TESTS COMPARED WITH GLOBAL RADIATION BEHIND WINDOW GLASS

The only (rather general) description of lamps in the ICH guideline for lamp options allows for using lamps with quite different spectral distributions — especially in the UV range of radiation. Some examples used for photostability tests on drugs in the past and those recommended in the ICH guideline are shown in Figure 5.17 and Figure 5.18 and compared with global solar radiation behind window glass. The graphs are normalized to 100% in the 300 to 800-nm (Figure 5.17) and 300- to 400-nm (Figure 5.18) wavelength regions, respectively.

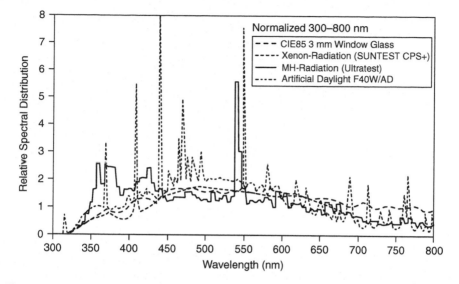

Figure 5.17 Relative spectral distribution of different laboratory light sources compared with CIE No. 85, Table 4 (UV and visible range).

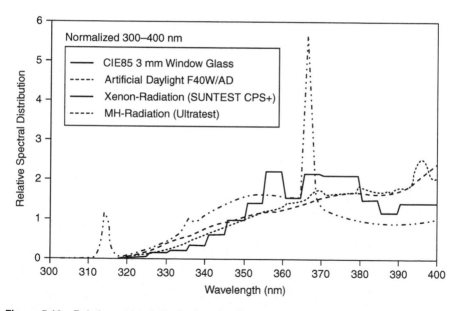

Figure 5.18 Relative spectral distribution of different laboratory light sources compared with CIE No. 85, Table 4 (UV wavelength range).

The spectral distribution over the visible wavelength range may be rather consistent for the different light sources, but there are significantly different spectral distributions in the UV range of radiation. This may be illustrated by comparing the

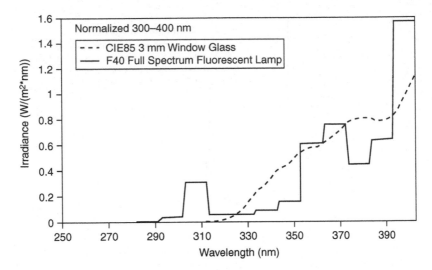

Figure 5.19 Spectral energy distribution of a "full spectrum" fluorescent lamp compared with CIE No. 85, Table 4 (UV region).

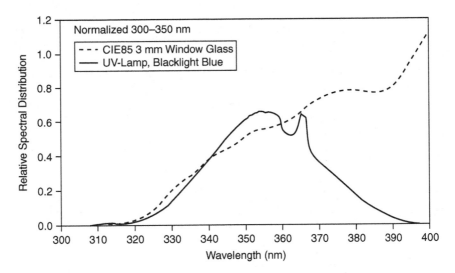

Figure 5.20 Relative spectral power distribution of a fluorescent UV-lamp (example) compared with CIE No. 85, Table 4, normalized to wavelength region 300 to 350 nm.

measured spectral energy distribution in the UV range of radiation of a so-called full-spectrum fluorescent lamp (Figure 5.19) with a specific fluorescent UV lamp (Figure 5.20). Another example for two fluorescent lamps used as daylight lamps for color-matching purposes is given in Figure 5.21. A similar situation is also demonstrated for metal halide lamps in general, including those types used in the past for

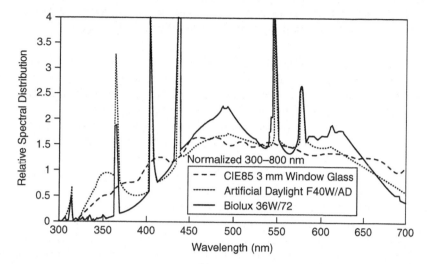

Figure 5.21   Relative spectral distribution of a fluorescent lamp (examples) compared with CIE No. 85, Table 4.

photostability testing of materials (Boxhammer, 1983), as well as of drugs (Thoma and Kerker, 1992). There is an enhanced amount of radiation in the UV wavelength range relative to the visible region (Figure 5.17). This does not correspond to the requirements on simulation of sunlight as specified today for solar simulation units (DIN 75 220) and that is covered by some types of available specific metal halide lamps (Boxhammer, 1998).

A rather uniform basic spectrum is given for appropriately filtered xenon arc radiation, at least as used in commercially available equipment.

The different spectral distributions of used laboratory light sources, especially in the UV range of radiation (Figure 5.18) but also in the visible wavelength range (Figure 5.17), may lead to test results that are not comparable because interactions between reactions caused by different wavelength bands of radiation may happen. This is especially true when using light sources with an emission only in the UV range of radiation or in the visible wavelength range when separate tests are performed. Discussions of light sources in the literature are in most cases on variations in spectral distributions based on data from lamp manufacturers. It should always be taken into account that the spectrum incident on the area or on a defined volume of exposure — and not the emission spectrum of the source — determines the processes that possibly may occur. This spectrum may be influenced (modified) by reflectors, mirrors, and the overall experimental design. Therefore, only specified and well-known (measured) spectral data under the conditions where the samples are exposed form a real basis for carrying out reliable tests.

Minor differences in the spectral distribution found in different kinds of equipment are mainly caused by a variation in filter elements for the infrared part (surface temperature) and the UV cutoff. The radiation level in the UV wavelength range may also vary between instruments.

5.7 LEVEL OF IRRADIANCE ON SAMPLE AREA AND RADIANT EXPOSURE IN PHOTOSTABILITY TESTING

The level of irradiance and the radiant exposure are two important parameters for conducting reliable photostability tests. The level of irradiance should be specified on sample level and the measurements (irradiance, radiant exposure) must be done on this sample level (where the drug, e.g., is exposed to the light source).

Assuming that the spectral distribution of an artificial light source does not contain radiation below the cut off of natural solar radiation behind window glass, acceleration of degradation processes is primarily determined by the irradiance level. The reaction rates may, however, be influenced by other factors such as temperature and humidity. The level of irradiance for accelerated photostability testing of technical materials is based on the highest value of global solar radiation of about 550 W/m^2 between 300 and 800 nm, using filtered xenon arc radiation. In special cases the samples can be exposed to lower or higher levels. For example, the very low illuminance level of 6 klx, equivalent to about 30 W/m^{-2} between 300 and 800 nm using filtered xenon arc radiation for light stability tests of color photographs (ISO 10 977), shows the wide range of recommended specified irradiances that exists, dependent on the type of materials and their application.

The levels of illuminance or irradiance proposed for photostability testing of drugs in the literature vary considerably: for fluorescent lamps between 10^3 and 10^6 lx, and for metal halide or xenon lamps between 200 and 750 W/m^2 in the wavelength range 300 to 800 nm. The conversion between photometric and radiometric units may be done by using values for some of the discussed radiation sources as listed in Table 5.5. Using quite different levels of irradiance as previously mentioned, the mechanism of material degradation may change compared with real-time exposure, even if the spectral energy distribution of radiation remains stable.

This must be considered thoroughly keeping in mind that, depending on the different lamp technologies and required size of exposure area that has a satisfactory uniformity of irradiance, a wide range of irradiances may be realized. The highest level of irradiance may be restricted when taking into account that the total irradiance determines the surface temperature of exposed samples at a given ambient temperature and velocity of air.

For confirmatory studies, two endpoint criteria have been specified in the ICH guideline as minimum exposure levels. These are expressed as luminent exposure (1.2 million lx h) based on indoor indirect daylight and radiant exposure (200 Wh/m^2) for the UV part of radiation. Assuming a thoroughly simulated indoor indirect daylight on the sample surface specimen plane, e.g., obtained by using appropriately filtered xenon radiation, these minimum endpoints (the photometric value converted to the equivalent radiometric quantity) will need quite different test durations. This must be taken into account in designing the stability test.

5.8 PERFORMANCE TESTING OF DRUGS

From the discussion of the previously specified technical requirements for photostability tests on materials other than pharmaceuticals and by comparison with the specified performance tests of drugs, it can be concluded that it is necessary to define more precisely the important factors determining correlation and variability of test results. The first step involves defining spectral energy distribution of radiation during real-time exposure as a real basis for simulation. Assuming that the different laboratory light sources are acceptable for use in accelerated photostability testing, the spectral distribution of these lamps, as well as levels of irradiance, should be defined as precisely as possible by specific test procedures. This should be done for the location where the samples are really exposed. Preference should be given to laboratory systems that closely meet the basic requirements for simulation. Requirements for measuring irradiance/radiant exposure on the sample area should be described.

The influence of temperature should be considered by specifying a maximum surface temperature level characterized by a black standard/white standard temperature, rather than an ambient temperature. In designing a test method that may replace section B (light sources) and section C (procedure) of the ICH guideline, the contents of the existing international standards as discussed in this chapter may be helpful. In general, it should clearly be stated that test results on samples from different test methods should not be compared unless correlation has already been established.

5.9 CONCLUSION

In the general field of material testing, the main objective of tests with laboratory light sources is to conduct tests under accelerated controlled conditions *vs.* natural exposure that result in good correlation as well as good repeatability and reproducibility of test results. Assuming that sunlight behind window glass represents the worst case concerning stress on drugs, available data for the global solar radiation have been discussed. Those data are an essential basis for designing methods for testing under simulated conditions using artificial light sources. Well-defined test methods prepared on international levels already exist. Focal points have been described to provide impetus for designing detailed test conditions on the basis of the existing ICH guideline.

By comparing existing photostability testing standards in the general field of material testing to the contents of the ICH guideline (section B [light sources] and section C [procedure]), a basic concept for a well-defined section concerning technical requirements for equipment is presented as a basis for a possible revision of these sections.

REFERENCES

Boxhammer, J., 1983, Einfluß unterschiedlicher Strahlungssysteme auf das Alterungsverhalten von Hochpolymeren, exemplarisch dargestellt anhand der Änderung des photochemischen Aufbaues einiger Polyäthylen–Werkstoffe, *Die angewandte Makromolekulare Chemie*, 114, 59–67.

Boxhammer, J., 1994, Current status of light and weatherfastness standards — new equipment technologies, operating procedures and application of standard reference materials, presentation, 2nd International Symposium on Weatherability, Tokyo, Sept. 27–29.

Boxhammer, J., 1995, Oberflächentemperaturen und Temperaturmeßtechnik in der Ebene exponierter Proben bei der zeitraffenden Bestrahlung/Bewitterung in Geräten, Seminar "Natürliches und Künstliches Bewittern Polymerer Werkstoffe," TA-Wuppertal, March 28–29, presentation.

Boxhammer, J., 1998, Design and validation characteristics of environmental chambers for photostability testing, in *Drugs — Photochemistry and Photostability*, Albini, A. and Fasani, E. (Eds.) The Royal Society of Chemistry, Cambridge, U.K.

Boxhammer, J., Kockott, D., and Trubiroha, P., 1993, Black standard thermometer, *Materialprüfung*, 35, 143–147.

Commission Internationale de L'Eclairage (CIE), 1972, Recommendations for Integrated Irradiance and Spectral Distribution of Simulated Solar Radiation for Testing Purposes, Report No. 20.

Commission Internationale de L'Eclairage (CIE), 1989, Technical Report: Solar Spectral Irradiance, Report No. 85.

Fischer, R.M. and Ketola, W.D., 1994, Surface temperatures of materials in exterior exposures and artificial accelerated tests; accelerated and outdoor durability, testing of organic materials, *ASTM, STP* 1202.

Kockott, D., 1999, How does modern technology contribute to the reliability of photostability tests of sunscreens? 27th Annual Meeting of the Society for Photobiology and 3rd International Meeting on the Photostability of Drugs and Drug Products (PPS 99), Washington, D.C., July 11–15.

Sandeep Nema, Washkuhn, R.J., and Beussink, D.R., 1995, Photostability testing: an overview, *Pharm. Technol.*, March, 170–185.

Searle, D., 1985, Spectral factors in photodegradation: activation spectra using the spectrograph and sharp cut filter techniques, 7th International Conference on Advances in Stabilization and Controlled Degradation of Polymers, Lucerne, May 22–24.

Thoma, K. and Kerker, R., 1992, Standardized photostability tests — a neglected field of drug safety, personal communication.

Thoma, K. and Kübler, N., 1994, Photostabilitätsprüfungen und andere Stabilitätsprüfungen, *Grundlagen der Körperpflegemittel*, internal report.

Trubiroha, P., 1987, Ermittlung der spektralen Empfindlichkeit der photochemischen Alterung durch monochromatisches Bestrahlen, International Symposium on Weathering, Essen, September 28–29, presentation.

Wendisch, P., 1966, Zum Temperaturverlauf in einer Platte bei Erwärmung durch intermittierende Infrarotbestrahlung, *Plaste und Kautschuk*, 13, 344–346.

CHAPTER 6

Photostability Testing: Design and Interpretation of Tests on New Drug Substances and Dosage Forms

N. H. Anderson and Stephen J. Byard

CONTENTS

0-415-30323-0/04/$0.00+$1.50

6.1 INTRODUCTION

Photostability testing is an essential part of product development and is required to ensure that satisfactory product quality is maintained during practical usage. A wide diversity of testing procedures has been employed by the pharmaceutical industry in the U.K. (Anderson et al., 1991) (Table 6.1) and in Germany (Thoma, 1996). This is not surprising, given (1) the absence of regulatory guidelines; (2) differing assumptions about practical product exposure (Table 6.2); and (3) the lack of published data on daylight (UV and visible) levels inside buildings. The principles had been discussed (Moore, 1987; Thoma and Kerker, 1992; Tønnesen, 1991) and the subject reviewed (Nema et al., 1995) before the 1996 publication of the International Conference on Harmonization (ICH) Photostability Guideline as an annex to the ICH Stability Guideline (ICH, 1996).

For unstable products, product quality is achieved through suitable labeling and use of a protective pack. This is in contrast to the situation with thermally unstable products, in which it is necessary for the manufacturer to determine the shelf-life

Table 6.1   Drug Product Photostability Testing: U.K. Practice

Light Source	
Accel. fluorescent	200–1200 lx
Exposure time	30–90 days
Total exposure	300–1080 klx days
Room fluorescent	1400–6000 lx
Exposure time	4–180 days
Total exposure	8–900 days
Xenon	150–180 klx
Exposure time	0.25–30 days
Total visible exposure	45–4500 klx days

Table 6.2   Assumptions about Practical Product Exposure

Products very rarely exposed to significant UV light
Blister pack left close to sunny window for up to 30 days
Accumulative exposure to visible light less than 1,000 klx h
Accumulative exposure to visible light up to 10,000 klx h

(i.e., period for which the product meets its quality specifications) under defined storage conditions.

Thus, although it is necessary to determine the *rate* of degradation for thermal stability testing, photostability testing only requires determining if the product photostability is sufficient to make a protective pack and warning label unnecessary; this is equivalent to a limit test. In order to determine the relative photostability of different product formulations, or the degree of protection afforded by different pack types, it may be necessary to conduct measurements of the photodegradation rate (Moore, 1990, 1996b; Tønnesen, 1991; Bunce, 1989). Such measurements are outside the scope of this chapter.

The methodology set forth in this chapter is largely based on the work of the ICH Photostability Working Group, which was formed to provide an international guideline for photostability testing of drug substances and drug products as part of the ICH process. The resulting ICH Guideline was published as a Step 4 document in November, 1996, and subsequently adopted by the Committee for Proprietary Medicinal Products on behalf of the European Agency for the Evaluation of Medicinal Products EMEA (1996); the U.S. Food and Drug Administration (1997); and the Japanese Ministry of Health and Welfare (1997). Photostability studies conducted for the purposes of product registration should be conducted according to this guideline. A useful technical guide and practical interpretation of the ICH Guideline has been published in two parts (Thatcher, 2001a, b).

UV irradiance is expressed in units of W m^{-2}, and UV exposure is defined as irradiance × time. The units Wh m^{-2} were chosen for practical convenience. Visible light is expressed as illuminance (lux) for practical convenience and also because indoor light levels are normally measured in lux; thus, the units for visible exposure are lux hours.

6.1.1 Basics of Photochemistry

In order to understand the basis of test design and the limitations of particular tests, it is important to understand the basic principles of photochemistry. Concepts such as the relationship between the photon flux of a source and its irradiance; quantum yield; and only absorbed photons causing photochemical reactions (Moore, 1987, 1996a) are fundamental. The spectral power distribution (SPD) of a source (i.e., the energy emitted as a function of wavelength and the absorption spectrum of the exposed product) is important in determining the nature and extent of the photochemical reaction.

It is also worth recalling that many covalent bonds are relatively strong and, unless an absorbed photon has the requisite energy, bond cleavage cannot occur. For example, a photon of 400 nm has an energy of 300 kJ mol^{-1}, which is less than that of most common single bonds in organic molecules; for example, C–C bond is 348 kJ mol^{-1}. In practice, UV irradiance is much more damaging to pharmaceutical products than visible radiation is because colorless (white) products absorb little or no visible radiation and UV photons have higher energy.

Table 6.3 Simulation of Light Levels

Exposure (klx h)	Days on Sunny Window Sill	Days 1 m from Window	Days under Artificial Light at 500 lx (24-h day)
1200	2–3	ca. 12	100
2400	4–6	ca. 24	200
3600	8–12	ca. 36	300
UV Exposure Wh m^{-2}			
200	1–2	ca. 50	
500	3–5	ca. 125	

6.2 OBJECTIVES OF TESTING — PHOTOSTABILITY UNDER PRACTICAL USAGE CONDITIONS

It is extremely difficult to establish the actual exposure of pharmaceutical products during practical usage. During discussions of the international harmonization of photostability testing of pharmaceutical products in which one of the authors (Anderson) participated, it was apparent that there were differences among Europe, the U.S., and Japan. In the U.S. and Japan, product exposure to glass-filtered daylight was believed to occur rarely, if at all, in hospitals or pharmacies (Riehl et al., 1995). In contrast, in Europe, it was recognized that products in hospital pharmacies or during distribution in hospitals, as well as possibly in the home, may be removed from their secondary (outer) carton and be exposed to glass-filtered daylight for periods of up to several days (Tønnesen and Karlsen, 1995). Therefore, it is important that the susceptibility of products to degradation by glass-filtered daylight or a source simulating glass-filtered daylight is determined.

The degree of visible and UV exposure to which products are subjected is not known with any certainty, and therefore any test of photostability is based on informed judgment, rather than fact. It is, however, possible to give approximate exposure values that correspond to time of exposure to indoor lighting or very close to a sunny window (Table 6.3). Thus, the results of photostability testing of pharmaceutical products should be regarded as essentially qualitative rather than quantitative.

6.3 FORCED PHOTODEGRADATION OF DRUG SUBSTANCE AND METHOD VALIDATION

The ICH Guideline states that the purpose of forced degradation studies is to evaluate the overall photosensitivity of the material for method development purposes and/or degradation pathway elucidation. No specific testing conditions are recommended. Clearly, in order to develop valid methods for determining the photostability of drug substances and products, a forced photodegradation study on the drug substance should be performed. It is easier to develop and largely validate methods for the product photodegradation using the drug substance. The "forced degradation" experiment on the drug substance should be conducted using a visible light and UV exposure in excess of that used for formal product testing (e.g., by a factor of three- to fivefold). Alternatively, exposure may be continued until significant degradation (up to 20%) has occurred.

These experiments are usually conducted on solid drug substance and also on a solution. The latter is more susceptible to degradation and is particularly relevant to the understanding of solution products' behavior. If no degradation is observed, no formal testing of drug substance photostability is required. The ICH Guideline on Validation of Analytical Procedures: Methodology (ICH, 1994) describes the principles of method validation.

According to the revised ICH Guideline on drug substance impurities (ICH, 2002) and the draft revised ICH Guideline on drug product impurities (ICH, 1999), which is currently at step 3, it is necessary to be able to report (i.e., quantify) impurities present in commercial drug substance as low as the 0.05% level (high-potency products ≤2 g/day) and the 0.03% level (low-potency products >2 g/day). For drug products, the current reporting limits are 0.1% for high potency products (≤1 g/day) and 0.05% for low potency products (>1 g/day) where the amount per day refers to the active ingredients. Thus, the methods developed should be capable of detecting photodegradants that could be formed in the formal tests at or above the ICH reporting levels. Because the structures of degradants are usually unknown, they are estimated by peak area relative to the drug substance, using HPLC with UV detection and assuming a response factor of 1 relative to drug substance. It is unnecessary to identify degradants below the identification thresholds given in the ICH Impurities Guidelines; practical details of testing are given in Section 6.5.

6.4 CONFIRMATORY TESTING (FORMAL TEST) FOR DRUG SUBSTANCE

6.4.1 Sample Presentation

The ICH Guideline gives useful practical advice on sample presentation, noting that the conditions should not alter the physical state of the drug substance and containers should allow full exposure of the sample. Thus, a thin layer of drug substance, typically 1 to 3 mm but not more than 3 mm, should be spread in a shallow dish and protected with a transparent cover such as a "cling" film or a thin film of polyethylene. The transmittance properties of the cover material may be determined by mounting a sample on a UV spectrophotometer cell holder and obtaining the absorption spectrum.

The extent of degradation is related to the sample surface area, which is related to particle size. Therefore, it is important that the sample of active ingredient used is representative of that to be used for making the clinical product. Because photo-degradation only occurs on the surface of solid samples, the surface area/weight ratio of the material exposed will affect the extent of observed degradation.

6.4.2   Light Sources

The ICH Guideline offers two sources: options 1 and 2. The merits and drawbacks of each of these sources are discussed in a separate section later.

6.4.3   Procedure

The ICH Photostability Guideline recommends that drug substance photostability testing be carried out under the same conditions as those used for product, in order for direct comparisons to be made. The recommended exposure is 1.2 million lx h and 200 Wh m^{-2} UV. During purification and manufacture of the drug substance, total exposure of the substance is extremely unlikely to exceed 100 klx h of visible light with no UV exposure. On this basis, the European Federation of Pharmaceutical Industries (EFPIA) Expert Working Group agreed informally that 100 klx h exposure was appropriate for the simulation of exposure during manufacture. Although such a test is not part of the ICH Guideline, it may be useful for internal control purposes.

The guideline recommends that samples be exposed side by side with a validated chemical actinometric system (e.g., quinine for near-UV region) to ensure that the specified exposure is obtained. Alternatively, exposure for the appropriate duration of time may be employed when conditions are monitored using calibrated radiometers/lux meters. The procedure used should ensure that any unevenness in the irradiance of sample area is taken into account. Any protected samples (e.g., wrapped in aluminum foil) used as dark controls to compensate for temperature effects should be placed alongside the authentic samples. However, foil-wrapped samples may, in fact, be at a lower temperature than unwrapped samples due to the reflecting properties of the foil.

6.4.4   Sample Analysis

At the end of the exposure period, the samples should be examined for any changes in physical properties (e.g., appearance, clarity, or color of solution) and for degradation by a validated method. Because any photodegradation is limited to the sample surface, it is important to ensure that a representative portion of solid drug substance is used in individual tests. Similar sampling considerations, such as homogenization of the entire sample, apply to other materials that may not be homogeneous after exposure. Analysis of the exposed sample should be performed concomitantly with that of any protected samples used as dark controls.

6.4.5   Evaluation of Results

The ICH Guideline states that

> Confirmatory studies should identify precautionary measures needed in manufacturing or in formulation of the drug product, and if light resistant packaging is needed. When evaluating the results of confirmatory studies to determine whether change due to exposure to light is acceptable, it is important to consider the results from the other

formal stability studies in order to assure that the drug will be within justified limits at the time of use (see the relevant ICH Stability and Impurity Guidelines).

This point is discussed in detail in Section 6.5.4 and Section 6.5.5.

6.5 TESTING PHARMACEUTICAL PRODUCTS

6.5.1 Sequential Product Testing

The ICH Guideline recommends a sequential approach as shown in Figure 6.1. For regulatory purposes, the product should first be tested unpacked. Alternatively, if necessary, a transparent container may be used for liquid/semisolid products. Unstable products are then further tested in primary and secondary (market) packs as necessary (shown in Figure 6.1). When products are stable in the primary pack but unstable without it, it is necessary to label products to prevent transfer into a less protective primary pack (for example, by a pharmaceutical wholesaler). It is unnecessary to conduct tests in containers completely impenetrable to light, such as metal tubes or cans, when these are used for direct dispensing to the patient. The product should be exposed to light using the conditions described for the drug substance in Section 6.4.3.

The design of in-use tests such as the stability of reconstituted lyophilized products, or products administered through an intravenous drip, is excluded from the ICH Guideline but should be based on the same principles and considerations as the formal product test. When high-volume infusion products are administered, they may be exposed to ambient light conditions for some hours and this could include exposure to glass-filtered sunlight.

It is normally only necessary to test one batch of drug substance and drug product during the development phase. Subsequently, the photostability characteristics of a single definitive pilot scale (or larger) batch is confirmed as described in the parent ICH Stability Guideline if the drug substance and product are clearly photostable or photolabile. If the results of the definitive studies are equivocal (borderline), confirmatory testing of up to two additional batches of drug substance and/or product should be conducted, or the material should be classified as unstable.

Figure 6.1   Sequential testing of product.

6.5.2 Sample Presentation

The same considerations apply as for the drug substance. For liquid products, exposure in a transparent container or a container with a transparent cover is recommended. All samples, whether packed or unpacked, should be presented in a way that maximizes exposure.

6.5.3 Sample Analysis

Normally the chromatographic method developed for the drug substance will also be valid for the product. However, if during photostability testing on product formulations, it becomes apparent that additional photodegradants are formed, it may be necessary to modify the chromatographic conditions to provide a suitable method. The ICH Guideline recommends that samples should be examined for changes in physical properties such as appearance and dissolution, as well as for degradation. The importance of using representative samples is emphasized.

6.5.4 Evaluation of Results

When evaluating the results of photostability studies to determine whether changes caused by exposure to light are acceptable, it is important to consider the results from other formal stability studies. The initial purity of the drug substance used to manufacture the product should also be taken into account. The point is discussed in the next section. When products are found to be unstable to light exposure, packaging and/or labeling and, when relevant, specific in-use instructions may be used to limit or prevent photodegradation.

6.5.5 Determination of "Acceptable" Change

For the drug substance, the recently revised ICH Guideline on Drug Substance Impurities (Q3A) describes how specification limits are set and how impurities may be qualified. Because the formation of photodegradants can be largely or completely prevented by suitable control measures, formal qualification should be only rarely needed. However, on occasions the same degradants in drug substance and/or product may be formed by thermal and photochemical degradation. Based on results of thermal and photochemical stability studies, a set of interrelated specification limits should be prepared following the procedure summarized in Figure 6.2.

1. Limit in drug substance at the time of manufacture. Unless the degradant is formed during purification of the drug substance, this limit should be very small.
2. Maximum limit in the drug substance to be used for product manufacture, corresponding to limit at the end of the retest period. This is the sum of the amounts of degradant formed under stability testing at 25°C/60% RH and the amount defined in the preceding point. No photodegradation of the drug substance should occur during storage and any occurring during packaging is taken into account in point 1.

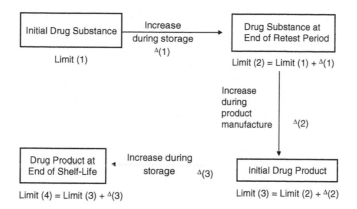

Figure 6.2 Setting degradant specification limits for a thermal/photodegradant.

3. Limit in the drug product at the time of manufacture. Some degradation may occur during product manufacture. Suitable controls during dispensing of the drug substance and product manufacture can be used to limit this degradation. The limit of degradant at the time of product manufacture is then the sum of the amount in the drug substance and that formed during the manufacture of the product.
4. Limit in the drug product at the end of shelf-life. The degradant limit at the end of shelf-life is the sum of the maximum amount present at the time of manufacture that formed through storage at 25°C/60% RH and should take into account degradant formed under formal photostability testing in the relevant packaging.

The limit at each stage is based on the limits at the previous stage plus any increase that may occur. For drug substances or drug products susceptible to photodegradation, it is up to the applicant to decide how much increase, if any, should be allowed for photodegradation. At each stage, specification limits should be justifiable and in accordance with the ICH Guideline.

6.6 LEVEL OF EXPOSURE AND LIGHT SOURCES

6.6.1 Exposure

During the course of ICH discussions on product testing, it was agreed that 1.2 million lx h, which represents 2 to 3 days' exposure close to a south-facing sunny window in the summer, is a suitable visible light exposure. A near-UV exposure of 200 Wh m^{-2} between 320 and 400 nm is recommended, corresponding to about 1 to 2 days close to a sunny window. This exposure recommendation is based on the assumption that products will be exposed to a mixture of glass-filtered natural light and indoor light, and not stored where they are exposed to sunlight for any length of time. When this is not the case, the UV exposure should be increased, for example, up to a maximum of 540 Wh m^{-2}, which is the highest figure considered by the ICH Working Group.

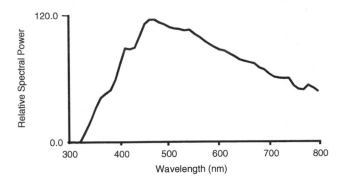

Figure 6.3 Standard indoor indirect daylight ID-65.

UV levels decline rapidly outside the region of direct sunlight inside a room; therefore, the ratio of UV/visible exposure recommended is less than that in "standard" glass-filtered daylight ID-65 (Clarke, 1979; ISO 10977:1993(E)) for which 1.2 million lx h corresponds to approximately 540 Wh m^{-2}. UV exposure below 320 nm is unnecessary because levels of radiation below this wavelength are negligible in glass-filtered daylight. The spectral distribution of standard glass-filtered daylight, ID-65, is shown in Figure 6.3.

6.6.2 Sources

The sources used should be comparable in spectral distribution to those to which products are exposed in practical use, namely, glass-filtered daylight and indoor lighting. A source closely simulating standard glass-filtered daylight (Clarke, 1979) will provide a relatively greater UV exposure than most practical situations. It is most important that the upper (360 to 400 nm) and lower (320 to 360 nm) UV ranges contain a significant percentage of the total UV irradiation (e.g., ≥25%) and that the UV irradiation essentially extends over the whole band. This is necessary to ensure that products that absorb only in a small part of the UV region — e.g., 320 to 340 nm or 380 to 400 nm — receive a meaningful UV exposure (Tønnesen and Karlsen, 1995). Some theoretical possibilities are illustrated in Figure 6.4.

The testing standard DIN 53 387 (DIN 53 387, 1989) requires the UV spectral distribution in the UV to be 3 ± 0.5% (320 to 360 nm) and 6 ± 1% (360 to 400 nm), where 100% is the total irradiance over 300 to 800 nm; this corresponds accurately to standard glass-filtered daylight. This distribution is not essential for pharmaceutical product testing; sources close to or meeting this standard can be used with confidence, but they will provide a greater UV exposure than necessary (ca. 540 Wh m^{-2} per 1.2 million lx h). When reproducibility of results between laboratories is important, the spectral distribution of the sources used should be as nearly identical as possible, irrespective of whether they conform to any particular specification.

The ICH Guideline recommends two options for sources. It also states that an appropriate control of temperature should be maintained to minimize the effect of localized temperature changes, or a dark control included in the same environment.

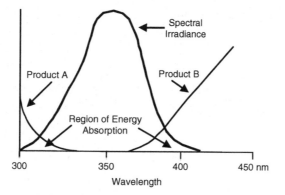

Figure 6.4 Energy absorption of hypothetical products from a near-UV tube.

For source options 1 and 2, one may rely on the spectral distribution specification provided by the light source manufacturer.

Option 1 — any light source designed to produce an output similar to the D65/ID65 emission standard, such as an artificial daylight fluorescent lamp combining visible and ultraviolet (UV) outputs, xenon, or metal halide lamp. D65 is the internationally recognized standard for outdoor daylight as defined in ISO 10977 (1993). ID65 is the equivalent indoor indirect daylight standard. For a light source emitting significant radiation below 320 nm, an appropriate filter may be fitted to eliminate such radiation.

Option 2 — the same sample should be exposed to:

1. A cool white fluorescent lamp designed to produce an output similar to that specified in ISO 10977 (1993)
2. A near-UV fluorescent lamp with a spectral distribution from 320 to 400 nm and maximum energy emission between 350 and 370 nm; a significant proportion of UV should be in bands of 320 to 360 nm and 360 to 400 nm

With option 2, testing may be performed sequentially on the same samples, or concurrently using a cabinet fitted with both sources. However, the testing period required to reach an exposure of 1.2 million lx h (typically $\geq$120 h) is longer than that required to reach the required UV exposure under a near-UV fluorescent lamp.

Advantages and disadvantages of option 1 and option 2 sources are summarized in Table 6.4 and discussed in more detail next. Apart from the advantages and disadvantages of these option sources, it is worth reiterating that, with a single source for visible and UV irradiance, the UV/visible ratio is fixed and the user cannot determine if the material under test is degraded by visible light alone. Therefore, it is advantageous to have a system that permits exposure to cool white fluorescent tubes. This system can also be used in combination with an option 1 source in a sequential test to provide the required ratio of UV/visible irradiation. For example, a xenon source could be used to provide the full UV exposure of 200 Wm^{-2} and the associated visible exposure of 0.444 million lx h. The balance of visible exposure, 0.756 million lx h could then be provided by the cool white fluorescent source. This

Table 6.4　Advantages and Disadvantages of Different Sources

	Advantages	Disadvantages
Option 1 sources		
Artificial daylight or full spectrum fluorescent lamp	Inexpensive Long life, >10,000 h Low heat output Large sample area	Not designed for scientific work — variability tube-to-tube and along tube length Uneven irradiation of sample area Long "burning in" period Long exposure needed Low emission over 380–420 nm Not suitable for use with quinine actinometer
Xenon or tungsten halide lamps	Designed for photostability testing/reproducible Good solar simulation Even irradiation of sample area Short exposure time Calibrated photometers usually available	Expensive Short life, <2000 h for xenon High heat output Small sample area
Option 2 sources		
Cool white fluorescent lamp	Inexpensive Long life, >10,000 h Low heat output Large sample area	Not designed for scientific work — variability tube to tube and along tube length Uneven irradiation of sample area Long "burning in" period Long exposure needed
Near UV fluorescent lamp	Inexpensive Long life, >10,000 h Low heat output Large sample area Suitable for quinine actinometry if SPD very similar to reference source used by FDA	Not designed for scientific work — variability tube to tube and along tube length Uneven irradiation of sample area Low emission over 380–420 nm Poor solar simulation

approach was discussed at a recent conference at which it was considered to be an acceptable alternative (Photostability, 1999).

Overall, it can be seen that each of the sources or combination of sources has its advantages and disadvantages that need to be taken into account when selecting the source or sources to use for testing. Although in terms of spectral power distribution the sources are different, it was not considered necessary to specify suitable sources more narrowly, at the time of publication of the ICH Guideline, because photostability testing is officially classified as a stress test.

Sources currently used or being evaluated by the pharmaceutical industry are:

1. Xenon lamps
2. Metal halide lamps
3. White fluorescent tubes
4. Artificial daylight tubes
5. "Full spectrum" daylight fluorescent tubes
6. Near UV fluorescent tubes

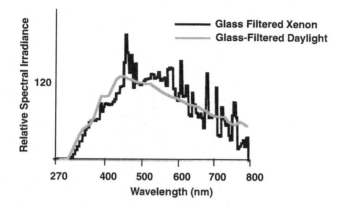

Figure 6.5 Glass-filtered xenon source and glass-filtered daylight.

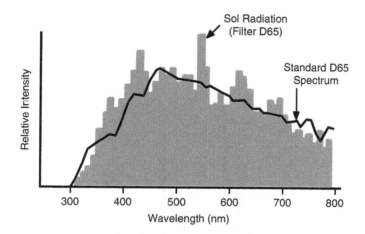

Figure 6.6 Metal halide lamp (Sol System).

The spectral distributions of these sources are shown in Figure 6.5 through Figure 6.10. The sources producing significant radiation below 320 nm are principally 1, 2, and 6; a window glass filter may then be used to remove such irradiation. Some lamp suppliers provide window glass filters. It should be noted that, for source 6, the quinine actinometric calibration has only been performed for the unfiltered source. Fluorescent tubes should be "burnt in" for 100 to 200 h or more because their spectral distribution changes during this period and their output declines noticeably. Output also varies from tube to tube and along the lengths of tubes. Therefore, the visible (and UV) levels across the sample area should be mapped to ensure samples are placed at points of equal irradiance. Tubes should be changed at defined intervals (e.g., 5000 or 10,000 h) or when the output has declined to the lowest level acceptable to the user.

The full spectrum True-Lite (U.K.)/Vita-Lite (U.S.) tube, manufactured by Aura (Sweden), emits 8% of its irradiance over 300 to 800 nm in the UV region. Only 0.5% of this is <320 nm, so a window glass filter is not necessary (Figure 6.9). The

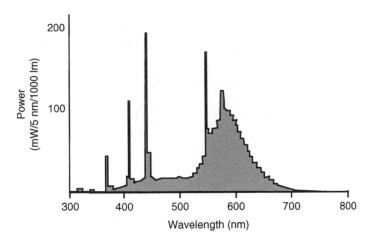

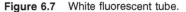

Figure 6.7   White fluorescent tube.

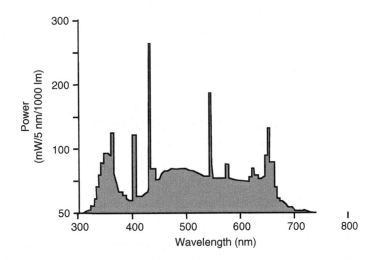

Figure 6.8   Artificial daylight fluorescent tube.

irradiance in the 380- to 400-nm region is relatively low (2.5%) compared to a glass-filtered xenon source (approximately 4%), and the total UV is also slightly lower (8% compared with 9.5% for glass-filtered xenon). However, this and similar full spectrum or artificial daylight tubes do provide an acceptable UV spectral distribution for photostability testing.

For tests simulating indoor light, for which no international standard exists, a typical cool white or white fluorescent tube with a color temperature of 3500 to 4300 K is recommended. For tests simulating glass-filtered daylight, artificial day-light tubes or full spectrum tubes from a specialist manufacturer of tubes designed for this purpose are preferable (see Figure 6.8 and Figure 6.9). Tests using white fluorescent and near-UV tubes sequentially or as a combination are more complicated

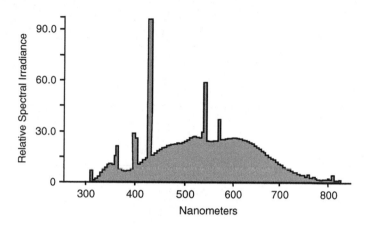

Figure 6.9 Vita-Lite "full spectrum" fluorescent tube.

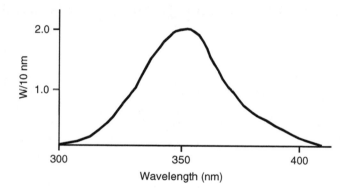

Figure 6.10 Near-UV fluorescent tube.

to conduct, and the spectral distribution is not as close to glass-filtered daylight (see Figure 6.7 and Figure 6.10).

Xenon lamps made by specialist suppliers are the most widely used daylight-simulation sources (Hibbert, 1991) because their spectral distribution is close to solar distribution; changes only to a very small extent during the lifetime of the lamp; and is not affected by voltage fluctuations. However, the small illuminated area (typically 25 × 25 cm for a small unit) and high heat output limit its usefulness, although a cooling unit can be fitted. A window glass filter is used to remove radiation below 320 nm. This will also absorb UV at lower wavelengths, particularly 320 to 360 nm. Specialist metal-halide lamps (with a glass filter) can give a spectral distribution (Figure 6.6) that is as representative of the solar distribution as that of the xenon source (Figure 6.5). The UV output declines to a small extent as the source ages. Heat output and the small sample area are again limitations.

If a combination of white fluorescent and near-UV tubes is used, it is important to check that the product does not absorb primarily in the 320- to 340-nm or 390- to 420-nm region, where it has been noted that this combination of sources may produce little or no output (Tønnesen and Moore, 1993; Tønnesen and Karlsen, 1995).

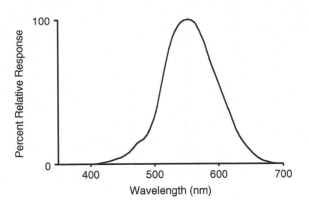

Figure 6.11   CIE photopic curve (Lux).

6.7 MEASUREMENT OF VISIBLE LIGHT AND UV IRRADIATION

6.7.1   Visible Light

A lux meter, calibrated at intervals prescribed by the manufacturers, is recommended. Because lux is a measurement of light as perceived by the human eye (Figure 6.11) rather than an absolute measurement, a lux meter is not suitable for comparing the output of different types of source. However, it is well suited for checking the evenness of illumination of the sample area and measuring changes in source output with time.

6.7.2   UV Irradiation

The ICH Working Party on Photostability sought to identify a simple and inexpensive means for the average pharmaceutical laboratory to ensure that samples receive the required level of UV exposure. Two methods are recommended in the ICH Guideline.

6.7.2.1   Radiometry

The UV filter radiometer comprises a detector with a UV wide-band filter, which allows UV irradiation of defined wavelengths to reach the sensor (see Figure 6.12). Unfortunately, no international standards have been established for filters, so the relative weighting given to the different UV wavelengths depends on the filters, i.e., meters from different manufacturers do not measure the same fraction of irradiance from a given source. Meters should be fitted with a cosine corrector to allow incident radiation from a wide angle of incidence to be measured. The geometrical relationship between the cosine corrector and the filter should be such that the radiation allowed to pass through the filter is not dependent on the angle of incidence.

UV filter radiometers may be calibrated to give an absolute measurement of irradiance for a particular type of source, e.g., xenon or fluorescent tube of a

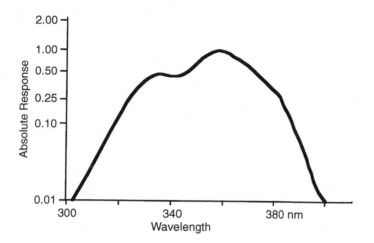

Figure 6.12 UV spectral response for 365-nm radiometer filter.

particular type. These meters provide a simple means of measuring evenness of irradiance across the sample area and changes with time for a particular source. Calibration requires specialized knowledge and equipment and should not be undertaken by unqualified personnel. Radiometers cannot be used to compare irradiance between sources unless calibrated specifically for each source. Recalibration should be conducted in accordance with the manufacturer's instructions. A spectroradiometer is a relatively expensive item of equipment that few pharmaceutical companies would wish to purchase; it is used to measure the absolute spectral irradiance of a source.

6.7.2.2 Quinine Actinometry

The principles of actinometry are well established, but there is no well-established actinometric system sensitive only to UV radiation (Kuhn et al., 1989; Piechocki and Wolters, 1993). Quinine was originally proposed by the Japanese National Institute of Health Sciences and the Japanese Pharmaceutical Manufacturer's Association (Yoshioka et al., 1994). The results of the work of the FDA and U.S. laboratories have confirmed its suitability (Drew et al., 1998). At the recommended concentration (2%), the quinine solution has an absorbance of ≥2 from 320 to 367 nm, so the UV irradiance over this range will be determined using this system. Quinine usually contains a number of alkaloid impurities, principally dihydroquinine. However, the level of the latter does not affect its performance as a UV actinometer.

The monohydrochloride dihydrate salt has been selected as having adequate solubility, and commercially sourced high-purity quinine monohydrochloride is suitable. Recent work by the FDA and U.S. industry has shown that during UV exposure, quinine solutions undergo a small fall in pH, causing an increase in absorbance at 400 nm, which is the wavelength recommended for actinometry. In order to make reliable measurements, it is therefore important that the vessels used are free from traces of acid, base, and buffer salts. The very steep slope of the quinine absorbance

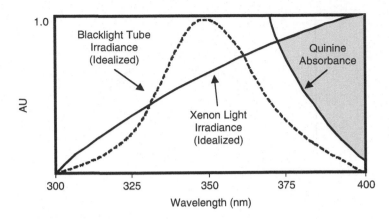

Figure 6.13   Possible limitation of quinine actinometer.

curve at 400 nm (>0.01 AU nm^{-1}) means that the monitoring wavelength of 400 nm must be set very accurately, preferably using a diode array spectrophotometer.

The quinine actinometric system described earlier will need to be calibrated for each type of light source used for photostability testing, using the same type of vessel (cuvette or ampule) to be used in actinometry. This is because the proportion of the UV irradiance absorbed by the quinine will depend on the spectral irradiance of the source (Figure 6.13). It can be seen that the shaded UV region of the xenon source is not absorbed by the quinine solution. The calibration will give the absorbance change at 400 nm equivalent to a defined UV exposure from a particular source.

For near-UV fluorescent tubes, the ICH Guideline indicates that an absorbance change of 0.5 AU can be taken as corresponding to 200 Wh m^{-2} UV irradiance for solution contained in a 1-cm quartz cell. This calibration relates to a Sylvania F20 T12/BLB lamp, but should be the same for other lamps of similar spectral power distributions. The consequences of using a near-UV lamp with a different SPD have been pointed out (Sequeira and Vozone, 2000). Exposure of ampules or cuvettes of quinine solution alongside test samples, or in a separate calibration experiment, can be used to ensure products receive the correct UV exposure.

More in-depth investigations of the behavior of quinine as an actinometric system have revealed some significant shortcomings. However, if careful attention is paid to certain practical points, it can be used to determine UV exposure. The first point to note is that the response of quinine to UV irradiation is temperature dependent, suggesting that the observed response is not the result of a purely photochemical process (Christensen et al., 2000, Baertschi, 1997). Therefore, the temperature of exposed solutions should be controlled to 20 to 25°C (Tønnesen and Karlsen, 1995; Sequeira and Vozone, 2000).

Although it has been suggested that the quinine response to UV irradiation is affected by the level of dissolved oxygen, this phenomenon has not been observed in Sanofi–Synthélabo's laboratories. The involvement of a nonphotochemical reaction in the change of quinine absorbance at 400 nm, after exposure to UV irradiation, is supported by observations that the absorbance at 400 nm continues to increase after the irradiation is complete (Kester et al., 1996; Christensen et al., 2000). This means that it is essential to analyze the quinine solutions immediately after the test is complete.

Quinine actinometry using solutions in cuvettes is slightly more inconvenient than UV radiometry for checking the evenness of irradiance across the sample area and for monitoring changes in source output with time. However, it has the advantage of never requiring recalibration unless the type or manufacturer of the source is changed.

6.7.2.3 Alternatives to Quinine

Because quinine does not meet the generally accepted requirements of a chemical actinometer (Bunce, 1989), possible alternatives have been investigated. Recent work (Allen et al., 2000) has confirmed that 2-nitro benzaldehyde (2NB) is converted to 2-nitroso benzoic acid on irradiation, following the original work (Pitts, 1966), and that the amount of acid formed can be readily estimated by titration. Alternatively, the amount of base corresponding to the acid liberated after the required exposure can be calculated and added to the 2NB solution before exposure. By monitoring the pH or adding phenolphthalein as an indicator, the end-point can readily be determined.

Allen et al. calculated the base equivalents for the required ICH exposure of 2.26×10^{-4} Einsteins (number of photons) for two types of near-UV tube (Toshiba FL40BL; Sylvania F20T/2/BLB); a full spectrum fluorescent tube (Vita-Lite, Duro-test); and a UV cool white tube (Sylvania FT20 T/2/CW). The two types of near-UV tube employed in this work are virtually identical, and the Sylvania F20T/2/BLB was used in the FDA's calibration of quinine. It should be noted that in the past, U.S. white fluorescent tubes emitted a significant level of UV radiation (demonstrated in the work of Allen et al.), whereas the European equivalents do not due to the use of iron-containing glass in the tube. Another alternative to quinine that has been considered is *trans*-2-nitrocinnamaldehyde. The photochemistry for this molecule is similar to that of 2-nitrobenzaldehyde, except that the material undergoes a *trans–cis* conversion as well as disproportionation to give 2-nitroso-cinnamic acid.

When exposure takes place in methanolic solution, the *cis* and *trans* acids are converted into the corresponding esters (Bovina et al., 1998). Similarly in ethanol, the ethyl esters are formed. Fortuitously, the *cis* and *trans* esters have similar extinction coefficients to the corresponding aldehydes at 440 nm, so, in spite of the complexity of the photochemistry, an approximately linear increase in absorbance at 440 nm is observed with time of exposure. However, the ideal actinometer should be based on a single photochemical change (Kuhn et al., 1989); therefore, the photochemical behavior of t-NCA in acetonitrile was examined (Anderson et al., 2001).

The absorbance change at 440 nm after an exposure of a 0.5% solution of t-NCA, in acetonitrile or methanol, to 200 Wh m^{-2} from a Heraeus Suntest unit (xenon source with window glass filter) was ca. 0.9 units. This is consistent with a change of ca. 0.75 observed (Bovina et al., 1998) after an exposure of 167 Wh m^{-2} using methanol as solvent. However, experiments have been completed to show that the linearity of the response obtained using acetonitrile as solvent is superior to that observed with methanol (Anderson et al., 2001). Similar absorbance changes, typically of 0.91, have been observed when exposing samples of t-NCA in acetonitrile to irradiation of 200Wh m^{-2} from a near-UV lamp.

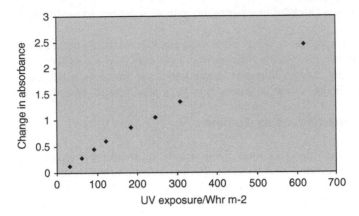

Figure 6.14 Change in absorbance of t-NCA in acetonitrile upon exposure to radiation from a near-UV source.

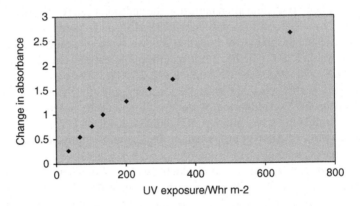

Figure 6.15 Change in absorbance of t-NCA in methanol upon exposure to radiation from a near-UV source.

As for experiments using the Heraeus Suntest, the linearity observed for t-NCA in acetonitrile was superior to that obtained for methanol over an exposure range extending to three times that recommended by the ICH. Coefficients of variation of 0.999 and 0.990 for t-NCA in acetonitrile and methanol, respectively, were obtained over a range extending up to 200 Wh m^{-2}. The data are shown in Figure 6.14 and Figure 6.15. This system is still under validation; however, data acquired to date indicate that it may be a good future alternative to quinine.

6.8 CONCLUSIONS

In order to provide a sound basis for making decisions with regard to labeling and the use of protective packs, pharmaceutical laboratories should determine the susceptibility of their products to simulated glass-filtered daylight, which can be optionally supplemented with visible light. The ICH Guideline on Photostability represents a realistic test, with an acceptable level of reproducibility to the regulatory agencies, and is reasonably straightforward to conduct in pharmaceutical R&D laboratories. The guideline represents a major step forward in harmonizing photostability testing of pharmaceutical products. Now that practical experience in the use of the guideline has been gained, opportunities for improving it may be identified. In particular, alternatives to quinine are now available, although a definitive study on the use of t-NCA in acetonitrile has yet to be published.

ACKNOWLEDGMENTS

Thanks are due to all the members of the ICH Photostability Working Party for their contributions to ICH Photostability Guideline. Informal discussions with Professor H.H. Tønnesen, Dr. J. Piechocki, and Dr. P. De Fillipis were appreciated. Also, the authors would like to acknowledge the work done by Mrs. P. Phillips and Mr. L. Stevens to assess the use of *trans*-2-nitrocinnamaldehyde as an actinometer.

REFERENCES

Allen, J.M., Allen, S.K., and Baertschi, S.W. (2000). 2-Nitrobenzaldehyde: a convenient UV-A and UV-B chemical actinometer for drug photostability testing. *J. Pharm. Biomed. Anal.*, 24, 167–178.

Anderson, N.H., Johnston, D., McLelland, M.A., and Munden, P. (1991). Photostability testing of drug substances and drug products in U.K. pharmaceutical laboratories. *J. Pharm. Biomed. Anal.*, 9, 443–449.

Anderson, N.H., Byard, S.J., Phillips, P., and Stevens, L. (2001). Unpublished observations.

Baertschi, S.W. (1997). Commentary on the quinine actinometry system described in the ICH draft guideline on photostability testing of new drug substances and products. *Drug Stability*, 1(4), 193–195

Bovina, E., De Filippis, P., Cavrini, V., and Ballardini, R. (1998) *trans*-2-Nitrocinnamaldehyde as a chemical actinometer for the UV-A range in photostability testing of pharmaceuticals. In Albini, A. and Fasani, E. (Eds.), *DRUGS Photochemistry and Photostability*, pp. 305–316, London: The Royal Society of Chemistry.

Bunce, N.J. (1989) Actinometry, in Sciano, J. (Ed.) *CRC Handbook of Organic Photochemistry*, pp. 241–260, Boca Raton, FL: CRC Press.

Christensen, K.L., Christensen, J.Ø., Frokjaer, S., Langballe, P., and Hansen, L.L. (2000). Influence of temperature and storage time after light exposure on the quinine monohydrochloride chemical actinometric system. *Eur. J. Pharm. Sci.*, 9, 317–321.

Clarke, F.J.J. (1979). *Proceedings of 19th Session of CIE*, Kyoto, Japan, p. 75.

DIN 53 387 (1989). Artificial weathering and ageing of plastics and elastomers by exposure to filtered xenon arc radiation. Deutsches Institut für Normung e.V., 10772 Berlin, Germany (E-mail postmaster@din.de). <www.din.de>

Drew, H.D., Brower, J.F., Juhl, W.E., and Thornton, L.K. (1998). Quinine photochemistry: a proposed chemical actinometer system to monitor UV-A exposure in photostability studies for pharmaceutical drug substance and drug product photostability studies. *Pharmacopoeial Forum*, 24, 6334–6346.

Hibbert, M. (1991). Shedding light on stability testing. *Manuf. Chemist*, 62, 32–33.

ICH (1994) Q2B, Validation of analytical procedures; methodology. <www.ich.org/ich5q.html> or ICH Secretariat c/o IFPMA, Geneva, Switzerland.

ICH (1996) Q1B, Photostability testing of new drug substances and products. <www.ich.org/ich5q.html> or ICH Secretariat c/o IFPMA, Geneva, Switzerland.

ICH (1999) Q3B(R), Impurities in drug products (revised guideline), <www.ich.org/ich5q.html> or ICH Secretariat c/o IFPMA, Geneva, Switzerland.

ICH (2002) Q3A(R), Impurities in new drug substances (draft revised guideline), <www.ich.org/ich5q.html> or ICH Secretariat c/o IFPMA, Geneva, Switzerland.

ISO 10977:1993(E) Photography — Processed photographic color films and paper prints — methods for measuring image stability. International Organization for Standardization, Case Postale 56, CH-1211 Geneve 20, Switzerland.

Kester, T.C., Zhan, Z., and Bergstrom, D.H. (1996). Quinine actinometry system described in the ICH draft guideline on photostability testing on new drug substances and products. Presentation given at the American Association of Pharmaceutical Sciences Meeting, Seattle, WA.

Kuhn, H.J., Braslavsky, S.E., and Schmidt, R. (1989). Chemical actinometry. *Pure Appl. Chem.*, 61, 187–210.

Moore, D.E. (1987). Principles and practice of drug photodegradation studies. *J. Pharm. Biomed. Anal.*, 5, 441–453.

Moore, D.E. (1990). Kinetic treatment of photochemical reactions. *Int. J. Pharm.*, 63, R5–R7.

Moore, D.E. (1996a). Photophysical and photochemical aspects of drug stability, in Tønnesen, H.H. (Ed.), *Photostability of Drugs and Drug Formulations*, pp. 9–38, London: Taylor & Francis.

Moore, D.E. (1996b). Standardization of photodegradation studies and kinetic treatment of photochemical reactions, in Tønnesen, H.H. (Ed.), *Photostability of Drugs and Drug Formulations*, pp. 63–82, London: Taylor & Francis.

Nema, S., Washkuhn, R.J., and Beussink, D.R. (1995). Photostability testing: an overview. *Pharm. Tech.*, 19(3), 170–185.

Piechocki, J.T. and Wolters, R.J. (1993). Use of actinometry in light-stability studies. *Pharm. Tech.*, 17(6), 46–52.

Pitts, J.N. (1966). Experimental methods in photochemistry, in Calvert, J.G. and Pitts, J.N. (Eds.), *Photochemistry*, pp. 783–788, New York: John Wiley & Sons.

Photostability (1999). 3rd International Conference on Photostability of Drug Substances and Drug Products, Washington, D.C., 10–14 July, 1999.

Riehl, J.P., Maupin, C.L., and Layloff, T.P. (1995). On the choice of a light source for the photostability testing of pharmaceuticals. *Pharmacopoeial Forum*, 21, 1654–1663.

Sequeira, S. and Vozone, C. (2000). Photostability of drug substances and products. *Pharm. Tech.*, 24(8), 30, 32, 34, 35.

Thatcher, S.R., Mansfield, R.K., Miller, R.B., Davis, C.W., and Baertschi, S.W. (2001a). Pharmaceutical photostability: a technical guide and practical interpretation of the ICH guideline and its application to pharmaceutical stability. I. *Pharm. Tech., N. Am.*, 25(3), 98, 100, 102, 104, 106, 108, 110.

Thatcher, S.R., Mansfield, R.K., Miller, R.B., Davis, C.W., and Baertschi, S.W. (2001b). Pharmaceutical photostability: a technical guide and practical interpretation of the ICH guideline and its application to pharmaceutical stability. II. *Pharm. Tech., N. Am.*, 25(3), 50, 52, 54, 56, 58, 60, 62.

Thoma, K. and Kerker, R. (1992). Photoinstabilität von Arzneimitteln. I. Mitteilung über das Verhalten von nur im UV-Bereich absorbierenden Substanzen bei der Tageslichtsimulation. *Pharmazeutische Industrie*, 54, 169–177.

Thoma, K. (1996). Photodecomposition and stabilization of compounds in dosage forms, in Tønnesen, H.H. (Ed.), *Photostability of Drugs and Drug Formulations*, pp. 136–137, London: Taylor & Francis Ltd.

Tønnesen, H.H. (1991). Photochemical degradation of components in drug formulations. I. An approach to the standardization of degradation studies. *Pharmazie,* 46, 263–265.

Tønnesen, H.H. and Moore, D.E. (1993). Photochemical degradation of components in drug formulations. *Pharm. Techn. Int.,* 5, 27–33.

Tønnesen, H.H. and Karlsen, J. (1995). Photochemical degradation of components in drug formulations. III. A discussion of experimental conditions. *Pharmeuropa,* 7, 137–141.

Yoshioka, S., Ishihara, Y., Terazono, T., Tsunakawa, N., Murai, M., Yasuda, T., Kitamura, T., Kunihiro, Y., Sakai, K., Hirose, Y., Tonooka, K., Takayama, K., Imai, F., Godo, M., Matsuo, M., Nakamura, K., Aso, Y., Kojima, S., Takeda, Y., and Terao, T. (1994). Quinine actinometry as a method for calibrating ultraviolet radiation intensity in light stability testing of pharmaceuticals. *Drug Dev. Ind. Pharm.,* 20, 2049–2062.

CHAPTER 7

The Questions Most Frequently Asked

Hanne Hjorth Tønnesen and Steve W. Baertschi

CONTENTS

7.1 INTRODUCTION

Stability studies are an integral part of the drug development process and are widely recognized as one of the most important areas in the registration of pharmaceutical products. A worldwide application of a single stability dosier for a new drug substance or product can be a significant challenge due to differences in requirements based on regional pharmacopoeia or local government regulations. As a step toward standardization of manufacturing and quality, the ICH Harmonized Tripartite Guideline on Stability Testing of New Drug Substances and Products (ICH Q1A) has been accepted by the largest pharmaceutical markets, i.e., the U.S., the European Union, Japan, and Canada. After January 1, 1998, it is obligatory to provide a stability database constructed according to this guideline for all new drug license applications filed in these regions. Within this parent guideline, it is stated that photostability testing should be an *integral part* of the studies undertaken to elucidate drug stability characteristics. Photostability testing is further addressed in a separate official ICH document (ICH Q1B).

Stability assessment begins with studies on the drug substance. A distinction is made between *forced degradation studies* (stress testing) and *confirmatory studies*. A forced degradation study is testing under extreme conditions used to characterize the drug substance or drug product; to determine degradation products and reaction mechanisms; and to develop appropriate methodology for quantitation (Chapter 10). Confirmatory studies apply to the drug substance and the drug product. Confirmatory studies involve testing under conditions designed to generate data to predict what might happen during storage under normal conditions, and to determine whether precautionary measures are needed during formulation, production, and storage. A confirmatory study can be regarded as a limit test, and should conclude with "acceptable" or "unacceptable" change. Stability studies of the drug product will be designed using the information obtained on the drug substance.

Despite implementation of the ICH stability and photostability guidelines (i.e., ICH Q1A and Q1B), issues remain that are not specifically covered in the documents and left to the applicant's discretion. The design of a testing protocol is of great importance because the results of the testing program can affect the chemistry, manufacturing, and controls part of drug development. The guideline on photostability testing allows for alternative approaches, assuming that these are scientifically sound. The aim of testing should be to demonstrate that exposure to irradiation does not result in an unacceptable change. The guideline does not address the evaluation of photostability characteristics during the product's actual period of use (i.e., in-use photostability testing). It is left to the applicant to establish how the product will be used and to undertake appropriate photostability studies. Photochemical reactions are far more complex than thermal processes and many questions must be considered in addressing photostability testing. This chapter is intended to assist nonexperts in this field in designing a test protocol by discussing the practical problems frequently encountered.

7.2 QUESTIONS RELATED TO RADIATION SOURCE

The term *light source* is used throughout the guideline. *Light* refers, however, to the photopic response, i.e., radiant energy acting on the retina. Visual perception is normally taken to cover a range of 400 to 800 nm with a maximum in response sensitivity around 550 nm. Although the term *light* often is recognized as having a broader meaning, the scientifically correct terms in this context are radiation, photon, or photolysis source. These will cover the UV regions and the visible light.

7.2.1 Why Combine UV Radiation and Visible Light?

Solar radiation below 300 nm is filtered by the ozone layer and does not reach the surface of the Earth. Normal window glass will not transmit radiation below 315 to 320 nm. UVA (320 to 400 nm) and visible light are therefore associated with daylight filtered through window glass or indoor fluorescent radiation while UVB (280 to 320 nm) is important when outdoor conditions are simulated.

Different practices might be encountered within Europe, the U.S., Japan, or other regions with respect to photoexposure of the drug substance or product. Routines for handling drug products seem to vary from continent to continent. In Europe, it is recognized that products may be removed from their secondary (outer) carton and thus be exposed to glass-filtered daylight for longer periods in hospitals or pharmacies (Tønnesen and Karlsen, 1995). In some cases, a drug product can even be exposed to direct sunlight. When the ICH photostability guideline was being developed, it was considered that in the U.S. and Japan, drug products would not be exposed to natural daylight and only indoor fluorescent radiation in combination with low amounts of window-filtered sunlight would need to be considered. Clearly, it is difficult to predict the amount of UV and visible (VIS) radiation to which the product may be exposed during the shelf-life. The spectral distribution and overall illumination used in a photostability study should provide a "worst-case" exposure and therefore include UV and visible radiation.

7.2.2 Why Two Options?

Again, the incorporation of two options into the ICH guideline reflects different practices in different continents for handling drug products. Option 1 radiation sources try to mimic daylight or indirect daylight filtered through window glass. These sources expose the samples to UVB, UVA, and visible light simultaneously. In practice, it is unlikely that the product will be exposed to direct sunlight for any length of time during manufacturing, shipping, and handling; therefore, a source providing glass-filtered daylight (corresponding to the ID65 standard as specified in the guideline) should be appropriate. Actually, option 1 offers choice among three different types of sources; glass-filtered daylight can be obtained by using a full spectrum or daylight fluorescent lamp, which has UV and VIS output, or by using a xenon or metal halide lamp in combination with appropriate filters (Chapter 5 and Chapter 6).

Option 2 radiation sources attempt to mimic indoor lighting conditions that (assuming that the room has a window) are composed of glass-filtered daylight and artificial radiation provided by white fluorescent lamps. Unlike natural daylight, indoor lighting conditions have no standard; the emission spectra of fluorescent tubes vary greatly. Option 2 allows the use of two separate lamps: one for UVA emission and one for visible light. The UV contribution is lower than would be obtained from lamps providing output meeting the D65 or ID65 emission standard. The lamps can be used in combination or sequentially. However, it is recommended to expose the sample simultaneously to UV and visible radiation unless previous results show that the sequence of irradiation (UV + visible, or visible + UV) is not important. The irradiation sequence may influence the results if the drug substance forms colored degradation products that can act as sensitizers or have a filter effect on the further degradation process (Tønnesen and Karlsen, 1995). It has been noted that a combination of sources according to option 2 may produce little or no output between 380 and 430 nm (Tønnesen and Moore, 1993). It is important to check that the sample does not absorb primarily in this region to exclude a "false" photostable formulation.

7.2.3   Will Options 1 and 2 Give the Same Results?

The choice of option can affect the extent of degradation and could therefore result in dramatically different results (Kerker, 1991; Thoma and Kerker, 1992a, b; Matsuo et al., 1996; Thoma and Kübler, 1996; Drew, 1998; Spilgies, 1998). From a scientific point of view, the two options given in the ICH guideline are very different (Piechocki, 1998). From a regulatory perspective, the two are regarded as equally sufficient in ascertaining the photostability of pharmaceuticals. Thus, no guidance is given in the ICH guideline for the choice of option 1 or 2. It is recommended that the same option be used for forced degradation studies and confirmatory studies to avoid a discrepancy in degradation rates and profiles. Information about the wavelength dependence of the photodegradation of the particular drug formulation would be of great help in order to select the right radiation source. Factors that may be considered during the selection of radiation source are outlined in Table 7.1.

7.3   QUESTIONS RELATED TO EXPERIMENTAL DESIGN

7.3.1   How to Obtain the Correct Overall Illumination?

For confirmatory studies, the ICH guideline recommends a total exposure of not less than 200 Wh m^{-2} in the UV range (320 to 400 nm) and 1.2 million lx h in the visible range (400 to 800 nm). Inside a room, the UV level decreases rapidly outside the region of direct sunlight (i.e., farther away from any windows). The minimum visible exposure levels proposed in the ICH guideline represent approximately 3 months of continuous exposure to artificial (visible) light in the pharmacy, warehouse, or home with the protective container removed from the product (Thatcher et al., 2001a, b). The recommended UV exposure (~320 to 400 nm) roughly corresponds to 1 to 2 days close to a sunny window (Chapter 6). The ratio of UV to VIS

Table 7.1 Questions to Be Addressed when Selecting Option 1 or Option 2

Parameters to be Considered	Preferred Option
Significantly increased heat level to be avoided	Option 1 or 2
Minimal exposure time	Option 1
Emission gap from 380 to 420 nm to be avoided	Option 1
Sequential exposure to be avoided	Option 1 or 2
Presentation problem due to cooling system (e.g., blowing of samples)	Option 2
Problem to eliminate UV overexposure	Option 2
Aim to mimic daylight conditions	Option 1
Aim to mimic indoor conditions	Option 2
Simple calibration procedure	Option 1
Large exposure area needed	Option 2
Low cost	Option 2
Easy to standardize	Option 1

that, according to the ICH guideline, is estimated to represent the "real" storage conditions differs from the UV/VIS ratio defined in the standard for indoor indirect daylight (ID65).

Because the spectral distribution of the selected radiation source should, according to the ICH guideline, correspond to the D65 or ID65 standard, this complicates the exposure procedure. For a lamp that fulfills the criteria given in option 1 (e.g., a spectral output similar to the ID65 standard), a total irradiance of 200 Wh m^{-2} in the UV region corresponds only to approximately 0.45 million lx h in the visible region. A test run with the end criterion of 1.2 million lux hours will therefore significantly exceed the minimum requirement 200 Wh m^{-2} by a factor of approximately 2.5 when a lamp that meets option 1 criteria is used.

At present no single source provides the combination of UV and VIS exposure levels desired by the ICH guideline without overexposure to the UV or the VIS. It is important to remember that because the recommended exposure is a minimum, exceeding the minimum exposure in the UV or VIS region is completely acceptable. Overexposure can be avoided by (1) excluding the excess of UV irradiation by use of filters after reaching the desired UV dose; (2) removing half of the samples after reaching the UV dose; or (3) running two separate tests — one for each criterion — with identical samples.

7.3.2 How to Determine Exposure Time?

The ICH guideline does not specify an irradiance level, only the overall illumination (i.e., end criterion). The irradiance level can therefore be adjusted according to individual requirements. Test conditions corresponding to the maximum output of the lamp will often be the first choice because the exposure time can thus be reduced. It is, however, important to realize that a high irradiance level can change the mechanisms of the degradation process even when the spectral distribution of the radiation source is kept constant. Differences can be observed because photodegradation processes can occur at rates competitive with thermal degradation processes. Thus, if the parent compound (or the initially formed photodegradation products,

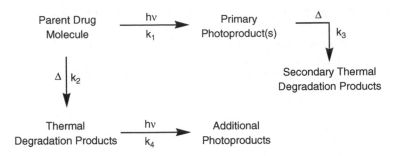

Figure 7.1 Thermal and photochemical processes likely to be included in the observed photodegradation profile.

which are often thermally unstable) has thermal instability that is observable over the time course of the photoexposure, the analytically observed degradation profile may vary as a function of time, temperature, and lamp intensity. Some consideration of these concepts is warranted.

The observed photodegradation profile from a photostability study will be a result of the thermal and photochemical processes outlined in Figure 7.1. This figure illustrates that any change to the k_1/k_2 or k_1/k_3 ratios can affect the observed degradation profile. As lamp intensity increases, the photodegradation rate (k_1) should increase linearly with the increased photon flux. In contrast, the thermal degradation rates (k_2 and k_3) will not increase "linearly" with increasing temperature, but will typically follow Arrhenius kinetics (Kennon, 1964):

$$k_2 = A * \exp - E_a/RT \tag{7.1}$$

where k_2 is the degradation rate constant; A is the "pre-exponential" constant; E_a is the energy of activation; R is the universal gas constant; and T is degrees in Kelvin. Because activation energies for most drug compounds fall between 12 to 24 kcal/mol (Kennon, 1964; Connors et al., 1986), the effect of temperature on thermal degradation rates can be estimated. If an E_a of 20 kcal/mol is assumed, k_2 will increase about fivefold if the temperature experienced by a sample is raised to 40°C from 25°C.

The concept can be illustrated by a practical example. Consider a photostability test performed with an option 2 lamp that has an intensity requiring 6 days (i.e., 144 h) to meet the required ICH minimum photoexposure, and the temperature is maintained at 25°C. Compare this to an Option 1 lamp (e.g., a xenon lamp) with an intensity that requires 8 h to achieve the minimum exposure but has a corresponding higher chamber temperature of 40°C. The thermal degradation rates at 40°C should be 2.6 to 5 times that of the rate at 25°C, assuming E_a values of 12 to 20 kcal/mol. The length of the photoexposure at 25°C is 18 times longer than the 40°C exposure. Therefore, more thermal degradation will occur during photoexposure at the lower temperature because of the increased length of time for the study. This illustrates that a higher intensity lamp causing higher temperatures in the chamber may cause less thermal degradation than a lower intensity lamp operating at a lower temperature (depending on the length of the exposure and the differences in temperatures). Summarizing (Figure 7.1):

- Any change to the k_1/k_2 or k_1/k_3 ratios can affect the observed degradation profile.
- If $k_1 \gg k_2$ or k_3, mostly primary photoproducts will be observed, especially at early time points.
- If $k_3 \gg k_1$, secondary thermal products may predominate.

There are implications for interpreting photostability study results. First, application of a high irradiance level compared to the "real" exposure may reduce the correlation of test results. That is, in long-term storage and handling, real-world photoexposures may be sporadic and time of storage may be sufficient for thermal reactions to contribute to the observed degradation profile. A short-term photostability study may form unstable photoproducts that, upon long-term storage, degrade into secondary thermal degradation products (from so-called "dark reactions") not observed in the short-term study. One approach to address such situations is to perform analytical evaluation of photostability samples immediately after photoexposure and then to expose the samples to elevated temperatures (e.g., accelerated stability conditions or stress conditions) for a period of time to assess the possibility of further thermal degradation processes.

A dark control should always be placed alongside the authentic samples to compensate for temperature and humidity effects. However, such a reference sample will only describe thermal reactions occurring from the ground state of the molecule and not take into account thermal reactions from its excited state, i.e., initiated by a photochemical process. A reference sample wrapped in aluminum foil would not necessarily experience the same surface heating effect as the exposed sample due to reflection (i.e., blocking of the surface heating). The consequence may be a difference in humidity between the two samples because an elevated surface temperature can lead to selective evaporation from water containing components within the exposed container.

7.4 PARAMETERS TO BE CONSIDERED IN OBTAINING OPTIMAL PRESENTATION OF SAMPLES WITHIN A TEST CHAMBER

The presentation of samples within the test chamber can have a significant effect on the outcome of the photostability study. Factors of importance are alignment of the samples relative to the irradiation source; sample form and layer thickness; and selection of protective material (Table 7.2).

7.4.1 Protection of Samples

Many commercial test chambers are equipped with a strong fan for temperature control. This can blow the drug substance or dosage form throughout the chamber unless a protecting material is used to cover the sample. The drug substance or product should be placed in a suitable container (e.g., a petri dish) of known transmittance for direct exposure. A transmission spectrum can be obtained by scanning (part of) the container in a UV/VIS spectrophotometer. Hygroscopic substances and low-melting or higly toxic compounds may require specially designed

Table 7.2   Important Issues for Presentation of Samples in Photostability Testing

Issue	Efforts
Protection of sample	Determine transmittance properties of container
Temperature increase	Low lamp intensity; make holes in cover; may be offset by shorter exposure time
Humidity increase, condensation of water	Humidity control of external facilities (room); temperature reduction
Thickness of sample layer	Dispensing into petri dish
Homogeneity, filter effect	Stirring (solution)
Uniform exposure of the samples	Mapping of test chamber
Appropriate dark control	Aluminum foil; temperature reduction
Number of batches	Results are consistent or equivocal; one batch if clear pass/fail result; two or more if results are borderline

containers (e.g., a sealed glass ampoule). Liquids may be exposed in clear glass vials or in quartz cuvettes.

7.4.2   Increase in Temperature

The temperature inside a petri dish or other suitable containers can easily exceed 40 to 50°C during exposure. This could lead to unwanted thermal reactions from the ground state or excited state of the drug molecule or its photoproducts; sublimation; melting; or evaporation. The heating effect can be reduced by a reduction in the lamp intensity (reduced irradiance). When possible, small holes can be made in the container (e.g., plastic film covering the petri dish). It is difficult to determine the exact surface temperature of individually exposed samples. A white standard thermometer can be used to predict the surface temperature of a reflecting surface, e.g., white tablets, while a black standard thermometer can be used to predict the highest possible surface temperature of a dark solid sample with poor thermal conductivity.

The surface temperature of colored samples ranges between that of a white reflecting surface and that of a fully absorbing (black) surface (Chapter 5). A temperature level strip (Thermax, Thermographic Measurements Ltd., South Wirral, England) may also be convenient in order to predict temperature of the exposed samples. The strip can be placed on the sample surface (e.g., surface of an infusion bag) or inside a container (e.g., inside a petri dish). A color change indicates that the rating quoted on the indicator has been reached. The maximum temperature during the exposure is thereby registered with good accuracy (±1%, according to the manufacturer).

7.4.3   Increase in Humidity and Water Condensation

An increase in sample temperature can result in a change of humidity inside the container as discussed previously. The evaporated water can further condense, forming water droplets inside the container. This often occurs when the lamp is turned off and the chamber cools down.

7.4.4 Thickness of Sample Layer

Solid drug substances should be evaluated using a very thin layer for forced degradation studies (preferentially <1 mm, although <3 mm is required by the guideline).
Intact tablets or capsules should be tested using a single layer. The drug substance
and solid dosage forms, as well as semisolid and liquid preparations that should be
tested outside the immediate container, could be placed in a petri dish. Thatcher
et al. (2001a, b) also describe a design for a single layer of tablets inside a bottle.

7.4.5 Filter Effect and Homogeneity of Samples

Liquid samples may have a high optical density, thus leading to a severe filter effect,
i.e., only a thin surface layer can absorb the radiation, thereby protecting the inner
part of the preparation from exposure. The filter effect should be avoided in kinetic
and mechanistic studies (Chapter 3). This can be obtained by diluting and stirring
the samples; the stirring must be standardized to obtain reproducible results. The
purpose of photostability testing according to the ICH guideline is to obtain information relevant for the conditions the preparation will experience under "real"
conditions. This will normally not involve prolonged stirring or shaking; thus,
avoiding stirring or shaking the samples during confirmatory studies is recommended.

7.4.6 Uniform Exposure of Samples

The irradiance may not be homogenous or uniform throughout a test chamber.
Samples placed in regions of low intensity may therefore be underexposed unless
appropriate time corrections are made. Mapping of the test chamber (i.e., measure
of irradiance at various locations in the chamber) should be performed every time
the photolysis source is replaced. The mapping should include the UV and visible
regions (see the following for calibration procedures).

7.4.7 Appropriate Dark Control

The dark control should be placed alongside the authentic sample. In cases in which
the sample is placed in a protective container, it is recommended to wrap each unit
individually in a protective material (e.g., aluminum foil) before placing it in the
container. For powder samples, the container may need to be wrapped in aluminum
foil. As discussed earlier, the aluminum foil can lead to temperature and humidity
conditions that are different from those of the exposed sample.

7.4.8 Number of Batches

The test should be carried out on one batch unless the results are equivocal. In
practice, this means that if a repeating experiment is likely to give a different outcome
(i.e., pass or fail), two additional batches could be tested before the final conclusion

on acceptable or unacceptable change is made. Lag time between exposure and quantitative analysis may influence the results and should be standardized for maximum reproducibility. Dark reactions initiated by a photochemical process can proceed slowly and may only be detected if the samples are analyzed after a certain period.

7.5 WHAT IS ACCEPTABLE AND UNACCEPTABLE CHANGE?

Photostability testing according to the ICH guideline will give an indication as to whether photochemical degradation of the drug substance or drug product is likely to occur during the shelf-life. Quantitative photostability results must be evaluated together with long-term stability results. Justification of impurity limits should be based on the ICH Drug Product Impurity Guidelines (ICH Q3A, Q3B). The photostability results are added to the definitive stability results for final evaluation of shelf-life (Thatcher et al., 2001a, b). When the combined results from photostability and thermal studies do not meet the specifications at the proposed expiry, the results must be considered as unacceptable unless a reduction in expiration date is an option. Appropriate labeling must also be defined. The product should be stored in its secondary container (e.g., a card box) if there is any risk of photodegradation.

7.6 WHAT INFORMATION CAN BE OBTAINED FROM RECOMMENDED CALIBRATION DEVICES?

The devices used for calibration of radiation sources and test chambers are discussed in Chapter 5 and Chapter 6. Neither the UV filter radiometer nor the luxmeter provides information on the spectral power distribution (SPD, the plot of radiation intensity vs. wavelength) of sources. Upon delivery, the device should have been calibrated by the manufacturer against a standard lamp and provided with a response curve. If the meters are used as received, they are well suited for measuring evenness of irradiance across the sample area and changes in total output with time.

Neither filter radiometers nor luxmeters can be used to obtain an absolute measurement of irradiance or to compare irradiance between sources unless they are calibrated *specifically for each source* (Tønnesen and Karlsen, 1997). A spectroradiometer is needed for a detailed estimate of the SPD but at present such equipment is not widely used on a regular basis because of cost and convenience. The total irradiance (i.e., actual number of photons) can be determined by chemical actinometry using a reaction of known photochemical efficiency (Chapter 3 and Chapter 6). The chemical actinometer listed in the ICH guideline (quinine hydrochloride) has its limitations and its suitability as actinometer has been questioned (Baertschi, 1997; Bovina et al., 1998; Drew, 1998). This actinometer is not suitable for calibration of option 1 radiation sources (Thatcher et al., 2001a, b). An alternative actinometer (2-nitrobenzaldehyde) has recently been discussed for calibration of option 1 (and other sources) for UV irradiance (Allen et al., 2000).

REFERENCES

Allen, J.M., Allen, S.K., and Baertschi, S.W. (2000) 2-Nitrobenzaldehyde: a convenient UV-A and UV-B chemical actinometer for drug photostability testing, *J. Pharm. Biomed. Anal.*, 24, 167–178.

Baertschi, S.W. (1997) Commentary on the quinine actinometry system described in the ICH draft guideline on photostability testing of new drug substances and products, *Drug Stab.*, 1, 193–195.

Bovina, E., De Filippis, P., Cavrini, V., and Ballardini, R. (1998) *trans*-2-Nitrocinnamaldehyde as chemical actinometer for the UV-A range in photostability testing of pharmaceuticals, in Albini, A. and Fasani, E. (Eds.) *Drugs: Photochemistry and Photostability*, 305–316 Cambridge: The Royal Society of Chemistry.

Connors, K.A., Gordon, A.L., and Valentino, S.J. (1986) *Chemical Stability of Pharmaceuticals: A Handbook for Pharmacists*, 19, 2nd ed., New York: Wiley Interscience.

Drew, H.D. (1998) Photostability of drug substances and drug products: a validated reference method for implementing the ICH photostability guidelines, in Albini, A. and Fasani, E. (Eds.) *Drugs: Photochemistry and Photostability*, 227–242, Cambridge: The Royal Society of Chemistry.

ICH Q1A (1994) Stability testing of new drug substances and products, *Fed. Reg.*, 59, 48754–48759.

ICH Q3A (1996) Impurities in new drug substances, *Fed. Reg.*, 61, 371–376.

ICH Q1B (1997) Photostability testing of new drug substances and products, *Fed. Reg.*, 62, 27115–27122.

ICH Q3B (1997) Impurities in new drug products, *Fed. Reg.*, 62, 27454–27461.

Kennon, L. (1964) Use of models in determining chemical pharmaceutical stability, *J. Pharm. Sci.*, 53, 815–818.

Kerker, R. (1991) Untersuchungen zur Photostabilität von Nifedipin, Glucocorticoiden, Molsidomin und ihren Zubereitungen. Dissertation, Munich: Ludwig–Maximilians–Universität.

Matsuo, M., Machida, Y., Furuichi, H., Nakamura, K., and Takeda, Y. (1996) Suitability of photon sources for photostability testing of pharmaceutical products, *Drug Stab.*, 1, 179–187.

Piechocki, J. (1998) Selecting the right source for pharmaceutical photostability testing, in Albini, A. and Fasani, E. (Eds.) *Drugs: Photochemistry and Photostability*, 247–271, Cambridge: The Royal Society of Chemistry.

Spilgies, H. (1998) Investigations on the Photostability of Cephalosporins and Beta-Lactamase Inhibitors, Dissertation, Munich: Ludwig–Maximilians–Universität.

Thatcher, S., Mansfield, R.K., Miller, R.B., Davis, C.W., and Baertschi, S.W. (2001a) Pharmaceutical photostability: a technical guide and practical interpretation of the ICH guideline and its application to pharmaceutical stability. I, *Pharm. Technol. U.S.*, 25(3), 98–110.

Thatcher, S., Mansfield, R.K., Miller, R.B., Davis, C.W., and Baertschi, S.W. (2001b) Pharmaceutical photostability: a technical guide and practical interpretation of the ICH guideline and its application to pharmaceutical stability. II, *Pharm. Technol. U.S.*, 25(4), 50–62.

Thoma, K. and Kerker, R. (1992a) Photoinstabilität von Arzneimitteln. I. Mitteilung über das Verhalten von nur im UV-Bereich absorbierenden Substanzen bei der Tageslichtsimulation, *Pharm. Ind.*, 54, 169–177.

Thoma, K. and Kerker, R. (1992b) Photoinstabilität von Arzneimitteln. II. Mitteilung über das Verhalten von im sichtbaren Bereich absorbierenden Substanzen bei der Tageslichtsimulation, *Pharm. Ind.*, 54, 287–293.

Thoma, K. and Kübler, N. (1993) Einfluss der Wellenlänge auf die Photozersetzung von Arzneistoffen, *Pharmazie*, 51, 660–664.

Tønnesen, H.H. and Moore, D.E. (1993) Photochemical degradation of components in drug formulations, *Pharm. Technol. Eur.*, 5, 27–33.

Tønnesen, H.H. and Karlsen, J. (1995) Photochemical degradation of components in drug formulations. III. A discussion of experimental conditions, *PharmEuropa*, 7, 137–141.

Tønnesen, H.H. and Karlsen J. (1997) A comment on photostability testing according to ICH guideline: calibration of light sources, *PharmEuropa*, 9, 735–736.

CHAPTER 8

Inconsistencies and Deficiencies in Current Official Regulations Concerning Photolytic Degradation of Drugs

Joseph C. Hung

CONTENTS

8.1 INTRODUCTION

Photolytic degradation is an important limiting factor in the stability of drugs that are sensitive to light. Some light-sensitive drugs are rapidly affected by nature's light (especially ultraviolet light) or artificial light (e.g., fluorescent light) and become discolored or cloudy in appearance, or develop precipitates. Others may slowly undergo photodegradation, which may not be visually apparent.

In a photochemical reaction, the light-sensitive drug molecules may be affected directly or indirectly by light, depending upon how the absorbing photon energy is transferred to the drug molecules. With a direct or indirect light-induced reaction, a drug can only undergo the photodegradation process if the absorbed energy exceeds a threshold. Because ultraviolet radiation has a higher energy level, it is the main cause of many degradation reactions of light-sensitive drugs. Colored-glass containers are the most commonly used method to protect these types of drugs. Yellow-green glass gives the best protection in the ultraviolet region; amber glass also offers consideration protection from ultraviolet light, but little protection from infrared light.

The photochemical reaction is a very complex process; many variables may be involved in the photolytic degradation kinetics. The velocity of the photochemical reaction may be affected not only by the light source, intensity, and wavelength of the light, but also by the size, shape, composition, and color of the container. To properly determine the deleterious or beneficial effects of light on the quality of a drug properly, standard light stability testing should consider all of the aforementioned variables. Once uniform standard light-stability testing procedures are instituted, proper packaging, storage environment, and expiration date for the light-sensitive drug can be established.

8.2 STANDARD LIGHT STABILITY TESTING

A pharmacopeia is an authoritative source for the stability testing of any drug substances and products. In the United States, the *United States Pharmacopeia and the National Formulary* (USP-NF) is recognized as an official compendium of drug standards. Unfortunately, no standard testing procedure originating from the *United States Pharmacopeia,* 27th revision, and *National Formulary,* 22nd edition (USP 27-NF 22) is available for evaluating the photosensitivity of drugs. The USP 27-NF 22 lists light transmission limits for testing light-resistant containers to ensure the proper protection of light-sensitive drugs (Light transmission, 2004); however, light-stability testing for drugs is not mentioned.

The official point of view from the U.S. Pharmacopeial Convention, Inc. (USPC), is that "it is not the intent of the USP to give specific guidelines for determining sensitivity to light for all compendial drugs and drug products" (Paul, 1992). The USPC feels that "the responsibility for determining the light stability of a particular product belongs to the manufacturer" (Paul, 1992). The USPC further states that the USP-NF has no specific guidelines for testing air, pH, moisture, trace metals, and commonly used excipients or solvents in active ingredients; however, the manufacturer must determine the effects of these conditions (Paul, 1992). Therefore, the USPC thinks that the primary goal of the USP-NF is to provide standards of the listed drugs and not to give specific assay and test procedures for determining compliance with every USP-NF standard. The organization also believes the obligation for determining the effect of the previously mentioned conditions on a drug's stability, including light sensitivity, is that of the drug manufacturer (Paul, 1992). Nevertheless, the USP 27-NF 22 does have test and assay procedures for pH and trace metals such as heavy metals, iron, and lead; see General Tests and Assays section of the USP 27-NF 22.

Each manufacturer is responsible for determining the light stability, pH, moisture, etc. for a given drug. However, because there are no standard evaluation procedures or specifications, different drug manufacturers may evaluate the effect of conditions (e.g., light) differently and may use varying guidelines for judging the results. As an example, if a drug or drug product is packaged in an opaque carton, and it takes 3 months at 200 lux before it discolors, should it carry a "protect from light" caution? If another drug or drug product is packaged in an amber glass container and loses 15% potency in 3 days at 200 lux, is "protect from light" adequate?

If the drug substance or drug product is susceptible to photolytic degradation, the official USP 27-NF 22 monograph for that drug or drug product will carry a cautionary statement such as "protect from light" or "preservation in a light-resistant container." With no standard or generally accepted light-stability testing method available, it is a puzzle how the USPC determined the light sensitivity for the drugs and drug products listed in the USP 27-NF 22.

The USPC believes that the U.S. Food and Drug Administration (FDA) "should be contacted for guidance and direction with regard to stability testing" because "the definition of stability testing is an important part of good manufacturing practice regulations used by FDA" (Paul, 1992). The FDA does have official guidelines in place for the evaluation of drug stability (Design and interpretation of stability studies, 1987; International harmonization, 1994). However, no detailed description regarding the device and procedures is officially recognized (e.g., light-source instruments, length of exposure to light, measurement of discoloration, intensity of the light source, and surrounding temperature conditions). In 1997 the FDA adopted a guideline titled "Guideline for the Photostability Testing of New Drug Substances and Products" (International Conference on Harmonisation, 1997), developed by the International Conference on Harmonisation of Technical Requirements for Registration of Pharmaceuticals for Human Use (ICH), in which a more detailed description of the recommended device and procedures for photostability testing is provided.

According to the FDA, this guideline is applicable to drug and biological products, and, as such, this guideline represents the FDA's current thinking on photostability testing on new drug substances and products (International Conference on Harmonisation, 1997). In addition, the FDA has also indicated that an alternative approach for testing photostability on drug and biological products may be used if such approach meets the requirements of the applicable statute, regulations, or both (International Conference on Harmonisation, 1997).

8.3 OFFICIAL REGULATIONS FOR LIGHT-SENSITIVE DRUGS

Currently in the U.S., the monographs in the USP 27-NF 22 and the package inserts for drugs are the only two official resources requiring cautionary statements regarding light sensitivity for drugs. However, the USP 27-NF 22, as well as the USP 27-NF 22 and the package inserts, has many inconsistencies and ambiguities in the legal requirements for the definitions, testing, packaging, storing, and warning labels.

8.3.1   Deficiencies and Inconsistencies in USP 27-NF 22

Opaque covering (enclosure). According to USP 27-NF 22, a light-resistant container protects light-sensitive drug contents from the effects of light by virtue of the specific properties of the material of which it is composed, including any light-resistant coating applied to it (Preservation, packaging, storage, and labeling, 2004). A container intended to provide protection from light or resistant to light must meet the requirements for light transmission (2004). A clear and colorless or translucent container made light resistant by means of an opaque covering is exempt from the requirements for light transmission (2004).

Because a container as defined in USP 27-NF 22 is not necessarily in direct contact with the drug (Preservation, packaging, storage, and labeling, 2004), the opaque enclosure would be considered a light-resistant container on the basis of the current USP 27-NF 22 definition. This implies that any drug substance or drug product prone to photolytic degradation can be packaged and stored in a clear and colorless or translucent container (e.g., ampoule, serum vial, or bottle) and maintained in a light-resistant carton or box, thereby fulfilling USP 27-NF 22 requirements (Preservation, packaging, storage, and labeling, 2004). However, the USP 27-NF 22 is not clear whether the paper carton or box is required to meet the light transmission limits. If it is required for such testing, the current testing procedure for light transmission does not have the standard preparation, procedure, and limits for the light-resistant paper container.

The USP 27-NF 22 does state, "Alternatively, a clear and colorless or a translucent container that is made light-resistant by means of an opaque enclosure is exempt from the requirements for light transmission" (Containers, 2004). Does this mean that the opaque enclosure that is supposed to provide protection and/or resistance from light effect is also exempt from light transmission testing in the light transmission section? Logical thinking would be that a clear and colorless or translucent container can be exempt from the requirements if it is made light resistant by using an opaque enclosure such as a paper carton or box. However, in order for the paper carton or box to claim light protection or resistance, it must be subject to the same testing procedures and requirements as stated in the light transmission chapter. Therefore, the USPC must update the light transmission testing section to include the preparation of specimen, procedure, and limits for the light-resistant paper material.

The recommended systematic approach to photostability testing on drug and biological products as listed in the *Guideline for the Photostability Testing of New Drug Substances and Products* (1997) includes tests on the drug product contained in the immediate pack and, if necessary, tests on the drug product in the marketing pack. The definition of the term "market pack," as per the *Guideline for the Photostability Testing of New Drug Substances and Products* (1997), is the combination of the immediate (primary) pack and other secondary packaging (e.g., a carton). As such, an opaque covering (enclosure) as stated in the USP 27-NF 22 is equivalent to secondary packaging of the market pack as described in the *Guideline for the Photostability Testing of New Drug Substances and Products* (1997), and therefore should be appropriately tested for its suitability in protecting the drug product from unacceptable changes that may occur due to light exposure.

Cautionary statement. When an opaque covering is used to make a clear and colorless or translucent container light resistant, the USP 27-NF 22 requires that "the label of the container bears a statement that the opaque covering is needed until the contents are to be used or administered" (Preservation, packaging, storage, and labeling, 2004). The USP 27-NF 22 is not clear which container (i.e., immediate container, outer container, or both) should have a cautionary statement for light sensitivity. It would be appropriate to have such a label on the immediate and outer containers in order to be consistent and thorough to alert the repackagers or end users regarding proper storage of the light-sensitive drug or drug products.

For any drugs that may be susceptible to photodegradation, the requirement in the USP 27-NF 22 for the proper packaging and storage of light-sensitive drugs reads, "preserve in light-resistant containers" or "protect from light." The USP 27-NF 22 further indicates that preservation in a light-resistant container is intended when the instruction to protect from light is given in an official monograph for any light-sensitive drug (Preservation, packaging, storage, and labeling, 2004). It appears that these two directions for packaging and storage of light-sensitive drugs are interchangeable in the USP 27-NF 22. However, it seems that one uniform statement would be more appropriate because it would help to avoid possible confusion and misinterpretation. Whenever one sees "protect from light" in an individual official monograph in the USP 27-NF 22 or in the package insert, the manufacturer must assure that the containers meet light transmission standards; the ultimate dispenser or the repackager needs to be aware that the original package must be retained or a suitable alternative must be used for proper protection of light-sensitive drugs or drug products.

The other alternative for cautionary labeling of the light-sensitive drug or drug product is to use symbols to show light sensitivity, for example, ☀, with a numeral, such as 0, 1, 2, and so on, to show the level of light sensitivity. Symbols are much more likely to be observed and followed than words. This use of symbols has been adopted worldwide for common instruction and for the rising non-English-speaking population of the U.S.; storage and cautionary symbols obviously meet a practical need.

Light-resistant container for single-use drug. Under the light-resistant container section of the USP 27-NF 22 (Preservation, packaging, storage, and labeling, 2004), "Where an article is required to be packaged in a light-resistant container and if the container is made light resistant by means of an opaque covering, a single-use, unit-dose container or mnemonic pack for dispensing may not be removed from the outer opaque covering prior to dispensing." This statement seems to suggest that single-use, unit-dose drugs or drugs packaged in a mnemonic pack that are susceptible to photolytic degradation must be stored in a clear or colorless or translucent container, using an opaque covering for light protection or resistance. However, the intention for such a requirement is not clear. Does this statement suggest that under the USP 27-NF 22 standards, a drug stored in a single-use, unit-dose container or a mnemonic pack susceptible to light degradation cannot be removed from the outer opaque covering at any given time? Would it be more appropriate to store a light-sensitive drug in an immediate light-resistant container, rather than depending upon the secondary light protection or resistance achieved through using an outer opaque covering?

In addition, it is not clear why the USP 27-NF 22 does not have the same restriction for the single-dose container, which is designed for drugs intended for parenteral administration only (Preservation, packaging, storage, and labeling, 2004). For parenteral drugs, it would seem to be more appropriate to store the drug solution injection in a clear and colorless prefilled syringe so that proper visual inspection can be performed to observe any color change, clouding appearance, or particulate matter in the drug solution. This type of drug would be more appropriately stored in a clear and colorless or translucent container that required an opaque enclosure to offer light protection or resistance.

Light exposure. Certain circumstances, such as compounding and/or dispensing of light-sensitive drugs or drug products in a light environment, would certainly not be in accordance with full compliance of the requirements, which state that the outer covering is not to be removed and discarded until the contents have been used (Preservation, packaging, storage, and labeling, 2004). Because the level of light sensitivity has never been defined, it is uncertain whether a light-sensitive drug substance or drug product can be exposed to light even for a short period of time during visual inspection, repackaging, compounding, or dispensing. In these situations, it is necessary to apply appropriate light-protection or resistance measures (e.g., dim the light resource, or use foil pouches to hold the dispensed drug) in order to minimize the photolytic effects. Short-term spikes due to opening an outer light-resistant container (e.g., paper carton or box) should be viewed as unavoidable and should be acceptable unless the drug substance or drug products are extremely sensitive to light exposure.

8.3.2   USP 27-NF 22 *vs.* Package Inserts

Light-sensitive cold kits and radiopharmaceuticals. USP 27-NF 22 is sometimes not in agreement with the light sensitivity cautionary statement listed in the package insert. Because the author's primary work is specialization in nuclear pharmacy, the examples mentioned in this chapter will be strictly related to the radiopharmaceuticals (i.e., radioactive drugs) used in nuclear pharmacy and nuclear medicine.

Several cold kits are sensitive to light exposure (Table 8.1). Although the package insert includes a cautionary statement with regard to light sensitivity for each of the cold kits listed in Table 8.1, the USP 27-NF 22 monographs stipulate only that technetium Tc 99m succimer injection (2004) and technetium Tc 99m tetrofosmin injection (2004) are to be protected from light. It is common practice in the nuclear pharmacy field to use reagent kits (a.k.a. cold kits; e.g., TechneScan MAG3®, Hepatolite®) and radioactive materials to prepare radiopharmaceuticals (e.g., technetium Tc 99m mertiatide injection, technetium Tc 99m disofenin injection).

Because USP 27-NF 22 lists radioactive drugs in names of the finished radiopharmaceutical products, it would therefore have the requirements and specifications for the final radioactive drugs. On the other hand, the package inserts for the cold kits contain the light-sensitivity information only for the cold kits. There is also no additional information, which could be determined from the package inserts of the light-sensitive cold kits (Table 8.1), regarding whether a radiopharmaceutical-prepared light-sensitive cold kit (Table 8.1) is also susceptible to light-induced decomposition.

If the reagents in the cold kit are light sensitive, is the radiopharmaceutical prepared from the light-sensitive cold kit also prone to photolytic degradation? If

Table 8.1   Light-Sensitive Cold Kits

Cold Kit Name	Manufacturer
AcuTect® (for preparation of technetium Tc 99m apecitide injection)	Berlex Laboratories, Wayne, NJ
DMSA (for preparation of technetium Tc 99m succimer injection)	Medi-Physics, Inc., Arlington Heights, IL
Hepatolite® (for preparation of technetium Tc 99m disofenin injection)	CIS-US, Inc., Bedford, MA
TechneScan MAG3® (for preparation of technetium Tc 99m mertiatide injection)	Mallinckrodt Inc., St. Louis, MO
Myoview™ (for preparation of technetium Tc 99m tetrofosmin injection)	Amersham Health Medi-Physics, Inc., Arlington Heights, IL
Neurolite® (for preparation of technetium Tc 99m bicisate injection) — vial A[a]	Bristol-Myers Squibb Medical Imaging, Inc., N. Billerica, MA
OctreoScan® (for preparation of indium In 111 pentetreotide injection) — reaction vial[b]	Mallinckrodt Inc., St. Louis, MO
UltraTag® RBC (for preparation of technetium Tc 99m red blood cells injection) — syringe I[c]	Mallinckrodt Inc., St. Louis, MO

[a] Vial A contains 0.9 mg bicisate dihydrochloride (ECD • 2 HCl); 0.36 mg edetate sodium, dihydrate; 24 mg mannitol; 83 µg stannous and stannic chloride, dihydrate.
[b] Reaction vial contains 10.0 µg pentetreotide; 2.0 mg gentisic acid; 4.9 mg trisodium citrate, anhydrous; 0.37 mg citric acid, anhydrous; 10.0 mg inositol.
[c] Syringe I contains 0.6 mg sodium hypochloride in 0.6 ml sterile water for injection.

so, what is the time period limit for light exposure that should apply to this type of radiopharmaceutical prepared from a light-sensitive cold kit?

Only two sets of cold kits and associated radiopharmaceuticals contain the cold kit and the reconstituted radiopharmaceutical claimed by the package insert and the USP 27-NF 22, respectively, to be light sensitive. These are DMSA (2000)/technetium Tc 99m succimer injection (2004) and Myoview™ (2003)/technetium Tc 99m tetrofosmin injection (2004). It is unclear whether the USPC considers the radiopharmaceuticals reconstituted from the light-sensitive cold kits listed in Table 8.1, other than DMSA and Myoview™, to be "light resistant" or whether the cautionary statement "protect from light" may have been inadvertently left out of each of the associated USP 27-NF 22 monographs (i.e., technetium Tc 99m apcitide injection, 2004; technetium Tc 99m disofenin injection, 2004; technetium Tc 99m mertiatide injection, 2004; technetium Tc 99m bicisate injection, 2004; indium In 111 pentetreotide injection, 2004; technetium Tc 99m red blood cells injection, 2004).

Interestingly, the product package insert for Choletec® (kit for the preparation of technetium Tc 99m mebrofenin injection) (Choletec®, 2000), which is an analogue of Hepatolite® (1999) (kit for the preparation of technetium Tc 99m disofenin injection), records no cautionary statement regarding light sensitivity in the USP 27-NF 22 or in the package insert for technetium Tc 99m mebrofenin injection (Choletec®, 2000).

8.3.3   Deficiencies of Package Inserts

The package insert for a drug is important official information regarding the drug's description, clinical pharmacology and indications, for example. The light sensitivity statement is usually listed under the "How Supplied" section or under the "Description" section of the package insert. It is interesting to note that although the final language state in the package insert of the drug must be approved by the FDA, it has not addressed the guidance and directions with regard to the light stability and light sensitivity standards.

Vague direction (i.e., use of the word "should" as stated in the package inserts for AcuTect® and Myoview™) (AcuTect®, 2001; Myoview™, 2003) must not be used in order to avoid any misinterpretation that the cautionary statement (i.e., protect from light) may be optional. The package insert for UltraTag® RBC (2000) contains the most straightforward and stern restrictions with regard to light sensitivity (i.e., "The syringe (I) must be protected from light to prevent degradation of the light-sensitive sodium hypochloride"). Other package inserts for cold kits such as DMSA (2000), Hepatolite® (1999), TechneScan MAG3® (2000), and Neurolite® (2001) provide a standard, but adequate, cautionary statement (e.g., protect from light). In addition, although the package insert for the U.S. product does include the cautionary statement for light sensitivity, it does not specify the light-sensitivity level of the drug.

The OctreoScan® cold kit was approved by the FDA for preparation of indium In 111 pentetreotide for imaging primary and metastatic neuroendocrine tumors bearing somatostatin receptors (OctreoScan®, 2000). The OctreoScan® reaction vial label, outer plastic container, and outer storage carton box bear the cautionary statement, "protect from light." Although the reaction vial is packaged appropriately, it is interesting to note that the package insert for the OctreoScan® kit does not include any warning statement regarding light sensitivity of the contents within the reaction vial of the OctreoScan® cold kit (OctreoScan®, 2000).

8.4 LABELING AND PACKAGING OF LIGHT-SENSITIVE COLD KITS

8.4.1 Labeling Light-Sensitive Cold Kits

Most of the light-sensitive cold kits listed in Table 8.1 comply with the USP 27-NF 22 requirement that labels on the immediate and outer containers must state "protect from light" (Preservation, packaging, storage, and labeling, 2004). However, two light-sensitive cold kits fail to comply with the USP 27-NF 22 labeling requirement (Preservation, packaging, storage, and labeling, 2004): AcuTect® and Myoview™.

Although each of the package inserts for AcuTect® (2001) and Myoview™ (2003) includes a cautionary statement for light sensitivity, the labels on the reagent vial and the outer cold kit box contain no warning statement regarding light sensitivity. In the previous edition of this book, this chapter cited two cold kits (i.e., Hepatolite® and Neurolite®) that violated the USP-NF labeling requirements because both outer box containers for the cold kits failed to include cautionary statements concerning the light sensitivity of Hepatolite® and vial A of Neurolite®. It is encouraging to learn that the aforementioned labeling deficiencies in these cold kits have since been properly rectified (Figure 8.1).

An additional note should be emphasized with regard to the storage of the Neurolite® cold kit. Each outer box contains two sets of cold kits for Neurolite®,

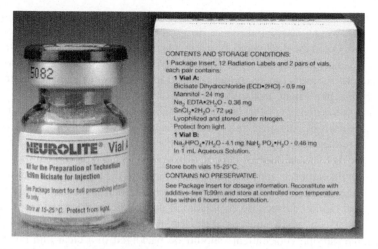

CONTENTS AND STORAGE CONDITIONS:
1 Package Insert, 12 Radiation Labels and 2 pairs of vials,
each pair contains:
   1 Vial A:
   Bicisate Dihydrochloride (ECD•2HCl) - 0.9 mg
   Mannitol - 24 mg
   Na₂ EDTA•2H₂O - 0.36 mg
   SnCl₂•2H₂O - 72 µg
   Lyophilized and stored under nitrogen.
   Protect from light.
   1 Vial B:
   Na₂HPO₄•7H₂O - 4.1 mg NaH₂ PO₄•H₂O - 0.46 mg
   In 1 mL Aqueous Solution.

Store both vials 15-25°C.
CONTAINS NO PRESERVATIVE.
See Package Insert for dosage information. Reconstitute with
additive-free Tc99m and store at controlled room temperature.
Use within 6 hours of reconstitution.

Figure 8.1   Left: vial A of the Neurolite® cold kits. "Protect from light" is stated on the vial label (lower right line); right: the back side of the outer box for the Neurolite® cold kits. Cautionary statement for light sensitivity now can be found on the labels of the outer box.

with each set consisting of dual reagent vials (i.e., vials A and B). Due to the light sensitivity of vial A's contents (Figure 8.1, left), each set of the Neurolite® cold kits must be kept inside the box during storage. The outer box cannot be removed and discarded until both sets of Neurolite® cold kits have been used.

8.4.2   Packaging Light-Sensitive Cold Kits

It is common practice for cold kit manufacturers to package light-sensitive drugs or drug products in clear and colorless glass vials or syringes. Although this is contrary to the standard practice of protecting light-sensitive drugs in light-resistant (e.g., amber, yellow-green, or blue) containers, a clear and colorless or translucent container does offer some advantages over a colored container.

Colored-glass or colored-plastic containers cost an average of approximately 25% more than the clear and colorless or translucent containers. This may be due to the more costly material of which they are composed, including any applied coating material, in order to make the containers light resistant. The higher cost of the light-resistant containers may also be due to the light transmission testing required by the USP 27-NF 22 (Light transmission, 2004). The colored containers must pass this test in order to qualify for use as light-resistant containers, whereas the clear and colorless or translucent container can be exempt from these requirements if it is made light resistant by an opaque enclosure (Preservation, packaging, storage, and labeling, 2004).

In addition, with colored glass or plastic, it is virtually impossible to observe a color change in a drug formulation, and it is difficult to examine an injectable drug solution for particulate matter. Therefore, injectable products raise the most difficult container selection problem for manufacturers because of the need to provide light protection (for light-sensitive drugs or drug products) as well as to allow for visual examination. Often the final inspection is made at the point of use by the dispensers. Many companies, in their attempt to balance these two factors, package their injectable products in clear glass containers placed in foil pouches or paper cartons as a secondary means of light protection.

Although the contents of light-sensitive injectable drugs can be transferred to a clear and colorless container (e.g., vial or syringe) to examine the clarity and particulate matter, this approach is not suitable for cold kit formulation in radiopharmaceuticals. The additional step of transferring the entire radioactive solution contained in the light-resistant vial into a colorless and clear syringe or vial increases the risk for degradation of any oxygen-sensitive radiopharmaceutical. The possibility for increased radiation exposure to personnel during the transfer and visual inspection is another concern. Therefore, the light-sensitive cold kits for the preparation of injectable radiopharmaceuticals should be packaged in a clear and colorless or translucent container. This container must be protected by a light-resistant opaque covering such as a paper carton or plastic box in order to comply with the USP 27-NF 22 requirements (Preservation, packaging, storage, and labeling, 2004).

It is also interesting to note that the Hepatolite® reaction vials are stored in a so-called "convenience pack" carton. This 30-vial convenience pack has six 12-mm diameter circular openings for checking for reorder point of the cold kit supply without the need to open the outer box (Figure 8.2). In addition, the convenience pack has an 11- × 3-cm round rectangular opening for retrieval of the Hepatolite®

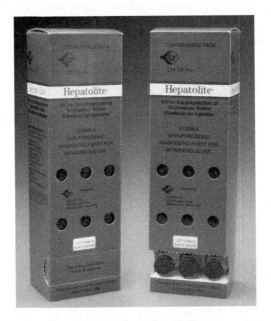

Figure 8.2   The convenience pack for Hepatolite® cold kits.

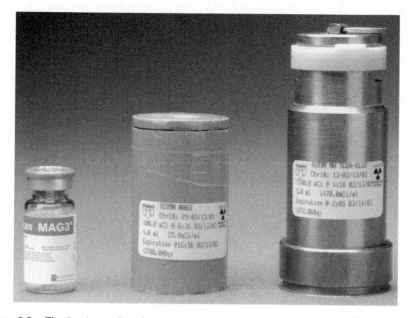

Figure 8.3   The lead container for the storage of technetium Tc 99m labeled radiopharmaceutical (middle) and the lead container for the storage of technetium Tc 99m pertechnetate injection (right) for the radiolabelling of the cold kit (left).

cold kits (Figure 8.2). These openings on the Hepatolite® carton do not provide an adequate light-resistant environment for the Hepatolite® cold kits, which are subject to photolytic degradation. The manufacturer should modify the storage box for the Hepatolite® cold kits in order to provide proper protection for the light-sensitive kit formulation. In the meantime, because the light-sensitive level for the Hepatolite® cold kit is not indicated, the openings on the Hepatolite® convenience pack should be properly covered to prevent light exposure of the cold kits.

Once the radiopharmaceutical is prepared from a light-sensitive cold kit, it is stored in a lead container to shield from radiation (Figure 8.3). As shown in Figure 8.3, the lead container completely encloses the radioactive vial inside the container, offering excellent protection from light exposure for the potentially light-sensitive radioactive drugs. Once the radiopharmaceutical is withdrawn into a syringe, the syringe, except the needle and a portion of the plunger, is completely surrounded with a lead syringe shield (Figure 8.4). The use of a lead syringe shield for radioactive drugs not only reduces radiation exposure to personnel, but also provides good protection from light exposure for the potentially light-sensitive radiopharmaceuticals. However, it remains unknown whether the lead glass or acrylic used in syringes and/or vial shields (most notably the 360° clear-view syringe/vial shields that are usually light yellow in color) (Figure 8.4) would meet USP 27-NF 22 light transmission testing requirements for providing adequate protection of light-sensitive radiopharmaceuticals from light-exposure degradation.

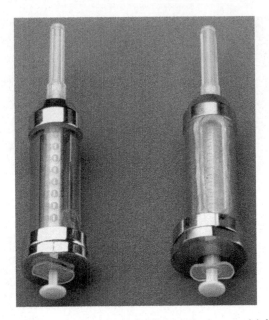

Figure 8.4   Right: lead syringe shield with the lead glass viewing panel; left: lead glass syringe with 360° syringe visibility.

8.5 RECOMMENDATIONS FOR OFFICIAL REGULATIONS FOR LIGHT-SENSITIVE DRUGS

The ICH/FDA guideline on photostability testing (Guideline for the Photostability Testing of New Drug Substances and Products, 1997) provides an in-depth and standardized basic testing of the intrinsic photostability characteristics of new drug substances and products. However, there are three major shortcomings in the aforementioned guideline that needed to be rectified:

1. Inclusion of two options for scientifically different irradiation sources, but no guidance is given as to the selection criteria for either option.
2. Failure to incorporate a design for in-use testing; i.e., taking into account the stability of reconstituted drug products (e.g., drug products administered through an i.v. drip or drug products that are repacked from a nontransparent container into a transparent unit-dose container).
3. Uncertainty as to whether those drug products that were rgistered before publication of the ICH/FDA guideline are also required to meet the requirements as stipulated in the new photostability testing guideline.

As stated in the Constitution and Bylaws (2004) of the USPC, the primary purpose of the USP 27-NF 22 is

to provide authoritative standards and specifications for materials and substances and their preparations that are used in health care or for the improvement or maintenance of health; they establish titles, definitions, descriptions, and standards for identity, quality, strength, purity, packaging, and labeling, and also, where practicable, bio-availability, stability, procedures for proper handling, storage, and shipment, methods for their examination, and formulas for their manufacture or preparation.

(Constitution and Bylaws, 2004)

It is clear that the main objectives of the USP 27-NF 22 are not only to establish the standards with which the listed drugs must comply, but also to give specific procedures and guidelines for determining the standards. The USP 27-NF 22 contains a section entitled "General Tests and Assays" (2004), which provides specific guide-lines and information for the performance of certain testing, such as sterility tests, oxygen determination, measurement of pH, and so on. Because light sensitivity is an important factor in the storage of drugs or drug products prone to photolytic degradation, it is absolutely necessary to include light-stability testing in the USP-NF.

The USPC should try to work with manufacturers and the FDA in order to update their official monographs for drugs and drug products and to standardize require-ments such as packaging and storage, labeling, and standard testing. Up-to-date and consistent official requirements among the USP-NF, FDA, and the package inserts would ensure better compliance from manufacturers, repackagers, and end users. Other suggestions to enhance compliance are to use uniform cautionary statements such as "protect from light" and/or symbols (e.g., ◐).

Once standard light-stability testing is set, using duration of light exposure, the USPC can assign several levels of increasing or decreasing sensitivity to light in a manner similar to that used for temperature and humidity (Quality of Biotechnolog-ical Products: Stability Testing of Biotechnological/Biological Products, 2004). The loss of specific drug potency; any specified degradant exceeding the drug's specifi-cation limit; or a drug or drug product exceeding its pH limit can be used as a gauge to determine the shelf-life of the light-sensitive drug or drug products. The USP-NF can then use a combination of symbols (e.g., ◐) and numerals such as 0, 1, and 2 to show the light sensitivity and level of the light sensitivity for the drugs and drug products. Depending on the levels of light sensitivity, the drug manufacturer can use proper precautions to protect the light-sensitive drug from light exposure occurring during compounding until filling into the final container and can select the proper light-resistant container to store the light-sensitive drugs or drug products. The end user and repackager need only be concerned with drugs or drug products that may become unacceptable for use within the time period limit permitted after the con-tainer is first opened or entered.

The USPC should define the parameters for storage conditions of light-sensitive drugs or drug products. Depending upon the light-sensitivity levels of the drugs or drug products, is it admissible to allow short-term spikes due to opening the light-resistant outer carton for light-sensitive drugs or drug products stored in a clear and colorless or translucent container? If the drug or drug products are very sensitive to light exposure, is it advisable to draw up the solution into a clear and colorless or translucent syringe, thereby causing possible degradation of the drug material? Should short-term spikes be considered unavoidable and acceptable for practical reasons? With regard to time limits for exposure to different light sources such as ultraviolet light or fluorescent light, it is important for the end user to exercise special caution for the storage, transfer, or compounding of light-sensitive drugs or drug products.

The USPC should modify the definitions for light-resistant container for single-use drug so that the definition and requirements have more practical meaning. A clear and colorless or a translucent container has several advantages over the light-resistant colored container. It is less expensive to manufacture (i.e., approximately 25% cheaper); it is exempt from meeting the light transmission testing requirements as stated in USP 27-NF 22 (Light transmission, 2004); and it enables visual inspection of the drug or drug products to examine for color change, clouding appearance, or precipitate formation.

If the light-sensitive drug or drug products must be stored in a light-resistant colored container (e.g., amber vial) to offer assurance of better protection from light sensitivity, should the cold kits for the preparation of the radioactive drugs and the single-dose syringes containing the light-sensitive drugs or drug products be exempted and allowed to be stored in a clear and colorless or translucent container? Although the contents of light-sensitive cold kits and radiopharmaceuticals can be transferred to such containers (e.g., vial or syringe) to examine the clarity and particulate matter, this approach is not suitable for cold kit formulations and radiopharmaceuticals because of possible oxidation effects on these drugs and the additional radiation exposure to personnel during the transfer. The reason for storing the injectable drugs in a predrawn single-dose syringe is the ease and convenience of drug administration. Transferring the drug content in the predrawn syringe to another clear and colorless container for the final inspection would defeat the purpose of the original idea for using the single-use syringe.

The USPC should modify its light transmission testing to include procedures and standard limits concerning use of the light-resistant paper carton or box. The labeling requirements for the secondary or outer light-resistant container, such as a carton or box, must be identical to the labels on the immediate container for the light-sensitive drugs or drug products. The outer light-resistant box or carton should have no openings in order to provide an adequate light-resistant environment for drugs or drug products susceptible to photolytic degradation. For a light-resistant outer container used to store multiple components of a drug or drug products, if even only one single component of the drug is light sensitive, the label on the outer container should still be required to bear a cautionary statement for light sensitivity so that the end user or repackager will be alerted to ensure the proper protection of the light-sensitive drug component.

ACKNOWLEDGMENTS

The author would like to thank Ms. Vicki S. Krage for her patience and assistance in the preparation of this chapter, which was presented in part in the following three articles: Photolytic degradation of drugs, *Am. J. Hosp. Pharm.*, 49, 2704–2705, 1992; Photochemical considerations of light-sensitive cold kits and radiopharmaceuticals,

J. Nucl. Med. Technol., 21, 90–91, 1993; The packaging of light-sensitive cold kits, *J. Nucl. Med. Technol.*, 23, 45–46, 1995; and Light-sensitive cold kits and the associated radiopharmaceuticals, *Ann. Nucl. Med. Sci.*, 16, 185–190, 2003.

REFERENCES

AcuTect® package insert, 2001, Berlex Laboratories, Inc., Wayne, NJ.

Choletec® package insert, 2000, Squibb Diagnostics, Inc., Princeton, NJ.

Constitution and Bylaws, 2004, in *United States Pharmacopeia,* 27th rev., and *National Formulary, 22*nd ed., The United States Pharmacopeial Convention, Inc., Rockville, MD, pp. 958–968.

Design and interpretation of stability studies, 1987, in *Guidelines for Submitting Documentation for the Stability of Human Drugs and Biologics,* Food and Drug Administration, Rockville, MD, pp. 8–9.

DMSA package insert, 2000, Medi-Physics, Inc., Arlington Heights, IL.

General Tests and Assays, 2004, in *United States Pharmacopeia,* 27th rev., and *National Formulary,* 22nd ed., The United States Pharmacopeial Convention, Inc., Rockville, MD, pp. 210–2402.

Hepatolite® package insert, 1999, CIS-US, Inc., Bedford, MA.

International harmonization; drift policy on standards; availability, 1994, *Fed. Reg.,* 59, 60870–60874.

International Conference on Harmonisation; Guideline for the Photostability Testing of New Drug Substances and Products; availability, 1997, *Fed. Reg.,* 62, 27116–27122.

Light transmission (General Chapter 661), 2004, in *United States Pharmacopeia,* 27th rev., and *National Formulary,* 22nd ed., The United States Pharmacopeial Convention, Inc., Rockville, MD, pp. 288–296.

Myoview™, package insert, 2003, Amersham Health Medi-Physics, Inc., Arlington Heights, IL.

Neurolite® package insert, 2001, Bristol-Myers Squibb Medical Imaging, Inc., N. Billerica, MA.

OctreoScan® package insert, 2000, Mallinckrodt Inc., St. Louis, MO.

Paul, W.L., 1992, Photolytic degradation of drugs (reply), *Am. J. Hosp. Pharm.,* 49, 2704–2705.

Preservation, packaging, storage, and labeling (General Notices), 2004, in *United States Pharmacopeia,* 27th rev., and *National Formulary,* 22nd ed., The United States Pharmacopeial Convention, Inc., Rockville, MD, pp. 9–11.

Quality of Biotechnological Products: Stability Testing of Biotechnological/Biological Products (General Chapter 1049), 2004, in *United States Pharmacopeia,* 27th rev., and *National Formulary,* 22nd ed., The United States Pharmacopeial Convention, Inc., Rockville, MD, pp. 481–484.

TechneScan MAG3® package insert, 2000, Mallinckrodt Inc., St. Louis, MO.

Technetium Tc 99m disofenin injection, 2004, in *United States Pharmacopeia,* 27th rev., and *National Formulary,* 22nd ed., The United States Pharmacopeial Convention, Inc., Rockville, MD, pp. 1771–1772.

Technetium Tc 99m succimer injection, 2004, in *United States Pharmacopeia,* 27th rev., and *National Formulary,* 22nd ed., The United States Pharmacopeial Convention, Inc., Rockville, MD, p. 1782.

Technetium Tc 99m tetrofosmin injection, 2004, in *United States Pharmacopeia,* 27th rev., and *National Formulary,* 22nd ed., The United States Pharmacopeial Convention, Inc., Rockville, MD, p. 1783.

UltraTag® RBC package insert, 2000, Mallinckrodt Inc., St. Louis, MO.

CHAPTER 9

Biological Effects of Combinations of Drugs and Light

Johan Moan and Petras Juzenas

CONTENTS

9.1 INTRODUCTION

The fact that some substances sensitize human skin to light has been known at least since the time of the Pharaohs. The Egyptians used plant extracts from *Ammi Majus*, which contains psoralens, and sunlight to treat skin disorders such as leukoderma (vitiligo). In one of India's sacred books, *Atharva Veda* (1400 B.C. or earlier), photochemotherapy of leukoderma with extracts of Psoralea corylifolia, which contains furocoumarins, is carefully described.

In our century, a new branch of medicine has developed: that of photomedicine, which is the science of how drugs, light, and biomolecules interact, mainly in human skin (Regan and Parrish, 1982; Spikes, 1997; Calzavara–Pinton et al., 2001). The skin is the largest organ of the human body, covering about 1.5 to 2.0 m², while its tissue volume is only about 50 cm³ and accounts for about 7 to 16% of total body weight. It protects against external exposures of toxic compounds and intruders and is among the most important sites of the immunological processes in the body (Shaw et al., 1991; Dowson, 1997).

Every minute a significant fraction of the blood flows through the skin and is exposed to small doses of light that can penetrate down to the dermis. At the borderline between dermis and epidermis, the proliferating basal cells, as well as most of the melanocytes, reside. Ultraviolet (UV) radiation can interact with these cells in such a way that they are transformed to cancer cells and give rise to melanoma or nonmelanoma skin cancer. In the UVB range (280 to 320 nm), this interaction is mainly a direct one, i.e., absorption of radiation directly in DNA. However, in the UVA (320 to 400 nm) and visible spectral range, photosensitized reactions are believed to play the major role in photocarcinogenesis (de Gruijl, 2002; Lacour, 2002). The photosensitizers involved have not been identified; they may be endogenously formed or derived from food ingredients, topically applied substances such as cosmetic products, or medicines.

Photosensitized reactions in skin can be beneficial, such as in photodynamic therapy of cancer (PDT) and use of psoralens and UVA radiation in PUVA therapy of psoriasis, or unwanted, such as in cases of photosensitivity induced by chlorpromazine, phenothiazine, tetracycline, and a large number of other drugs. These reactions can be classified as photoallergic or phototoxic. Some photosensitized reactions are dependent on oxygen while others are not. Oxygen-dependent reactions, which occur via the formation of singlet oxygen, are termed photodynamic reactions or photosensitized reactions of type II (Foote, 1991; Ochsner, 1997). All other reactions, oxygen-dependent as well as oxygen-independent ones, are termed type I reactions. These reactions mainly occur via the formation of free radicals.

9.2 PENETRATION OF VISIBLE LIGHT AND UV RADIATION INTO LIVING TISSUE

Proteins and nucleic acids are the main absorbers of UVB radiation in human tissues. In the UVA and visible range, melanin and hemoglobin are the main absorbers. All human tissues are inhomogeneous and contain high concentrations of light-scattering particles and boundaries. The concentration and type of absorbers and scatterers vary widely from tissue to tissue; therefore, the optical penetration depth also varies within the range of 0.2 to 4 mm. The optical penetration depth (δ) is defined as the distance over which the space irradiance decreases by a factor of $e = 2.718$ (Eichler et al., 1977; Moan et al., 1998a).

About 5% of the incident fluence of optical radiation is reflected from the stratum corneum, and, due to backscattering, the space irradiance increases down to about 0.1 mm. Further down, the space irradiance below a skin surface exposed to a wide beam of parallel light decays approximately exponentially (Moan et al., 1998a; Juzenas et al., 2002a). The thickness of the stratum corneum is typically 10 to 150 μm and that of the dermis is 1 to 5 mm. The role of melanin as absorber in the

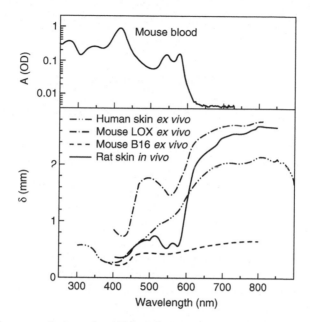

Figure 9.1   Upper panel: absorption of blood measured with a spectrophotometer. The blood was sampled from a hairless female BALB/c mouse and diluted in PBS (1:100) immediately before the measurements. Lower panel: the penetration depth (δ) of light as a function of the wavelength measured in various biological tissues. Human skin excised from an abdominal region of a female patient was provided by Dr. J.F. Evensen. Amelanotic melanoma (LOX) or pigmented melanoma (B16-F10) cells were injected subcutaneously (approximately 1.0 to 2.0·10⁶ cells per mouse) into a flank of a hairless female BALB/c mouse and the tumors were excised after 10 and 20 days, respectively. Measurements in skin *in vivo* were performed in a living female hairless RWT rat.

skin is demonstrated by the fact that about 50% of the incident UVA may reach the basal layer of white skin while only 5 to 15% may reach the same layer in black skin (Parrish et al., 1978; Jablonski and Chaplin, 2000; Moan, 2001; Ortonne, 2002).

The penetration depth of visible light from a halogen lamp in normal skin and in tumors was measured as a function of wavelength by means of a conventional fluorescence spectrometer equipped with a fiber-optic system (Moan et al., 1998a; Juzenas et al., 2002a). The absorption of hemoglobin plays a significant role in the wavelength regions around 420 nm and 520 to 580 nm. The dips of the penetration curves practically correspond to the peaks of the blood absorption (Figure 9.1).

Furthermore, the light penetration in the skin *ex vivo* is different from that *in vivo*, where oxygenated hemoglobin is prevalent. Oxyhemoglobin has distinct absorption peaks at around 420, 540, and 580 nm, while deoxyhemoglobin has two peaks at around 440 and 560 nm (Voet and Voet, 1995; Yaroslavsky et al., 2002). In pigmented melanoma, the light penetration is reduced by melanin absorption. Generally, the light penetration increases with increasing wavelength up to about 700 nm (Figure 9.1). Thus, dyes absorbing in the red part of the spectrum can be excited more efficiently in tissue than dyes absorbing at shorter wavelengths. This is taken advantage of when porphyrins, chlorins, phthalocyanines, and naphthalocyanines are used as sensitizers in PDT. These sensitizers absorb at wavelengths greater than 600 nm, i.e., outside the main bands of absorption of hemoglobin.

9.3 TYPE I AND TYPE II REACTIONS

In type I reactions, electrons or hydrogen atoms are transferred between sensitizer molecules and substrate or solvent molecules. Oxygen may participate in subsequent reactions. The probability that a type I reaction will occur increases with decreasing oxygen concentration and with increasing substrate concentration. The use of psoralens and UVA radiation in the PUVA treatment of psoriasis and vitiligo is the best known application of a type I reaction (Musajo et al., 1974; Pathak et al., 1974; Harber et al., 1982; Gasparro, 2000). Psoralens intercalate in DNA even without light exposure.

Light exposure within the absorption spectrum of the psoralens (300 to 400 nm) leads to covalent binding of the psoralen to pyrimidine bases in DNA. Monofunctional psoralens, such as angelicines, bind to only one of the DNA strands; bifunctional ones (linear psoralens) bind to both strands, and therefore cross-link one strand to the other. This will slow down the rate of cell division, which is abnormally high in psoriatic skin. Psoralens may also act on membranes in a type II process.

It is not known which of the two processes plays the main role in treatment. PUVA can also increase the number and size of melanocytes in skin. This is the basis for its use in the treatment of vitiligo. It is debatable whether PUVA pigmentation protects DNA in the basal cells. Experiments carried out in our laboratory show that pigmentation generated by PUVA protects skin slightly against UV-induced DNA strand breaks (Kinley et al., 1997). However, it seems that PUVA treatment may be associated with a significant carcinogenic risk (Lindelof et al., 1991; Lindelof, 1999; Seidl et al., 2001).

Photophoresis is a variant of PUVA treatment (Edelson et al., 1987; Horio, 2000). Leukocytes from patients with cutaneous T-cell lymphoma or autoimmune disorders (rheumatoid arthritis and systemic sclerosis) are exposed to UVA radiation in the presence of a psoralen and given back to the patients. It is believed that PUVA changes some surface markers of T-cells and thus triggers the immune system. Thus, this treatment can be regarded as a beneficial application of a photoallergic reaction.

Photoimmunotherapy was introduced 20 years ago (Mew et al., 1983). Photosensitizer immunoconjugates have been constructed to improve the specificity of PDT (Mew et al., 1983). A major problem in this field has been to synthesize and purify conjugates. The use of benzoporphyrin derivative monoacid ring A (BPD) or Verteporfin has been applied in the treatment of immunologically related diseases (Jiang et al., 1990; Levy et al., 1994). Recently a new method was reported to prepare BPD immunoconjugates, which photodynamically kill tumor cells that overexpress the epidermal growth factor receptor (EGFR) via specific binding to this receptor (Savellano and Hasan, 2003). Because BPD absorbs light in the wavelength region of 650 to 700 nm, where penetration depth into tissue is large (Figure 9.1), transcutaneous light application may be efficient.

In type II reactions (photodynamic reactions), singlet oxygen is formed by energy transfer from sensitizers in the lowest excited triplet state to oxygen molecules in the ground state (Wasserman and Murray, 1979; De Rosa and Crutchley, 2002). Thus, type II reactions cannot occur in the absence of oxygen. For porphyrins, photoinactivation of cells is reduced by 50% at an oxygen concentration of about 15 μM, which corresponds to 1% partial pressure, compared to the yield in air-saturated medium with oxygen concentration of more than 70 μM (5%) (Moan and Sommer, 1985). This oxygen dependency is quite similar to the oxygen dependency of gamma- and x-ray radiation (Hall, 2000; Harrison et al., 2002). Cell sensitivity to radiation is reduced at oxygen levels less than about 0.5 to 2.0% (Pass, 1993; Hall, 2000). Many sensitizers have a very high quantum yield of singlet oxygen, typically 0.2 to 0.7 (Redmond and Gamlin, 1999). The lifetime of singlet oxygen is about 10 to 30 times longer in D_2O than in H_2O (Weldon et al., 1999; Girotti, 2001) — a fact often used to check whether singlet oxygen is involved in a photosensitized reaction (Foster et al., 1991; Sharman et al., 2000). However, if singlet oxygen is generated from lipophilic sensitizers in cell membranes, only a negligible D_2O effect is expected because singlet oxygen usually reacts while diffusing in the membrane and before it reaches the aqueous phase (Kochevar and Redmond, 2000; Lavi et al., 2002).

When two dyes were used in different concentrations, one generating singlet oxygen during light exposure and the other being degraded by singlet oxygen, the lifetime of singlet oxygen in cells was estimated to be about 10 to 40 ns, corresponding to a diffusion radius of $d = (6Dt)^{\frac{1}{2}} = 0.01$ to 0.02 μm (Moan and Berg, 1991), where D is the diffusion coefficient of oxygen and the lifetime $t = 10$ to 40 ns. This radius is very small compared with a typical cell diameter of around 10 μm. Even the size of cell organelles (~0.5 to 1 μm) is at least an order of magnitude larger than the diffusion radius of singlet oxygen. It was therefore concluded that photodynamic damage to cells is always induced close to cellular locations where the concentration of sensitizer is high. This is in agreement with the observations: most of the lipophilic sensitizers used in photodynamic therapy of cancer are not taken up by cell nuclei and sensitize photodamage only to a small fraction of the DNA localized close to the nuclear membrane (Kvam et al., 1992). In fact, using a method based on this observation, it was possible to estimate the average length of DNA between each attachment point to the nuclear membrane in interphase cells to about 180 kilobases (Kvam et al., 1992), which correspond to approximately 80 μm of the 30 nm chromatin fiber (Henegariu et al., 2001).

To prove that a photosensitized reaction is a type II reaction is not easy. Several scavengers, such as β-carotene, react efficiently with singlet oxygen and can be used to indicate its involvement in a reaction. A hydroperoxide formed when singlet oxygen reacts with cholesterol can be used as an indicator (Suwa et al., 1978; Girotti, 2001; Min and Boff, 2002). However, no test is really conclusive because singlet oxygen may be formed in the membranes of cells (where no water is available to manifest the D_2O effect) or in the aqueous phase of the cytoplasm (where there is no cholesterol and the lipophilic β-carotene does not localize) (Sharman et al., 2000). Furthermore, all compartments of cells contain large concentrations of molecules that react with singlet oxygen, shorten its lifetime, and reduce its steady state concentration to below levels at which its 1260-nm luminescence can be detected (Krasnovsky, Jr. et al., 1983, Gorman and Rodgers, 1992, Niedre et al., 2002).

Singlet oxygen is a very reactive species that efficiently degrades a number of biomolecules. The most important ones are guanine, tryptophan, tyrosine, histidine, cysteine, cystine, and cholesterol. Singlet oxygen causes protein cross links and DNA strand breaks.

9.4 PHOTOSENSITIZING AND PHOTOLABILE DYES IN MEDICINE

In 1900 Oscar Raab reported in *Zeitschrift für Biologie* that paramecia were killed by exposure to acridine and light (Spikes, 1997; Calzavara–Pinton et al., 2001). Neither acridine alone or light alone had any harmful effect on the organisms. A few years later von Tappeiner and Jodlbauer (1904) demonstrated that oxygen was required for the biological response to acridine and light, terming the process "photodynamic action" (von Tappeiner and Jodlbauer, 1904). These are the first scientific reports of phototoxic reactions. Later it was shown that some phototoxic reactions, notably those sensitized by psoralens, do not require oxygen.

Phototoxic reactions in humans occur exclusively in skin exposed to light. Their morphology and clinical symptoms may vary; in some cases a burning and painful sensation is felt during light exposure, while in others reactions such as erythema, edema, and vesiculation occur later. On the cellular level, the phototoxic reactions may involve DNA, proteins, lipids, lysosomes, mitochondria, and the plasma membrane. Phototoxic reactions can be caused by a number of drugs. Among the most common are (Harber et al., 1982):

- Aminobenzoic acid derivatives
- Anthiaquinone dyes
- Chlorothiazides
- Chlorpromazine
- Anthracine
- Acridine
- Pyrene
- Nalidixic acid
- Phenothiazine
- Protriptyline
- Psoralens
- Sulfanilamide
- Tetracycline

The phototoxic effects of chlorpromazine may be partly due to its metabolites (see references in Harber et al., 1982). Photodynamic and nonphotodynamic processes have been demonstrated for this drug. Stable photoproducts of the drug may account for some of its effects. The action spectrum for chlorpromazine phototoxicity is red-shifted (max at 330 nm) compared with its absorption spectrum (max at 305 nm). This may be due to the shape of the penetration spectrum of light into tissue and/or to metabolic production of compounds that absorb at longer wavelengths. Also, the action spectrum for tetracycline phototoxicity seems to be red-shifted to a position with a peak value at about 400 nm, and demethylchlorotetracycline is more potent than the parent compound tetracycline.

On the other hand, the action spectrum for phototoxicity caused in human skin by sulphanilamide correlates well with the absorption spectrum of the drug (see references in Harber et al., 1982). A photo-oxidation product, p-hydroxylaminobenzene, seems to have toxic effects. Sulphanilamide induces not only phototoxic, but also photoallergic reactions; the latter require immunologic responses. The general scheme for such a reaction is that antigens that activate lymphocytes are produced by photosensitized reactions in proteins. The sensitizer in the skin is photoconverted to a product that reacts with proteins to form antigens. In contrast to phototoxic reactions, photoallergic reactions are systemic and can result in symptoms in irradiated and unirradiated tissue. Most photoallergic reactions can be considered to involve delayed-type hypersensitivity immunological mechanisms. Photoallergic reactions can be caused by low sensitizer concentrations, while larger sensitizer concentrations are usually needed to elicit phototoxicity. The most widely known sensitizers of photoallergy include (Harber et al., 1982):

- Aminobenzoic acids
- Bithionol
- Chlorpromazine
- Chlorpropamide
- Fentichlor
- Salicylanilides
- 6-Methylcoumarin
- Promethazine
- Sulfanilamide
- Thiazides

Among the drug-induced photosensitivity diseases, different variants of the porphyrias are best known (Rimington, 1989; Thunell, 2000). Porphyria cutanea tarda (PCT) can be induced or aggravated by chlorophenols, estrogens, ethanol, hexachlorobenzene, and 2-benzyl-4,6-dichlorophenol.

ALA (5-aminolevulinic acid synthetase) is a rate-limiting enzyme in the biosynthesis of heme. A number of chemicals, including hypnotics, sedatives, analgesics, antirheumatics, antibiotics, hypoglycemic agents, and sex steroids, induce ALA synthetase (Harber et al., 1982; Thunell, 2000). This enzyme seems to be controlled by an operator gene that can be regulated by an aporepressor. Certain chemicals can inhibit this repressor and thus result in uncontrolled ALA synthetase activity and elevated porphyrin production. In heme biosynthesis, hydrophilic and lipophilic porphyrinogens are involved. When overproduced, the porphyrinogens are easily oxidized to photoactive porphyrins.

As a general rule hydrophilic porphyrins sensitize photodamage to lysosomes, while the lipophilic ones sensitize photodamage to mitochondrial membranes and other membrane systems (Sandberg, 1981; Sandberg and Romslo, 1982; Dougherty et al., 1998). As discussed later, application of ALA can induce overproduction of protoporphyrin IX (PpIX) in cells (Malik and Lugaci, 1987; Peng et al., 1987) — a phenomenon applied in PDT (Charlesworth and Truscott, 1993; Peng et al., 1997a). In addition to porphyrias, lupus erythematosus and pellagra can be induced by photosensitization.

9.5 TREATMENT OF ADVERSE PHOTOSENSITIZED REACTIONS IN HUMANS

The main treatment of the diseases mentioned earlier is, whenever possible, to stop exposure to photosensitizing substances or substances that initiate endogenous production of photosensitizers. In cases in which this is impractical or impossible, quenchers of singlet oxygen, such as β-carotene, can help if photodynamic processes are involved (Poh–Fitzpatrick, 2000). Other treatments of porphyrias, lupus erythematosus, and pellagra are described in textbooks of dermatology.

9.6 THERAPIES BASED ON USE OF PHOTOSENSITIZING DRUGS

The use of psoralens in the treatment of psoriasis and vitiligo has been discussed earlier. Psoriasis is a very common disease and photochemotherapy is a frequently used treatment (Tremblay and Bissonnette, 2002). Due to the carcinogenic effect of the PUVA treatment (Lindelof, 1999; Seidl et al., 2001), photosensitizers other than DNA intercalating psoralens are sought. It seems that PpIX induced by ALA may offer a good alternative (Bissonnette et al., 2002).

PDT of cancer is a rapidly developing field (Oleinick et al., 2002; Morton et al., 2002) and the treatment has a low carcinogenic risk (Fuchs et al., 2000). Since the introduction of porphyrin derivatives in clinical PDT in the late 1970s, a number of sensitizers have been proposed: phthalocyanines, chlorins, benzoporphyrins, porphyrins, and PpIX induced by ALA or its derivatives (Sharman et al., 1999; Konan et al., 2002). Some of these sensitizers are used in the clinic. The most widely used one, a mixture of hematoporphyrin derivatives called Photofrin, is approved for PDT of cancer in a number of European and Asian countries and in North America (Dougherty et al., 1998; Sharman et al., 1999). When injected in tumor-bearing animals (or humans), these sensitizers are taken up and/or retained in tumors with some selectivity with respect to normal tissues such as skin, muscles, and brain (Dougherty, 1987; Moan and Berg, 1992; Vrouenraets et al., 2003).

The reasons for the tumor selectivity are only partly understood. One explanation may be that these drugs, notably the lipophilic ones, tend to bind to lipoproteins in serum. Because tumors have a high concentration of receptors for low-density lipoproteins, drugs that bind to these lipoproteins may also be directed to the tumors. Many of the photosensitizers mentioned aggregate in aqueous solutions. Tumors usually contain many macrophages that may take up aggregates by endocytosis, so aggregation may play a role in tumor selectivity. The low lymphatic drainage and the leaky vascular system of tumors are other factors to consider. Finally, because tumors have a low pH, drugs that change from hydrophilic to lipophilic upon lowering of the pH, e.g., from 7.5 to 6.5, are expected to be tumor selective. Injecting glucose in tumor-bearing animals has shown that the low tumor pH plays an important role for selective uptake of photosensitizers. Glucose injection selectively lowers the tumor pH and increases the uptake of Photofrin and meso-tetrahydroxyphenyl-chlorin (mTHPC) in the tumors (Peng et al., 1991; Moan et al., 1999).

Recently, ALA–PDT has become widely used in the treatment of skin tumors, notably basal cell carcinomas (BCC) (Peng et al., 1997b; Kelty et al., 2002). Application of excess ALA leads to an accumulation of PpIX with some tumor selectivity (Kennedy et al., 1990, 1996). After exogenous application of ALA, the last step in heme synthesis, the action of the enzyme ferrochelatase, which adds ferrous iron to PpIX to form heme, is not efficient enough under such conditions and excess PpIX accumulates. In contrast to heme, PpIX is a very efficient photosensitizer (Kennedy and Pottier, 1992). The tumor selectivity may have several reasons, such as a higher activity of porphobilinogen deaminase (PBGD) and a lower activity of ferrochelatase (FC) in tumors than in normal tissues (el Sharabasy et al., 1992; van Hillegersberg et al., 1992; Gibson et al., 1998), and a low concentration of iron in tumors (Rittenhouse–Diakun et al., 1995; Gibson et al., 1998).

The main barrier for the penetration of topically applied drugs is principally attributed to the stratum corneum (Scheuplein and Blank, 1971; Moser et al., 2001). The tumor selectivity may partly be due to the fact that the skin overlaying tumors often is more permeable than normal skin (Peng et al., 1997b; Moan et al., 2001). ALA–PDT can be made more efficient by:

- Gently scraping the skin overlaying the tumors to remove the stratum corneum (Peng et al., 1997b; van den Akker et al., 2000; Soler et al., 2000)
- Adding penetration enhancers such as dimethylsulfoxide (DMSO) (Malik et al., 1995; Peng et al., 1997b; De Rosa et al., 2000) and azacycloalkane derivatives like HPE-101 (Yano et al., 1993; van den Akker et al., 2002)
- Applying iron chelators such as ethylendiaminetetraacetic acid (EDTA); 1,2-diethyl-3-hydroxypyridin-4-one (CP94); and deferoxamine (or desferrioxamine, DF) (Berg et al., 1996; Chang et al., 1997; Peng et al., 1997b; Ninomiya et al., 1999; Soler et al., 2000)
- Adding drugs that promote PBGD activity (Gibson et al., 1998; Hilf et al., 1999) or inactivate FC (Iinuma et al., 1994)
- Applying ALA ester derivatives that are more lipophilic than ALA (Gaullier et al., 1997; Kloek et al., 1998; Uehlinger et al., 2000; Ninomiya et al., 2001)

The ALA esters are less polar than ALA but are converted to ALA by esterases present in the tissues (Kloek and Beijersbergen van Henegouwen, 1996; Washbrook and Riley 1997; Tunstall et al., 2002). Ultrasound can also enhance the penetration of ALA (Ma et al., 1998). Thus, even nodular BCCs with depths of more than 2 mm can be treated, and ALA–PDT gives very good cosmetic results (Fritsch et al., 1998; Soler et al., 2000; Peng et al., 2001).

So far, attempts to treat cutaneous malignant melanomas by PDT have not been successful (Fritsch et al., 1998; Kalka et al., 2000; Morton et al., 2002), supposedly because the light absorption by melanin reduces the penetration depth of light into the melanomas (Ito et al., 1992; Lui and Anderson, 1993; Losi et al., 1993). Nevertheless, promising results were recently reported on the use of PDT for treatment of melanomas (Haddad et al., 1998; Woodburn et al., 1998; Busetti et al., 1999; Juzenas et al., 2002b).

9.7 ACTION MECHANISMS OF PDT ON THE CELLULAR LEVEL

Depending on the conditions, PDT destroys tumors by a direct effect on tumor cells or by an effect on cells of the vascular system (Moan and Berg, 1992; Dougherty et al., 1998; Morton et al., 2002). The endothelial cells lining the vascular walls seem to take up large concentrations of photosensitizers after intravenous, as well as intraperitoneal, administration.

Cells exposed to light in the presence of photosensitizers can be destroyed through a number of mechanisms. Large light exposures of cells incubated with lipophilic sensitizers lead to immediate cell inactivation caused mainly by destruction of membranes. Mitochondrial membranes and the plasma membrane are destroyed. This damage is seen immediately after PDT: blebs are formed on the plasma membrane and the cells swell and disintegrate. Lipids and proteins in the membranes are destroyed and the singlet oxygen photoproduct of cholesterol is often found. Under anoxic conditions, such as in the central parts of poorly vascularized tumors or after prolonged light exposure, free radical mechanisms (type I reactions) may play a role. An indication of this is that photoprotoporphyrin is formed in small yields in ALA–PDT. The lower the oxygen concentration is the more important this photoreaction is expected to be (Harel et al., 1976).

At low-light exposures, a major cause of cell inactivation is destruction of free tubulin (Berg et al., 1992). After such destruction, fewer microtubules are formed in the cells and the cells are unable to perform mitosis. This is seen as an accumulation of cells in the metaphase after PDT (Berg et al., 1992). Furthermore, it seems that PDT may trigger apoptosis (Zaidi et al., 1993; Dougherty et al., 1998; Bissonnette et al., 2002). Cells surviving low exposures may possess damaged proteins on their surfaces. *In vivo*, such damaged proteins may be recognized by the immune system (Dougherty et al., 1998). This may explain the efficiency of PDT (and that of photophoresis) as well as the low metastatic potential of tumors after PDT (Canti et al., 1983; Edelson et al., 1987; Sindelar et al., 1991; Rousset et al., 1999; Horio, 2000; Morton et al., 2002).

9.8 PHOTODEGRADATION OF DYES DURING PDT

Most photosensitizers are degraded during light exposure (Moan, 1986; Bonnett and Martinez, 2001). The photolabilities of dyes differ widely (Figure 9.2). Under anoxic conditions, the quantum yield of degradation is usually smaller than in the presence of oxygen, and other products are formed. From porphyrins, for instance, chlorine-type photoproducts are formed under reducing conditions (Harel et al., 1976; Moan, 1986; Rotomskis et al., 1997; Bonnett and Martinez, 2001).

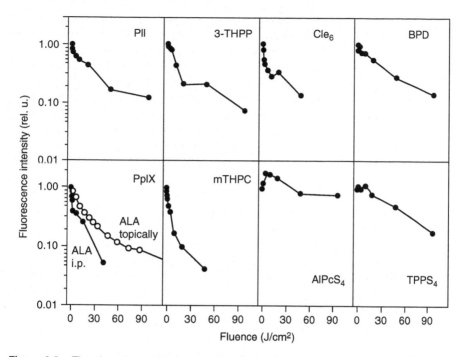

Figure 9.2   The photodegradation of a number of dyes in mouse skin (hairless female BALB/c mice), measured fluorimetrically during laser exposure *in vivo* at a fluence rate of 100 mW/cm² and at the specified wavelengths. The mice were injected i.p. with the dyes 24 h before the laser exposure. PII — Photofrin (10 mg/kg, Quadra Logic, λ = 630 nm); 3-THPP — 3-tetrahydroxyphenyl porphine (Porphyrin Products, λ = 645 nm); Cle$_6$ — chlorin e$_6$ (10 mg/kg, Porphyrin Products, λ = 630 nm); BPD — benzoporphyrin derivative monoacid ring A (10 mg/kg, Quadra Logic, λ = 690 nm); PpIX — protoporphyrin IX (generated *in vivo* from 5-aminolevulinic acid, Sigma, 100 mg/kg or topically 20% w/w cream, λ = 635 nm); mTHPC — mesotetra-hydroxyphenyl chlorin (1 mg/kg, Scotia, λ = 670 nm); AlPcS$_4$ — aluminum phthalocyanine tetrasulfonate (10 mg/kg, Porphyrin Products, λ = 670 nm); TPPS$_4$ — tetraphenylporphine tetrasulfonate (10 mg/kg, Porphyrin Products, λ = 645 nm).

An example of photoproduct is the formation of photoprotoporphyrin (PPp) during light exposure of PpIX (Inhoffen et al., 1969; Cox et al., 1982; Krieg and Whitten, 1984a; Konig et al., 1993; Robinson et al., 1998; Juzenas et al., 2001). PPp is a type of chlorin (Zheng et al., 2000) and its formation is a photooxygenation process (Cox et al., 1982; Ericson et al., 2003). Because oxygen is consumed in many photosensitized reactions, the reaction pathways may change during light exposure under conditions in which the oxygen supply is limited, such as in a tumor. Chlorin-type photosensitizers absorb light at longer wavelengths than porphyrins and they are good photosensitizers (Spikes, 1990; Stewart et al., 1998; Brault et al., 2001). Thus, one should pay attention to the previously mentioned photoconversion phenomenon, notably when the optimal wavelength band for therapy is sought.

For PpIX, one should use light of wavelengths between 600 and 700 nm to excite the porphyrin and the chlorine-type photoproduct (Moan et al., 2000). Photoprotoporphyrin is more water soluble than PpIX and therefore has different binding and retention properties in cells and tissues. It has been shown that photoprotoporphyrin formed during PDT leaves the cells and diffuses out into remote sites *in vivo* (Juzenas et al., 2001). Photosensitizing efficiency and photodegradation of photosensitizers depend on environment (Kongshaug et al., 1990; Lavi et al., 2002). Generally, many sensitizers are photodegraded much faster in the presence of amino acids and proteins than in pure buffer solutions (Krieg and Whitten, 1984a; Pogue et al., 2001). Type I reactions are more likely to occur in the presence of proteins, and protein photo-oxidation products formed in reactions between singlet oxygen and proteins may react with and degrade the photosensitizers. For sulfur-containing amino acids and proteins the latter reaction pathway is important (Krieg and Whitten, 1984b).

In tissue, of course, this is of utmost importance. Light exposure may result in degradation of the binding sites of sensitizers on proteins (Moan et al., 1988, 1997). Thus, the photosensitizer is liberated and may move to other binding sites that are possibly more vital for cell survival (Moan et al., 1997). It has been shown that, during light exposure of red cells in patients with erythropoietic protoporphyria (EPP), PpIX molecules move from their binding sites on globin to the cell membrane and further to surrounding cells in contact with the red cells (Brun et al., 1990). Such "photomovement" may explain some of the skin pathogenesis seen in EPP patients. Similarly, light exposure of cells containing hydrophilic sensitizers in their lysosomes leads to a permeabilization of the lysosomes, followed by a transfer of the sensitizers from the lysosomes to the cell nucleus (Moan et al., 1989, 1994). This transfer can take place in cells after sublethal PDT.

When sensitizers leak out of the lysosomes, they become diluted. This dilution leads to an increase in the fluorescence quantum yield as well as an increase in the quantum yield for cell inactivation (Berg et al., 1991; Berg and Moan, 1994). Such lysosomal rupture is demonstrated *in vivo* by the increase of fluorescence of tetra-phenylporphine tetrasulfonate (TPPS$_4$) or aluminum phthalocyanine tetrasulfonate (AlPcS$_4$) in tumors during PDT (see Figure 9.2). Under other conditions, a transient decrease of fluorescence is seen during light exposure. This is an indication of a transient transfer of the fluorescing molecules to binding sited in the tissue with a lower quantum yield of fluorescence (Moan and Anholt, 1990; Moan et al., 1998b). Changes in the quantum yield of photoinactivation of cells have been observed also for extralysosomally localized sensitizers and are probably due to a relocalization of the dye during PDT (Moan, 1988). Small light exposures of tumors at a time when a significant amount of a sensitizer is in the circulation sometimes lead to an increase in the tumor uptake of the sensitizer (Ma et al., 1992).

9.9 ADVANTAGES AND DISADVANTAGES OF USING PHOTOLABILE DYES IN PDT

Photosensitizers are widely different with respect to photolability (Figure 9.2). The disadvantage of using a photolabile dye for PDT is that one needs to use considerably larger fluences of light to achieve the same photodynamic dose than would be needed in the case of a photostable photosensitizer (Mang et al., 1987; Moan, 1988; Rotomskis et al., 1998). However, PDT with photolabile dyes may contribute to a larger therapeutic ratio. For most cancer treatments, normal tissue damage limits the therapeutic dose that can be applied. For PDT, prolonged cutaneous photosensitivity, which may last for several weeks after drug application, is the main adverse effect (Roberts et al., 1989; Peng et al., 1997b). If one uses low drug doses, the photosensitizer in skin and muscles surrounding the tumor may be photodegraded to non-phototoxic levels before unacceptable photodamage occurs (Boyle and Potter, 1987). The tumor, however, may still be destroyed, since it contains 2 to 10 times more of the photosensitizer than the normal tissues (Moan et al., 1996, 2000).

Moreover, photoproducts formed during the photodegradation of photosensitizers are photodynamically active and may contribute to the efficacy of PDT (Cox et al., 1982; Konig et al., 1990; Giniunas et al., 1991). Chlorin-type photoproducts have an absorption peak at about 660 to 670 nm and extinction coefficients larger than that of the long-wavelength peak of the Q-band of porphyrins (Brault et al., 2001; Theodossiou and MacRobert, 2002). The red-shift and larger extinction coefficient of the Q-band of chlorins and chlorin-type photoproducts are features that make the photoproducts attractive as second-generation photosensitizers for PDT. The action spectra for the cell inactivation with HpD and PpIX peak at about 630 and 635 nm, respectively (Konig et al., 1990; Szeimies et al., 1995), while those for cell inactivation with the photoproducts of HpD and PpIX have maxima at around 640 to 650 and 670 nm, respectively (Konig et al., 1990; Bagdonas et al., 2000). Moreover, in the case of PpIX *in vitro*, the cell inactivating effect was slightly larger when the 670-nm exposure was given after the 635-nm exposure compared to that when 670-nm exposure was given before 635-nm exposure (Bagdonas et al., 2000).

Because the photoproducts display photosensitizing properties, it has been proposed that light sources for PDT should contain some light with wavelengths at around 650 to 700 nm in addition to 630 to 640 nm (Konig et al., 1993; Dietel et al., 1997; Moan et al., 2000). It should be noted that the penetration of light into tissue increases in this wavelength region (Figure 9.1). However, the yields of photoproducts are low *in vivo*, photosensitive, and can be photobleached by 670 nm-light exposure (Bagdonas et al., 2000). It seems that the photodynamic activity of the photoproducts *in vivo* is smaller than that of the parent photosensitizers: photofrin; hematoporphyrin derivative (Konig et al., 1990; Giniunas et al., 1991); and PpIX (Ma et al., 2001). However, one comparative study *in vitro* was recently reported in which PPp showed higher phototoxicity than PpIX at its optimal activation wavelength of 670 nm and at 635 nm (Theodossiou and MacRobert, 2002). This may be related to lipophilicity because the quantum yield of cell inactivation tends to decrease with increasing water solubility of a photosensitizer (Moan, 1988; Strauss et al., 1997). However, in some prokaryotes, accumulation of hydrophilic copro- or uroporphyrin appears to increase cell photoinactivation compared with accumulation of lipophilic porphyrins like PpIX (Ramstad et al., 1997).

ACKNOWLEDGMENTS

The present work was supported by the Association for International Cancer Research (AICR) and the Norwegian Cancer Society (DNK). The authors are indebted to

Vladimir Iani, Li-Wei Ma, Asta Juzeniene, and Silje Stakland for help with the experiments.

REFERENCES

Bagdonas, S., Ma, L.W., Iani, V., Rotomskis, R., Juzenas, P., and Moan, J. (2000) Phototransformations of 5-aminolevulinic acid-induced protoporphyrin IX *in vitro*: a spectroscopic study, *Photochem. Photobiol.*, 72: 186–192.

Berg, K., Madslien, K., Bommer, J.C., Oftebro, R., Winkelman, J.W., and Moan, J. (1991) Light induced relocalization of sulfonated meso-tetraphenylporphines in NHIK 3025 cells and effects of dose fractionation, *Photochem. Photobiol.*, 53: 203–210.

Berg, K. and Moan, J. (1994) Lysosomes as photochemical targets, *Int. J. Cancer*, 59: 814–822.

Berg, K., Steen, H.B., Winkelman, J.W., and Moan, J. (1992) Synergistic effects of photoactivated tetra(4-sulfonatophenyl)porphine and nocodazole on microtubule assembly, accumulation of cells in mitosis and cell survival, *J. Photochem. Photobiol. B Biol.*, 13, 59–70.

Berg, K., Anholt, H., Bech, O., and Moan, J. (1996) The influence of iron chelators on the accumulation of protoporphyrin IX in 5-aminolevulinic acid-treated cells, *Br. J. Cancer*, 74: 688–697.

Bissonnette, R., Tremblay, J.F., Juzenas, P., Boushira, M., and Lui, H. (2002) Systemic photodynamic therapy with aminolevulinic acid induces apoptosis in lesional T-lymphocytes of psoriatic plaques, *J. Invest Dermatol.*, 119: 77–83.

Bonnett, R. and Martinez, G. (2001) Photobleaching of sensitizers used in photodynamic therapy, *Tetrahedron*, 57: 9513–9547.

Boyle, D.G. and Potter, W.R. (1987) Photobleaching of photofrin II as a means of eliminating skin photosensitivity, *Photochem. Photobiol.*, 46: 997–1001.

Brault, D., Aveline, B., Delgado, O., and Martin, M.T. (2001) Chlorin-type photosensitizers photochemically derived from vinyl porphyrins, *Photochem. Photobiol.*, 73: 331–338.

Brun, A., Western, A., Malik, Z., and Sandberg, S. (1990) Erythropoietic protoporphyria: photodynamic transfer of protoporphyrin from intact erythrocytes to other cells, *Photochem. Photobiol.*, 51: 573–577.

Busetti, A., Soncin, M., Jori, G., and Rodgers, M.A. (1999) High efficiency of benzoporphyrin derivative in the photodynamic therapy of pigmented malignant melanoma, *Br. J. Cancer*, 79: 821–824.

Calzavara–Pinton, P.-G., Szeimies, R.-M., and Ortel, B. (2001) *Photodynamic Therapy and Fluorescence Diagnosis in Dermatology*, Elsevier, Amsterdam.

Canti, G., Ricci, L., Cantone, V., Franco, P., Marelli, O., Andreoni, A., Cubeddu, R., and Nicolin, A. (1983) Hematoporphyrin derivative photoradiation therapy in murine solid tumors, *Cancer Lett.*, 21: 233–237.

Chang, S.C., MacRobert, A.J., Porter, J.B., and Bown, S.G. (1997) The efficacy of an iron chelator (CP94) in increasing cellular protoporphyrin IX following intravesical 5-aminolevulinic acid administration: an *in vivo* study, *J. Photochem. Photobiol. B: Biol.*, 38: 114–122.

Charlesworth, P. and Truscott, T.G. (1993) The use of 5-aminolevulinic acid (ALA) in photodynamic therapy (PDT), *J. Photochem. Photobiol. B: Biol.*, 18: 99–100.

Cox, G.S., Bobillier, C., and Whitten, D.G. (1982) Photo-oxidation and singlet oxygen sensitization by protoporphyrin IX and its photo-oxidation products, *Photochem. Photobiol.*, 36: 401–407.

de Gruijl, F.R. (2002) Photocarcinogenesis: UVA *vs.* UVB radiation, *Skin Pharmacol. Appl. Skin Physiol.*, 15: 316–320.

DeRosa, F.S., Marchetti, J.M., Thomazini, J.A., Tedesco, A.C., and Bentley, M.V. (2000) A vehicle for photodynamic therapy of skin cancer: influence of dimethylsulphoxide on 5-aminolevulinic acid *in vitro* cutaneous permeation and *in vivo* protoporphyrin IX accumulation determined by confocal microscopy, *J. Control Release*, 65: 359–366.

De Rosa, M.C. and Crutchley, R.J. (2002) Photosensitized singlet oxygen and its applications, *Coord. Chem. Rev.*, 233–234: 351–371.

Dietel, W., Fritsch, C., Pottier, R., and Wendenburg, R. (1997) 5-Aminolevulinic-acid-induced formation of different porphyrins and their photomodifications, *Lasers Med. Sci.*, 12: 226–236.

Dougherty, T.J. (1987) Photosensitizers: therapy and detection of malignant tumors, *Photochem. Photobiol.*, 45: 879–889.

Dougherty, T.J., Gomer, C.J., Henderson, B.W., Jori, G., Kessel, D., Korbelik, M., Moan, J., and Peng, Q. (1998) Photodynamic therapy, *J. Natl. Cancer Inst.*, 90: 889–905.

Dowson, D. (1997) Tribology and the skin surface, in *Bioengineering of the Skin: Skin Surface Imaging and Analysis*, Wilhelm, K.P., Elsner, P., Berardesca, E., and Maibach, H.I. (Eds.), pp. 159–180, CRC Press, Boca Raton, FL.

Edelson, R., Berger, C., Gasparro, F., Jegasothy, B., Heald, P., Wintroub, B., Vonderheid, E., Knobler, R., Wolff, K., and Plewig, G. (1987) Treatment of cutaneous T-cell lymphoma by extracorporeal photochemotherapy. Preliminary results, *N. Engl. J. Med.*, 316: 297–303.

Eichler, J., Knof, J., and Lenz, H. (1977) Measurements on the depth of penetration of light (0.35 to 1.0 µm) in tissue, *Radiat. Environ. Biophys.*, 14: 239–242.

el Sharabasy, M.M., el Waseef, A.M., Hafez, M.M., and Salim, S.A. (1992) Porphyrin metabolism in some malignant diseases, *Br. J. Cancer*, 65: 409–412.

Ericson, M.B., Grapengiesser, S., Gudmundson, F., Wennberg, A.M., Larko, O., Moan, J., and Rosen, A. (2003) A spectroscopic study of the photobleaching of protoporphyrin IX in solution, *Lasers Med. Sci.*, 18: 56–62.

Foote, C.S. (1991) Definition of type I and type II photosensitized oxidation, *Photochem. Photobiol.*, 54: 659.

Foster, T.H., Murant, R.S., Bryant, R.G., Knox, R.S., Gibson, S.L., and Hilf, R. (1991) Oxygen consumption and diffusion effects in photodynamic therapy, *Radiat. Res.*, 126: 296–303.

Fritsch, C., Goerz, G., and Ruzicka, T. (1998) Photodynamic therapy in dermatology, *Arch. Dermatol.*, 134: 207–214.

Fuchs, J., Weber, S., and Kaufmann, R. (2000) Genotoxic potential of porphyrin type photosensitizers with particular emphasis on 5-aminolevulinic acid: implications for clinical photodynamic therapy, *Free Radic. Biol. Med.*, 28: 537–548.

Gasparro, F.P. (2000) The role of PUVA in the treatment of psoriasis. Photobiology issues related to skin cancer incidence, *Am. J. Clin. Dermatol.*, 1: 337–348.

Gaullier, J.M., Berg, K., Peng, Q., Anholt, H., Selbo, P.K., Ma, L.W., and Moan, J. (1997) Use of 5-aminolevulinic acid esters to improve photodynamic therapy on cells in culture, *Cancer Res.*, 57: 1481–1486.

Gibson, S.L., Cupriks, D.J., Havens, J.J., Nguyen, M.L., and Hilf, R. (1998) A regulatory role for porphobilinogen deaminase (PBGD) in δ-aminolevulinic acid (δ-ALA)-induced photosensitization, *Br. J. Cancer*, 77: 235–243.

Giniunas, L., Rotomskis, R., Smilgevicius, V., Piskarskas, A., Didziapetriene, J., Bloznelyte, L., and Griciute, L. (1991) Activity of hematoporphyrin derivative photoproduct in photodynamic therapy *in vivo*, *Lasers Med. Sci.*, 6: 425–428.

Girotti, A.W. (2001) Photosensitized oxidation of membrane lipids: reaction pathways, cytotoxic effects, and cytoprotective mechanisms, *J. Photochem. Photobiol. B: Biol.*, 63: 103–113.

Gorman, A.A. and Rodgers, M.A. (1992) Current perspectives of singlet oxygen detection in biological environments, *J. Photochem. Photobiol. B: Biol.*, 14: 159–176.

Haddad, R., Blumenfeld, A., Siegal, A., Kaplan, O., Cohen, M., Skornick, Y., and Kashtan, H. (1998) *In vitro* and *in vivo* effects of photodynamic therapy on murine malignant melanoma, *Ann. Surg. Oncol.*, 5: 241–247.

Hall, E.J. (2000) The oxygen effect and reoxygenation, in *Radiobiology for the Radiologist*, John, J.-R., Sutton, P., and Marino, D. (Eds.), pp. 91–111, Lippincott, Williams, & Wilkins, Philadelphia.

Harber, L.C., Kochevar, I.E., and Shalita, A.R. (1982) Mechanisms of photosensitization to drugs in humans, in *The Science of Photomedicine*, Regan, J.D. and Parrish, J.A. (Eds.), pp. 323–347, Plenum Press, New York.

Harel, Y., Manassen, J. and Levanon, H. (1976) Photoreduction of porphyrins to chlorins by tertiary amines in the visible spectral range. Optical and EPR studies, *Photochem. Photobiol.*, 23: 337–341.

Harrison, L.B., Chadha, M., Hill, R.J., Hu, K., and Shasha, D. (2002) Impact of tumor hypoxia and anemia on radiation therapy outcomes, *Oncologis*, 7: 492–508.

Henegariu, O., Grober, L., Haskins, W., Bowers, P.N., State, M.W., Ohmido, N., Bray-Ward, P., and Ward, D.C. (2001) Rapid DNA fiber technique for size measurements of linear and circular DNA probes, *BioTechniques*, 31, 246–250.

Hilf, R., Havens, J.J., and Gibson, S.L. (1999) Effect of δ-aminolevulinic acid on protoporphyrin IX accumulation in tumor cells transfected with plasmids containing porphobilinogen deaminase DNA, *Photochem. Photobiol.*, 70: 334–340.

Horio, T. (2000) Indications and action mechanisms of phototherapy, *J. Dermatol. Sci.*, 23 (Suppl. 1): S17–S21.

Iinuma, S., Farshi, S.S., Ortel, B., and Hasan, T. (1994) A mechanistic study of cellular photodestruction with 5-aminolevulinic acid-induced porphyrin, *Br. J. Cancer*, 70: 21–28.

Inhoffen, H.H., Brockmann, H., and Bliesener, K.M. (1969) Photoprotoporphyrine und ihre Umwandlung in Spirographis-sowie Isospirographis-Porphyrin, *Liebigs Ann. Chem.*, 730: 173–185.

Ito, A.S., Azzellini, G.C., Silva, S.C., Serra, O., and Szabo, A.G. (1992) Optical absorption and fluorescence spectroscopy studies of ground state melanin–cationic porphyrin complexes, *Biophys. Chem.*, 45: 79–89.

Jablonski, N.G. and Chaplin, G. (2000) The evolution of human skin coloration, *J. Hum. Evol.*, 39: 57–106.

Jiang, F.N., Jiang, S., Liu, D., Richter, A., and Levy, J.G. (1990) Development of technology for linking photosensitizers to a model monoclonal antibody, *J. Immunol. Methods*, 134: 139–149.

Juzenas, P., Iani, V., Bagdonas, S., Rotomskis, R., and Moan, J. (2001) Fluorescence spectroscopy of normal mouse skin exposed to 5-aminolevulinic acid and red light, *J. Photochem. Photobiol. B: Biol.*, 61: 78–86.

Juzenas, P., Juzeniene, A., Kaalhus, O., Iani, V., and Moan, J. (2002a) Noninvasive fluorescence excitation spectroscopy during application of 5-aminolevulinic acid *in vivo*, *Photochem. Photobiol. Sci.*, 1: 745–748.

Juzenas, P., Juzeniene, A., Stakland, S., Iani, V., and Moan, J. (2002b) Photosensitizing effect of protoporphyrin IX in pigmented melanoma of mice, *Biochem. Biophys. Res. Commun.*, 297: 468–472.

Kalka, K., Merk, H., and Mukhtar, H. (2000) Photodynamic therapy in dermatology, *J. Am. Acad. Dermatol.*, 42: 389–413.

Kelty, C.J., Brown, N.J., Reed, M.W., and Ackroyd, R. (2002) The use of 5-aminolevulinic acid as a photosensitizer in photodynamic therapy and photodiagnosis, *Photochem. Photobiol. Sci.*, 1: 158–168.

Kennedy, J.C., Pottier, R.H., and Pross, D.C. (1990) Photodynamic therapy with endogenous protoporphyrin IX: basic principles and present clinical experience, *J. Photochem. Photobiol. B: Biol.*, 6: 143–148.

Kennedy, J.C. and Pottier, R.H. (1992) Endogenous protoporphyrin IX, a clinically useful photosensitizer for photodynamic therapy, *J. Photochem. Photobiol. B: Biol.*, 14: 275–292.

Kennedy, J.C., Marcus, S.L., and Pottier, R.H. (1996) Photodynamic therapy (PDT) and photodiagnosis (PD) using endogenous photosensitization induced by 5-amino-levulinic acid (ALA): mechanisms and clinical results, *J. Clin. Laser Med. Surg.*, 14: 289–304.

Kinley, J.S., Brunborg, G., Moan, J., and Young, A.R. (1997) Photoprotection by furocou-marin-induced melanogenesis against DNA photodamage in mouse epidermis *in vivo*, *Photochem. Photobiol.*, 65: 486–491.

Kloek, J. and Beijersbergen van Henegouwen, G.M.J. (1996) Prodrugs of 5-aminolevulinic acid for photodynamic therapy, *Photochem. Photobiol.*, 64: 994–1000.

Kloek, J., Akkermans, W., and Beijersbergen van Henegouwen, G.M.J. (1998) Derivatives of 5-aminolevulinic acid for photodynamic therapy: enzymatic conversion into proto-porphyrin, *Photochem. Photobiol.*, 67: 150–154.

Kochevar, I.E. and Redmond, R.W. (2000) Photosensitized production of singlet oxygen, *Methods Enzymol.*, 319: 20–28.

Konan, Y.N., Gurny, R., and Allemann, E. (2002) State of the art in the delivery of photosen-sitizers for photodynamic therapy, *J. Photochem. Photobiol. B: Biol.*, 66: 89–106.

Kongshaug, M., Rimington, C., Evensen, J.F., Peng, Q., and Moan, J. (1990) Hematoporphyrin diethers — V. Plasma protein binding and photosensitizing efficiency, *Int. J. Biochem.*, 22: 1127–1131.

Konig, K., Felsmann, A., Dietel, W., and Boschmann, M. (1990) Photodynamic activity of HPD-photoproducts, *Studia Biophysica*, 138: 219–228.

Konig, K., Schneckenburger, H., Ruck, A., and Steiner, R. (1993) *In vivo* photoproduct formation during PDT with ALA-induced endogenous porphyrins, *J. Photochem. Photobiol. B: Biol.*, 18: 287–290.

Krasnovsky, A.A., Jr., Kagan, V.E., and Minin, A.A. (1983) Quenching of singlet oxygen luminescence by fatty acids and lipids, *FEBS Lett.*, 155: 233–236.

Krieg, M. and Whitten, D.G. (1984a) Self-sensitized photo-oxidation of protoporphyrin IX and related porphyrins in erythrocyte ghosts and microemulsions: a novel photo-oxidation pathway involving singlet oxygen, *J. Photochem.*, 25: 235–252.

Krieg, M. and Whitten, D.G. (1984b) Self-sensitized photo-oxidation of protoporphyrin IX and related free-base porphyrins in natural and model membrane systems. Evidence for novel photo-oxidation pathways involving amino acids, *J. Am. Chem. Soc.*, 106: 2477–2479.

Kvam, E., Stokke, T., Moan, J., and Steen, H.B. (1992) Plateau distributions of DNA fragment lengths produced by extended light exposure of extranuclear photosensitizers in human cells, *Nucleic Acids Res.*, 20: 6687–6693.

Lacour, J.P. (2002) Carcinogenesis of basal cell carcinomas: genetics and molecular mechanisms, *Br. J. Dermatol.*, 146 (Suppl. 61): 17–19.

Lavi, A., Weitman, H., Holmes, R.T., Smith, K.M., and Ehrenberg, B. (2002) The depth of porphyrin in a membrane and the membrane's physical properties affect the photosensitizing efficiency, *Biophys. J.*, 82: 2101–2110.

Levy, J.G., Chowdhary, R., Ratkay, L., Waterfield, D., Obochi, M., Leong, S., Hunt, D., and Chan, A. (1994) Immune modulation using transdermal photodynamic therapy, in *Photodynamic Therapy of Cancer II, Proc. SPIE 2325*, Brault, D., Jori, G., Moan, J., and Ehrenberg, B. (Eds.), pp. 155–165, The International Society for Optical Engineering, Bellingham, WA.

Lindelof, B., Sigurgeirsson, B., Tegner, E., Larko, O., Johannesson, A., Berne, B., Christensen, O.B., Andersson, T., Torngren, M., and and Molin, L. (1991) PUVA and cancer: a large-scale epidemiological study, *Lancet*, 338: 91–93.

Lindelof, B. (1999) Risk of melanoma with psoralen/ultraviolet A therapy for psoriasis. Do the known risks now outweigh the benefits? *Drug Saf.*, 20: 289–297.

Losi, A., Bedotti, R., Brancaleon, L., and Viappiani, C. (1993) Porphyrin-melanin interaction: effect on fluorescence and non-radiative relaxations, *J. Photochem. Photobiol. B: Biol.*, 21: 69–76.

Lui, H. and Anderson, R.R. (1993) Photodynamic therapy in dermatology: recent developments, *Dermatol. Clin.*, 11: 1–13.

Ma, L.W., Moan, J., and Peng, Q. (1992) Effects of light exposure on the uptake of photofrin II in tumors and normal tissues, *Int. J. Cancer*, 52: 120–123.

Ma, L.W., Moan, J., Peng, Q., and Iani, V. (1998) Production of protoporphyrin IX induced by 5-aminolevulinic acid in transplanted human colon adenocarcinoma of nude mice can be increased by ultrasound, *Int. J. Cancer*, 78: 464–469.

Ma, L.W., Bagdonas, S., and Moan, J. (2001) The photosensitizing effect of the photoproduct of protoporphyrin IX, *J. Photochem. Photobiol. B: Biol.*, 60: 108–113.

Malik, Z. and Lugaci, H. (1987) Destruction of erythroleukaemic cells by photoactivation of endogenous porphyrins, *Br. J. Cancer*, 56: 589–595.

Malik, Z., Kostenich, G., Roitman, L., Ehrenberg, B., and Orenstein, A. (1995) Topical application of 5-aminolevulinic acid, DMSO and EDTA: protoporphyrin IX accumulation in skin and tumors of mice, *J. Photochem. Photobiol. B: Biol.*, 28: 213–218.

Mang, T.S., Dougherty, T.J., Potter, W.R., Boyle, D.G., Somer, S., and Moan, J. (1987) Photobleaching of porphyrins used in photodynamic therapy and implications for therapy, *Photochem. Photobiol.*, 45: 501–506.

Mew, D., Wat, C.K., Towers, G.H., and Levy, J.G. (1983) Photoimmunotherapy: treatment of animal tumors with tumor-specific monoclonal antibody-hematoporphyrin conjugates, *J. Immunol.*, 130: 1473–1477.

Min, D.B. and Boff, J.M. (2002) Chemistry and reaction of singlet oxygen in foods, *Compr. Rev. Food Sci. Food Saf.*, 1: 58–72.

Moan, J. and Sommer, S. (1985) Oxygen dependence of the photosensitizing effect of hematoporphyrin derivative in NHIK 3025 cells, *Cancer Res.*, 45: 1608–1610.

Moan, J. (1986) Effect of bleaching of porphyrin sensitizers during photodynamic therapy, *Cancer Lett.*, 33: 45–53.

Moan, J., Rimington, C., and Malik, Z. (1988) Photoinduced degradation and modification of Photofrin II in cells *in vitro*, *Photochem. Photobiol.*, 47: 363–367.

Moan, J. (1988) A change in the quantum yield of photoinactivation of cells observed during photodynamic treatment, *Lasers in Med. Sci.*, 3: 93–97.

Moan, J., Berg, K., Kvam, E., Western, A., Malik, Z., Ruck, A., and Schneckenburger, H. (1989) Intracellular localization of photosensitizers, *Ciba Found. Symp.*, 146: 95–107.

Moan, J. and Anholt, H. (1990) Phthalocyanine fluorescence in tumors during PDT, *Photochem. Photobiol.*, 51: 379–381.

Moan, J. and Berg, K. (1991) The photodegradation of porphyrins in cells can be used to estimate the lifetime of singlet oxygen, *Photochem. Photobiol.*, 53: 549–553.

Moan, J. and Berg, K. (1992) Photochemotherapy of cancer: experimental research, *Photochem. Photobiol.*, 55: 931–948.

Moan, J., Berg, K., Anholt, H., and Madslien, K. (1994) Sulfonated aluminum phthalocyanines as sensitizers for photochemotherapy. Effects of small light doses on localization, dye fluorescence and photosensitivity in V79 cells, *Int. J. Cancer*, 58: 865–870.

Moan, J., Iani, V., Ma, L.W., and Peng, Q. (1996) Photodegradation of sensitizers in mouse skin during PCT, in *Photochemotherapy: Photodynamic Therapy and Other Modalities, Proc. SPIE 2625*, Ehrenberg, B., Jori, G., and Moan, J. (Eds.), pp. 187–193, The International Society for Optical Engineering, Bellingham, WA.

Moan, J., Streckyte, G., Bagdonas, S., Bech, O., and Berg, K. (1997) Photobleaching of protoporphyrin IX in cells incubated with 5-aminolevulinic acid, *Int. J. Cancer*, 70: 90–97.

Moan, J., Peng, Q., Sorensen, R., Iani, V., and Nesland, J.M. (1998a) The biophysical foundations of photodynamic therapy, *Endoscopy*, 30: 387–391.

Moan, J., Iani, V., and Ma, L.W. (1998b) *In vivo* fluorescence of phthalocyanines during light exposure, *J. Photochem. Photobiol. B: Biol.*, 42: 100–103.

Moan, J., Ma, L.W., and Bjorklund, E. (1999) The effect of glucose and temperature on the *in vivo* efficiency of photochemotherapy with meso-tetra-hydroxyphenyl-chlorin, *J. Photochem. Photobiol. B: Biol.*, 50: 94–98.

Moan, J., Juzenas, P., and Bagdonas, S. (2000) Degradation and transformation of photosensitizers during light exposure, in *Recent Research Developments in Photochemistry and Photobiology*, Pandalai, S.G. (Ed.), pp. 121–132, Transworld Research Network, Trivandrum.

Moan, J. (2001) Visible light and UV Radiation, in *Radiation at Home, Outdoors and in the Workplace*, Brune, A., Hellborg, R., Persson, B.R.R., and Pääkkönen, R. (Eds.), pp. 69–85, Scandinavian Science Publisher, Oslo.

Moan, J., van den Akker, J.H.T.M., Juzenas, P., Ma, L.W., Angell–Petersen, E., Gadmar, Ø.B., and Iani, V. (2001) On the basis for tumor selectivity in the 5-aminolevulinic acid-induced synthesis of protoporphyrin IX, *J. Porphyr. Phthalocyanines*, 5: 170–176.

Morton, C.A., Brown, S.B., Collins, S., Ibbotson, S., Jenkinson, H., Kurwa, H., Langmack, K., McKenna, K., Moseley, H., Pearse, A.D., Stringer, M., Taylor, D.K., Wong, G., and Rhodes, L.E. (2002) Guidelines for topical photodynamic therapy: report of a workshop of the British Photodermatology Group, *Br. J. Dermatol.*, 146: 552–567.

Moser, K., Kriwet, K., Naik, A., Kalia, Y.N., and Guy, R.H. (2001) Passive skin penetration enhancement and its quantification *in vitro*, *Eur. J. Pharm. Biopharm.*, 52: 103–112.

Musajo, L., Rodighiero, F., Caporale, G., Dall'Acqua, F., Marciani, S., Bordin, F., Baccichetti, F., and Bevilaqua, R. (1974) Photoreactions between skin-photosensitizing furocoumarins and nucleic acids, in *Sunlight and Man*, Fitzpatrick, T.B., Pathak, M.A., Harber, L.C., Seiji, M., and Kukita, A. (Eds.), pp. 369–387, University of Tokyo Press, Tokyo.

Niedre, M., Patterson, M.S. and Wilson, B.C. (2002) Direct near-infrared luminescence detection of singlet oxygen generated by photodynamic therapy in cells *in vitro* and tissues *in vivo*, *Photochem. Photobiol.*, 75: 382–391.

Ninomiya, Y., Itoh, Y., Henta, T., and Ishibashi, A. (1999) Photodynamic diagnosis of basal cell carcinoma on the lower eyelid using topical 5-aminolevulinic acid and desferrioxamine, *Br. J. Dermatol.*, 141: 580–581.

Ninomiya, Y., Itoh, Y., Tajima, S., and Ishibashi, A. (2001) *In vitro* and *in vivo* expression of protoporphyrin IX induced by lipophilic 5-aminolevulinic acid derivatives, *J. Dermatol. Sci.*, 27: 114–120.

Ochsner, M. (1997) Photophysical and photobiological processes in the photodynamic therapy of tumors, *J. Photochem. Photobiol. B: Biol.*, 39: 1–18.

Oleinick, N.L., Morris, R.L., and Belichenko, I. (2002) The role of apoptosis in response to photodynamic therapy: what, where, why, and how, *Photochem. Photobiol. Sci.*, 1: 1–21.

Ortonne, J.P. (2002) Photoprotective properties of skin melanin, *Br. J. Dermatol.*, 146 (Suppl. 61): 7–10.

Parrish, J.A., Anderson, R.K., Urbach, F., and Pitts, D. (1978) UV-A, *Biological Effects of Ultraviolet Radiation with Emphasis on Human Responses to Longwave Ultra Violet*, Plenum Press, New York and London.

Pass, H.I. (1993) Photodynamic therapy in oncology: mechanisms and clinical use, *J. Natl. Cancer Inst.*, 85: 443–456.

Pathak, M.A., Kramer, D.M., and Fitzpatrick, T.B. (1974) Photobiology and photochemistry of furocoumarins (psoralens), in *Sunlight and Man*, Fitzpatrick, T.B., Pathak, M.A., Harber, L.C., Seiji, M., and Kukita, A. (Eds.), pp. 335–368, University of Tokyo Press, Tokyo.

Peng, Q., Evensen, J.F., Rimington, C., and Moan, J. (1987) A comparison of different photosensitizing dyes with respect to uptake C3H-tumors and tissues of mice, *Cancer Lett.*, 36: 1–10.

Peng, Q., Moan, J., and Cheng, L.S. (1991) The effect of glucose administration on the uptake of photofrin II in a human tumor xenograft, *Cancer Lett.*, 58: 29–35.

Peng, Q., Berg, K., Moan, J., Kongshaug, M., and Nesland, J.M. (1997a) 5-Aminolevulinic acid-based photodynamic therapy: principles and experimental research, *Photochem. Photobiol.*, 65: 235–251.

Peng, Q., Warloe, T., Berg, K., Moan, J., Kongshaug, M., Giercksky, K.E., and Nesland, J.M. (1997b) 5-Aminolevulinic acid-based photodynamic therapy. Clinical research and future challenges, *Cancer*, 79: 2282–2308.

Peng, Q., Soler, A.M., Warloe, T., Nesland, J.M., and Giercksky, K.E. (2001) Selective distribution of porphyrins in skin thick basal cell carcinoma after topical application of methyl 5-aminolevulinate, *J. Photochem. Photobiol. B: Biol.*, 62: 140–145.

Pogue, B.W., Ortel, B., Chen, N., Redmond, R.W., and Hasan, T. (2001) A photobiological and photophysical-based study of phototoxicity of two chlorins, *Cancer Res.*, 61: 717–724.

Poh–Fitzpatrick, M.B. (2000) Porphyrias: photosensitivity and phototherapy, *Methods Enzymol.*, 319: 485–493.

Ramstad, S., Futsaether, C.M., and Johnsson, A. (1997) Porphyrin sensitization and intracellular calcium changes in the prokaryote Propionibacterium acnes, *J. Photochem. Photobiol. B: Biol.*, 40: 141–148.

Redmond, R.W. and Gamlin, J.N. (1999) A compilation of singlet oxygen yields from biologically relevant molecules, *Photochem. Photobiol.*, 70: 391–475.

Regan, J.D. and Parrish, J.A. (1982) *The Science of Photomedicine*, Plenum Press, New York and London.

Rimington, C. (1989) Haem biosynthesis and porphyrias: 50 years in retrospect, *J. Clin. Chem. Clin. Biochem.*, 27: 473–486.

Rittenhouse–Diakun, K., van Leengoed, H., Morgan, J., Hryhorenko, E., Paszkiewicz, G., Whitaker, J.E., and Oseroff, A.R. (1995) The role of transferrin receptor (CD71) in photodynamic therapy of activated and malignant lymphocytes using the heme precursor δ-aminolevulinic acid (ALA), *Photochem. Photobiol.*, 61: 523–528.

Roberts, W.G., Smith, K.M., McCullough, J.L., and Berns, M.W. (1989) Skin photosensitivity and photodestruction of several potential photodynamic sensitizers, *Photochem. Photobiol.*, 49: 431–438.

Robinson, D.J., de Bruijn, H.S., van der Veen, N., Stringer, M.R., Brown, S.B., and Star, W.M. (1998) Fluorescence photobleaching of ALA-induced protoporphyrin IX during photodynamic therapy of normal hairless mouse skin: the effect of light dose and irradiance and the resulting biological effect, *Photochem. Photobiol.*, 67: 140–149.

Rotomskis, R., Streckyte, G., and Bagdonas, S. (1997) Phototransformation of sensitizers 2. Photoproducts formed in aqeous solutions of porphyrins, *J. Photochem. Photobiol. B: Biol.*, 39: 172–175.

Rotomskis, R., Bagdonas, S., Streckyte, G., Wendenburg, R., Dietel, W., Didziapetriene, J., Ibelhauptaite, A., and Staciokiene, L. (1998) Phototransformation of sensitizers: 3. Implications for clinical dosimetry, *Lasers Med. Sci.*, 13: 271–278.

Rousset, N., Vonarx, V., Eleouet, S., Carre, J., Kerninon, E., Lajat, Y., and Patrice, T. (1999) Effects of photodynamic therapy on adhesion molecules and metastasis, *J. Photochem. Photobiol. B: Biol.*, 52: 65–73.

Sandberg, S. (1981) Protoporphyrin-induced photodamage to mitochondria and lysosomes from rat liver, *Clin. Chim. Acta*, 111: 55–60.

Sandberg, S. and Romslo, I. (1982) Phototoxicity of uroporphyrin as related to its subcellular localization in rat livers after feeding with hexachlorobenzene, *Photobiochem. Photobiophys.*, 126: 318–324.

Savellano, M.D. and Hasan, T. (2003) Targeting cells that overexpress the epidermal growth factor receptor with polyethylene glycolated BPD Verteporfin photosensitizer immunoconjugates, *Photochem. Photobiol.*, 77: 431–439.

Scheuplein, R.J. and Blank, I.H. (1971) Permeability of the skin, *Physiol Rev.*, 51: 702–747.

Seidl, H., Kreimer–Erlacher, H., Back, B., Soyer, H.P., Hofler, G., Kerl, H., and Wolf, P. (2001) Ultraviolet exposure as the main initiator of p53 mutations in basal cell carcinomas from psoralen and ultraviolet A-treated patients with psoriasis, *J. Invest Dermatol.*, 117: 365–370.

Sharman, W.M., Allen, C.M., and van Lier, J.E. (1999) Photodynamic therapeutics: basic principles and clinical applications, *Drug Discov. Today*, 4: 507–517.

Sharman, W.M., Allen, C.M., and van Lier, J.E. (2000) Role of activated oxygen species in photodynamic therapy, *Methods Enzymol.*, 319: 376–400.

Shaw, J.E., Prevo, M., Gale, R., and Yum, S.I. (1991) Percutanous absorption, in *Physiology, Biochemistry, and Molecular Biology of the Skin*, Goldsmith, L.A. (Ed.), pp. 1447–1479, Oxford University Press, New York.

Sindelar, W.F., Dehany, T.F., Tochner, Z., Thomas, G.F., Dachowski, L.J., Smith, P.D., Frianf, W.F., Cole, J.M., and Glatstein, E. (1991) Techniques of photodynamic therapy for disseminated intraperitoneal malignant neoplasm, *Arch. Surg.*, 126: 318–324.

Soler, A.M., Angell–Petersen, E., Warloe, T., Tausjo, J., Steen, H.B., Moan, J., and Giercksky, K.E. (2000) Photodynamic therapy of superficial basal cell carcinoma with 5-aminolevulinic acid with dimethylsulfoxide and ethylendiaminetetraacetic acid: a comparison of two light sources, *Photochem. Photobiol.*, 71: 724–729.

Spikes, J.D. (1990) Chlorins as photosensitizers in biology and medicine, *J. Photochem. Photobiol. B: Biol.*, 6: 259–274.

Spikes, J.D. (1997) Photodynamic action: from paramecium to photochemotherapy, *Photochem. Photobiol.*, 65S: 142S–147S.

Stewart, F., Baas, P., and Star, W. (1998) What does photodynamic therapy have to offer radiation oncologists (or their cancer patients)? *Radiother. Oncol.*, 48: 233–248.

Strauss, W.S., Sailer, R., Schneckenburger, H., Akgun, N., Gottfried, V., Chetwer, L., and Kimel, S. (1997) Photodynamic efficacy of naturally occurring porphyrins in endothelial cells *in vitro* and microvasculature *in vivo*, *J. Photochem. Photobiol. B: Biol.*, 39: 176–184.

Suwa, K., Kimura, T., and Schaap, A.P. (1978) Reaction of singlet oxygen with cholesterol in liposomal membranes. Effect of membrane fluidity on the photo-oxidation of cholesterol, *Photochem. Photobiol.*, 28: 469–473.

Szeimies, R.M., Abels, C., Fritsch, C., Karrer, S., Steinbach, P., Baumler, W., Goerz, G., Goetz, A.E., and Landthaler, M. (1995) Wavelength dependency of photodynamic effects after sensitization with 5-aminolevulinic acid *in vitro* and *in vivo*, *J. Invest Dermatol.*, 105: 672–677.

Theodossiou, T. and MacRobert, A.J. (2002) Comparison of the photodynamic effect of exogenous photoprotoporphyrin and protoporphyrin IX on PAM 212 murine keratinocytes, *Photochem. Photobiol.*, 76: 530–537.

Thunell, S. (2000) Porphyrins, porphyrin metabolism and porphyrias. I. Update, *Scand. J. Clin. Lab Invest*, 60: 509–540.

Tremblay, J.F. and Bissonnette, R. (2002) Topical agents for the treatment of psoriasis, past, present and future, *J. Cutan. Med. Surg.*, 6: 8–11.

Tunstall, R.G., Barnett, A.A., Schofield, J., Griffiths, J., Vernon, D.I., Brown, S.B., and Roberts, D.J. (2002) Porphyrin accumulation induced by 5-aminolevulinic acid esters in tumor cells growing *in vitro* and *in vivo*, *Br. J. Cancer*, 87: 246–250.

Uehlinger, P., Zellweger, M., Wagnieres, G., Juillerat–Jeanneret, L., van den Bergh, H., and Lange, N. (2000) 5-Aminolevulinic acid and its derivatives: physical chemical properties and protoporphyrin IX formation in cultured cells, *J. Photochem. Photobiol. B: Biol.*, 54: 72–80.

van den Akker, J.T., Iani, V., Star, W.M., Sterenborg, H.J., and Moan, J. (2000) Topical application of 5-aminolevulinic acid hexyl ester and 5-aminolevulinic acid to normal nude mouse skin: differences in protoporphyrin IX fluorescence kinetics and the role of the stratum corneum, *Photochem. Photobiol.*, 72: 681–689.

van den Akker, J.T., Iani, V., Star, W.M., Sterenborg, H.J., and Moan, J. (2002) Systemic component of protoporphyrin IX production in nude mouse skin upon topical application of aminolevulinic acid depends on the application conditions, *Photochem. Photobiol.*, 75: 172–177.

van Hillegersberg, R., van den Berg, J.W., Kort, W.J., Terpstra, O.T., and Wilson, J.H. (1992) Selective accumulation of endogenously produced porphyrins in a liver metastasis model in rats, *Gastroenterology*, 103: 647–651.

Voet, D. and Voet, J.G. (1995) Hemoglobin: protein function in microcosm, in *Biochemistry*, Rose, N. (Ed.), pp. 215–250, John Wiley & Sons, New York.

von Tappeiner, H. and Jodlbauer, A. (1904) Über die wirkung der photodynamischen (fluorescierenden) Stoffe auf Protozoen und Enzyme, *Dtsch. Arch. Klin. Med.*, 80: 427–487.

Vrouenraets, M.B., Visser, G.W., Snow, G.B., and van Dongen, G.A. (2003) Basic principles, applications in oncology and improved selectivity of photodynamic therapy, *Anticancer Res.*, 23: 505–522.

Washbrook, R. and Riley, P.A. (1997) Comparison of δ-aminolevulinic acid and its methyl ester as an inducer of porphyrin synthesis in cultured cells, *Br. J. Cancer*, 75: 1417–1420.

Wasserman, H.H. and Murray, R.W. (1979) *Singlet Oxygen*, Academic Press, New York.

Weldon, D., Poulsen, T.D., Mikkelsen, K.V., and Ogilby, P.R. (1999) Singlet sigma: the "other" singlet oxygen in solution, *Photochem. Photobiol.*, 70: 369–379.

Woodburn, K.W., Fan, Q., Kessel, D., Luo, Y., and Young, S.W. (1998) Photodynamic therapy of B16F10 murine melanoma with lutetium texaphyrin, *J. Invest Dermatol.*, 110: 746–751.

Yano, T., Higo, N., Fukuda, K., Tsuji, M., Noda, K., and Otagiri, M. (1993) Further evaluation of a new penetration enhancer, HPE-101, *J. Pharm. Pharmacol.*, 45: 775–778.

Yaroslavsky, A.N., Priezzhev, A.V., Rodriguez, J., Yaroslavsky, I.Y., and Battarbee, H. (2002) Optics of blood, in *Handbook of Optical Biomedical Diagnostics*, Tuchin, V. (Ed.), pp. 169–216, The International Society for Optical Engineering, SPIE Press, Bellingham, WA.

Zaidi, S.I., Oleinick, N.L., Zaim, M.T., and Mukhtar, H. (1993) Apoptosis during photodynamic therapy-induced ablation of RIF-1 tumors in C3H mice: electron microscopic, histopathologic and biochemical evidence, *Photochem. Photobiol.*, 58: 771–776.

Zheng, G., Shibata, M., Dougherty, T.J., and Pandey, R.K. (2000) Wittig reactions on photoprotoporphyrin IX: new synthetic models for the special pair of the photosynthetic reaction center, *J. Org. Chem.*, 65: 543–557.

CHAPTER **10**

In Vitro Screening of the Photoreactivity of Antimalarials: A Test Case

Hanne Hjorth Tønnesen and Solveig Kristensen

CONTENTS

10.1 INTRODUCTION

Preformulation studies play a significant part in anticipating formulation problems and identifying logical paths in liquid- and solid dosage form technology. By investigating the intrinsic stability of the drug, it is possible to advise on formulation approaches and indicate types of excipient, specific, protective additives and packaging likely to improve the integrity of the drug and product. A pharmaceutical product should retain potency above 95% throughout shelf-life under recommended storage conditions, and it should look and perform as it did when newly manufactured.

Possible interaction between drugs and irradiation is an important issue for the quality of pharmaceuticals. Irradiation has two main effects on drugs. The first is the influence of light exposure on the stability of the drug substances and drug formulations. In order to provide data for determining the photostability of drug substances and products, a forced degradation study should be performed. According to the ICH Guideline on photostability testing, the design of the prescribed forced degradation studies is left to the manufacturer. In most cases it will be appropriate to irradiate the sample with a high-intensity photon source and to terminate the experiments when extensive decomposition is reached or an appropriate exposure level has been used. However, this will not necessarily provide a complete strategy for assessing the light sensitivity of a drug product as was further discussed in Chapter 6.

The second aspect of drug–light interactions is the biological effects caused by the reaction of drugs, photoproducts, or metabolites of drugs with light and biomolecules, resulting in drug-induced photosensitivity. A better understanding of molecular mechanisms involved in photosensitivity reactions will aid the new drug development process by providing means to assess the potential of a new compound to provoke adverse effects. In order to understand the photochemical processes fully and thereby establish the extent of the light-sensitivity problem *in vitro* and *in vivo*, it may be necessary to perform an extensive screening of the drug.

An appropriate *in vitro* assay should be designed to specify the photoreactivity* of the compound in order to understand implications for lack of efficacy, changes in physicochemical properties, or possible photosensitizing activity *in vivo*. Such a screening represents an important field of interdisciplinary research involving photochemistry, photophysics, photomedicine, photobiology, and pharmacokinetics. The photoreactivity of a molecule can change dramatically when the substance is introduced into a biological system. The selection of experimental conditions (including the photon source) is therefore critical in order to obtain a correlation between *in vitro* and *in vivo* photoreactivity. This often seems to be neglected in the discussion of the results obtained from *in vitro* experiments.

In vitro screening of drug photoreactivity has been carried out for a series of compounds within various therapeutical groups (e.g., non-steroidal anti-inflammatory drugs, diuretic agents, antipsychotic agents) (Moore and Tamat, 1980; Moore and Burt, 1981; Moore and Chappuis, 1988; Moore et al., 1990, 1998; Moore, 1998; Moore and Wang, 1998). In the authors' laboratory, an *in vitro* assay has been applied in the evaluation of the photoreactivity of antimalarial drugs. The photoreactivity of structural analogues can be difficult to predict, as was discussed in Chapter 4, and thus it might be necessary to investigate each drug candidate within a therapeutic group. As mentioned earlier, the photodecomposition of a drug cannot only result

---

* In this context, the term "photoreactivity" is used to describe how a compound responds to irradiation *in vitro* and *in vivo*. This includes degradation reactions; other processes like radical formation, energy transfer, and luminescence; and reactions with endogenous substances.

in a loss of potency of the product or a change in physicochemical properties but also has the potential to cause phototoxic reactions *in vivo*. The aim of this work was to study the processes involved in the photodegradation of antimalarial compounds and to apply the data to explain the underlying cause of ocular and cutaneous side effects resulting from treatment with these substances.

Changes in skin pigmentation, bleaching of the hair, corneal opacity, cataracts in the lens, and visual disturbances are frequently observed during medication with these drugs. However, irreversible retinal damage (retinopathy) and blindness are the most serious adverse effects, observed after long-term medication or after high accumulative doses of some of these compounds (Fraunfelder and Meyer, 1989; Moore and Hemmens, 1982; Tanenbaum and Tuffanelli, 1980). The antimalarial substances investigated are mainly structural analogues (Figure 10.1). They fulfill almost every criterion needed to subject a drug to a complete photoassay (Chapter 1):

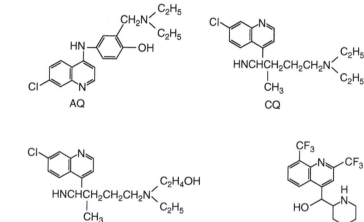

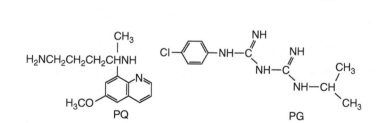

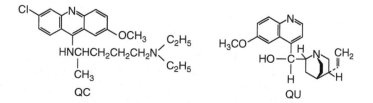

Figure 10.1 Chemical structures of the antimalarials investigated.

- *Accumulation in skin, eye, and hair.* Several of the antimalarial compounds possess a large distribution volume, an extremely long half-life, and strong interaction with melanin *in vitro*. These drugs are known to accumulate in melanin-rich areas of the body (eye, skin, hair) (Tanenbaum and Tuffanelli, 1980). Due to low turnover of melanin in the eye, a long-term retention, even for years, may be observed for drugs with high melanin affinity (Lindquist, 1986).
- *Administration at a high accumulative dosage or used in long term medication.* The antimalarial drugs are often administered over a long period of time. Some of the drugs also possess anti-inflammatory effects although in much larger doses than those employed in prophylaxis and treatment of malaria (Webster, 1990). Accumulative doses of 200 to 300 g can be reached in certain cases (Ehrenfeld et al., 1986).
- *Photolabile in vitro.* Several of the drugs have been demonstrated to decompose in solution during photoirradiation (Kristensen et al., 1993; Laurie et al., 1986, 1988; McHale et al., 1989; Nord et al., 1991; Taylor et al., 1990; Tønnesen et al., 1988; Tønnesen and Grislingaas, 1990).
- *Forms photolabile degradation products or in vivo metabolites.* This is demonstrated for some of the drugs (Laurie et al., 1988; Nord et al., 1991; Tønnesen et al., 1988).
- *Molecules contain essential functionalities* (aromatic moiety, amino group, halogen atoms, etc.).
- *Environmental criteria.* Prophylactic use of antimalarial drugs is usually combined with a life in areas with high solar irradiance.

The antimalarial compounds are clearly qualified to undergo a complete evaluation according to the strategy demonstrated in Figure 10.2. The *in vitro* photoassay used in this study is illustrated in Figure 10.3. Such an assay should include

- Evaluation of absorption and emission characteristics
- Determination of photodegradation for the drug in different media under the exposure to UV radiation or visible light, including determination of rate constants
- Identification of products formed
- Detection of reactive intermediates generated during irradiation
- Quenching and sensitizing properties of the compounds
- Determination of quantum yields for the various processes
- Evaluation of action spectra for selected processes
- Studies of physicochemical changes (e.g., change in color, viscosity) induced by exposure

10.2 PHOTOREACTIVITY STUDIES

10.2.1 Absorption Characteristics

The photoreactivity of a pure drug substance will follow the basic law of photochemical absorption, i.e., a photochemical (or subsequent photobiological) reaction cannot occur unless nonionizing radiation is absorbed by the compound. Determination of the absorption spectrum of a drug will immediately establish whether the substance can absorb the optical radiation present in natural daylight or indoor

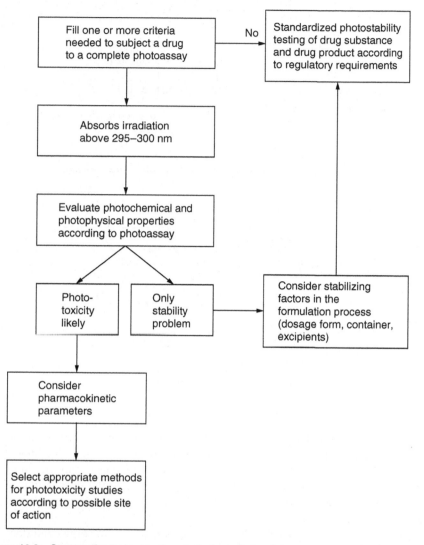

Figure 10.2 Strategy for the evaluation of photoreactivity of drugs as a part of the preformulation work.

illumination; irradiation penetrating the viable layers of the skin; or various segments of the eye. The fact that a drug molecule absorbs radiation in the UV or visible region of the spectrum means that it absorbs energy sufficient to break a bond in the molecule.

The absorption property is a first indication of a possible photochemical process leading to decomposition or other changes in a drug formulation. It is also an *in vitro* estimation of the drug's potential to photosensitize. However, in most cases, the absorption spectrum will not reveal the presence of other compounds in the preparation that can influence the photoreactivity of the drug. Such compounds can be impurities from the synthesis or (photo)degradation products formed during the

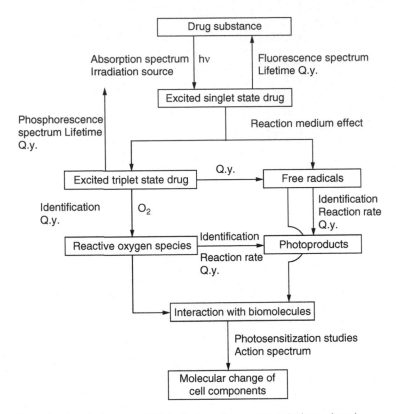

Figure 10.3   *In vitro* photoassay of biologically active compounds (e.g., drugs).

development process. These products will, in most cases, have absorption characteristics close to the main compound and be present only in small amounts. They are therefore difficult to detect unless a separation technique is applied. Although they are present only in trace amounts, they may sensitize the degradation of the parent compound; participate in photoinduced redox processes; or cause phototoxic reactions *in vivo*.

Irradiation below approximately 295 nm is cut off by the cornea and does not reach the lens (Bachem, 1956); irradiation above 300 nm can reach the dermis with its capillary blood vessels (Megaw and Drake, 1986). Thus, a drug must absorb irradiation above these wavelengths in order to damage the lens and the skin, respectively, through photosensitized reactions. All the antimalarials investigated absorb irradiation above 295 to 300 nm.

10.2.2   Reaction Medium and Photon Source

The reaction medium and photon source must be selected with care in order to get a realistic description of the photoreactivity of a certain drug. In practice, the solubility of the substance can be a limiting factor in the experimental design. Phosphate buffer, pH 7.4, is often used to "mimic" biological conditions. The aqueous medium can contain micelles, liposomes, cyclodextrines, or organic co-solvents to improve drug solubility. It is important to realize that even minor amounts of such additives may alter the photoreactivity of the drug (Chapter 16). For instance, adding 6 5mM carbon tetrachloride to a sample of 6-methoxy-2-naphthylacetic acid prepared in acetonitrile caused the photodegradation rate to increase nearly 10 times (Boscá et al., 2000).

Photolysis studies in polar and apolar organic media can provide additional information about fragmentation pattern and reaction mechanisms. Such information may be useful for predicting drug photoreactivity in lipophilic environments. However, one should keep in mind that organic solvents can participate in the reaction and lead to formation of "nonphysiological" degradation products. This is clearly demonstrated by the photolysis of chloroquine in isopropanol (Nord et al., 1991). The photochemical decomposition of chlorquine was apparently favored in isopropanol compared to water; therefore, it was chosen as reaction medium for the isolation of the degradation products. Isopropanol forms several photodecomposition products of low molecular weight (Pacakova et al., 1985) under the actual conditions. Several condensation products formed from fragments of chloroquine and the reaction medium were identified.

The majority of pharmaceutical formulations is stored under air and thereby in contact with oxygen during the shelf-life. Furthermore, most photosensitized reactions in biological systems require the involvement of molecular oxygen. It is therefore natural to apply oxygen-containing media in photoreactivity studies. Anaerobic conditions can provide valuable information in mechanistic studies of photodegradation, however. In some exceptions oxygen-independent processes have an important place in the overall picture of photosensitization.

It has previously been shown that chlorpromazine can photosensitize the hemolysis of erythrocytes under anaerobic conditions (Kochevar and Lamola, 1979) and that furocoumarins bind covalently to DNA upon irradiation in the absence of oxygen (Rodighiero and Dal'Acqua, 1976). Photoreactivity studies carried out under anaerobic conditions should therefore be considered as a part of the in vitro assay. A correlation between results obtained in deaerated organic solvents and in aqueous medium at physiological pH does not always exist, however. Dechlorination of chloroquine and hydroxychloroquine plays an important role in the photolysis of these compounds under anaerobic conditions (Moore and Hemmens, 1982), while the chlorine atom seems to be intact in all the photodegradation products identified under aerobic conditions (Nord et al., 1991; Tønnesen et al., 1988). In such a case, the in vivo photoreaction pathway may be strongly dependent on the localization of the drug molecule within the body tissues. This example emphasizes the value of applying more than one set of experimental conditions in the in vitro assay.

Many drug molecules are weak acids or weak bases. The pH of the reaction medium may in that case strongly influence the results obtained. The photosensitized oxidation of mefloquine and other antimalarials shows a clear pH dependency (Tønnesen and Moore, 1991). Valuable information can be lost if experiments are carried out only in one medium, i.e., organic solvent that cannot differentiate between protonated and deprotonated forms of the molecules, or at nonphysiological pH. This is clearly demonstrated for the antimalarials of 4-aminoquinoline structure

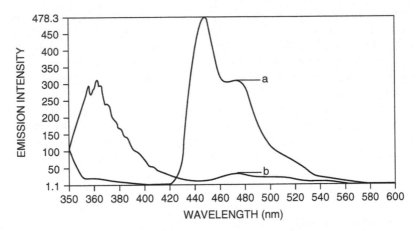

Figure 10.4 The observed luminescence spectra at 77 K of desethylhydroxychloroquine when changing the solvent from EPA to EPA + 0.5% 1 *M* HCl at excitation wavelengths 328 and 346 nm, respectively. Peaks between 350 and 430 nm correspond to the fluorescence emission and peaks between 430 and 600 nm correspond to the phosphorescence emission. The relative areas under the two uncorrected phosphorescence emission spectra are 1:10. a: desethylhydroxychloroquine in EPA + 0.5% 1 *M* HCl (i.e., the heterocyclic N is ionized); b: desethylhydroxychloroquine in EPA (i.e., the heterocyclic N is deionized).

(Nord et al., 1994). These compounds are strongly fluorescent in the deprotonated form; phosphorescence dominates when the compounds are protonated, leading to different photoreactivity in different media (Figure 10.4). This would suggest that phototoxicity exhibited through triplet state energy transfer to form singlet oxygen may be more pronounced if the drug exists in hydrophilic regions of the cell such as the cytosole or is incorporated in acid vesicles.

It is of great importance to obtain data on the spectral distribution and intensity of the photon source applied in drug photoreactivity studies (see Chapter 2, Chapter 3, and Chapter 6). Many forced degradation studies previously reported in the literature did not take into account the need for using a standardized irradiation source. Kinetic data reported in the literature therefore often have one thing in common: the lack of standardized experimental conditions, leading to discrepancy in the results. Typical statements are that the sample was placed "0.5 m from a north-facing window so that constant daylight illumination was obtained" (Kirk, 1987) or "0.9 m from a continuously illuminated fluorescent bulb (30 W)" (Newton et al., 1981). These expressions ignore the fact that the intensity of "daylight" is dependent on factors like altitude; latitude; season; time of the day and the year; and that the spectral distribution and intensity of the photon source vary with the type and age of the lamp.

As a result, the half-life reported in the literature for a given compound can differ by several hundred percent (Bosanquet, 1986). The precautions to be taken in a hospital ward or in a hospital pharmacy in order to protect a certain drug formulation against light will consequently show great variations. When an *in vitro* photoassay is performed, it would be necessary to include exposure to physiologically important wavelengths like UV-A, UV-B, and visible light that also cover likely in-use conditions (Tønnesen and Karlsen, 1995). Exposure to irradiation below 280 nm (UV-C) can be convenient for qualitative purposes due to a faster degradation rate, but the observed reactions may not necessarily be relevant under normal storage conditions or in biological systems.

Only irradiation *absorbed* by a sample can cause photochemical reactions. The overlap between the absorption spectrum of the product and the emission spectrum of the irradiation source will determine photoreactions that can occur in the sample. The emission intensity at the absorbed wavelengths will determine the rate at which the products will form. To ensure the formation of all possible degradation products, including products formed in sensitized reactions, the sample must be irradiated at all absorbing wavelengths. This can be achieved by using a broad spectrum irradiation source. The qualitative and quantitative aspects of the process will be changed if the emission spectrum of the lamp only partly overlaps with the absorption spectrum of the sample, as may be the case if a combination of irradiation sources is used. In the authors' laboratory, a sun-simulating irradiation source (i.e., xenon lamp in combination with appropriate filters) would normally be included in the photoreactivity studies to cover "real" conditions.

10.2.3 Photolysis and Identification of Degradation Products

Studies of drug photolysis will provide valuable information for making decisions with regard to selection of dosage form, excipients, and containers. Pathways for the formation of free radicals and reactive oxygen species (ROS) can be proposed from photodegradation studies. Formation of free radicals and ROS can strongly influence the physicochemical properties of the formulation (e.g., viscosity, color, melting point) and further induce phototoxic and photoallergic effects after medication.

Adverse effects may also occur as a consequence of the formation of toxic or photoreactive degradation products. For instance, when photolysis of the drug molecule occurs on or near the epidermal layer, lipophilic degradation products would be expected to partition into cell membranes and be cleared only slowly from the system. If the chromophore is retained, the generation of phototoxic species (e.g., free radicals, ROS) may be observed even when the parent compound is fully degraded. For many compounds, it is found that phototoxic degradation products may be formed by photodecomposition of the parent compound or by metabolic pathways. Isolation and identification of degradation products combined with photoreactivity studies of these compounds and of known metabolites are therefore important (Kristensen et al., 1997).

Photodegradation studies have been carried out for several antimalarial compounds (Kristensen et al., 1993; Nord et al., 1991; Tønnesen et al., 1988; Tønnesen and Grislingaas, 1990). The main degradation products of chloroquine and hydroxychloroquine correspond to their *in vivo* metabolites. These degradation products are also demonstrated to be photolabile (unpublished results). Some of the photodegradation products of hydroxychloroquine seem to be more potent as photosensitizers than the parent drug (Kristensen et al., 1994a).

10.2.4 Degradation Rate

Although great effort should be taken to stabilize the formulation in such a way that the shelf-life becomes independent on the storage conditions, it might still be necessary to take precautions to exclude or minimize the amount of irradiation reaching the formulation. The precautions taken are normally based on a stipulated photochemical half-life of the drug substance within the formulation. As discussed earlier, a kinetic approach to the photochemical degradation of drugs is without any value unless the light source is specified. For the absolute determination of the extent of a photoreaction, the spectral distribution and intensity of the light source must be known. Determination of reaction quantum yield using monochromatic light in combination with an actinometer provides the *exact kinetic data* for a specific photochemical reaction and is usually obtained at the absorption maximum of the compound tested.

The *observed* decomposition rate of a drug substance or drug product can be obtained by using an irradiation source simulating indoor light or sunlight. The intensity of the irradiation source must be related to the likely exposure or to the actual intensity of sunlight for calculating the accelerating effect of the lamp. This calculation can be based on a defined value for the irradiance of sunlight at a specific location or as a global average (Tønnesen and Moore, 1993). The order of the reaction can only be determined if the reaction is followed up to more than 50% degradation (Sande and Karlsen, 1993). The kinetic treatment of photochemical reactions is further discussed in Chapter 3.

Temperature and humidity conditions in the test chamber should be kept at a level to ensure that the effects of changes such as sublimation, evaporation, or melting are minimized in the physical state of the samples. Rate constants for photochemical reactions depend on the temperature because of secondary thermal reactions of the parent compound or primary products. Thermal stability of the material should independently be determined through accelerated stability testing. The ambient temperature and the temperature of the samples during irradiation are related to the photon source used for testing and the intensity and distance of the sample from the photon source.

Infrared (IR) radiation has a significant effect on heating of materials. The amount of IR radiation will affect the ambient temperature and the temperature of the sample; the latter is also strongly influenced by the color of the material. An increase in surface temperature is likely to be observed in dark samples with low thermal conductivity. The use of foil-wrapped dark control samples will normally be adequate. For products with low thermal stability, it may be necessary to determine the surface temperature of the light-exposed samples and place the dark control samples in a chamber under temperature and humidity conditions corresponding to the surface temperature and humidity of the drug during irradiation. The surface temperature of a light sample (e.g., white tablet) and a dark sample can be compared with the temperature of a white standard thermometer and a black standard thermometer, respectively, at equal distance from the photon source. The importance of a dark control is often neglected in the discussion of photochemical stability data. In most cases, the photolytic rate constants are much larger than the rate constants of the thermal reactions. For some compounds, e.g., primaquine, the influence of the dark reactions cannot be ignored (Kristensen, to be published).

10.2.5 Changes in Physicochemical Properties

Changes in physicochemical properties of the drug substance (e.g., color, crystal modification) may take place upon irradiation. Efforts should be made to observe such changes during the *in vitro* assay. The sample absorption spectrum should be recorded before and after irradiation; surface color of solid samples should be evaluated by appropriate methods; and the identity of the sample crystal modification should be confirmed when the drug is irradiated in the solid state. The humidity in the test chamber can influence the photochemical stability of certain solid samples, as demonstrated for mefloquine. The photoinduced yellowing of uncoated meflo-quine tablets is accelerated by an increase in humidity. These tablets are mainly used in tropical countries and the real in-use conditions will include high relative humidity. In such cases, the influence of the humidity on the photostability must be taken into account (Tønnesen et al., 1997).

In cases in which the results from the assay have revealed that the drug is likely to form free radicals or ROS, one should keep in mind that the drug then has the potential to induce a sensitized degradation of commonly used excipients like poly-meric thickening agents (Baldursdottir et al., 2003a, b). The result is a change in the viscosity and rheological properties of the pharmaceutical product.

10.2.6 Fluorescence and Phosphorescence

The back transition to the ground state from the excited singlet or triplet state of a molecule may be accompanied by the emission of photons (fluorescence and phos-phorescence, respectively). The emission characteristics of a molecule can therefore provide direct information about excited state properties. Qualitative and quantitative changes in emission spectra resulting from a change of reaction medium provide information about interaction between the excited states and the environment. Such information can be of great importance in explaining the photoreactivity of a drug molecule in a certain formulation (e.g., in the presence and absence of surfactants) or in a biological system (e.g., protein bound *vs.* free molecule).

Most photosensitized reactions originate from the triplet state of the molecules (Roberts, 1988) but an apparent relation also seems to exist between fluorescence quantum yield *in vitro* and photosensitizing ability *in vivo* (Moore and Hemmens, 1982). The production of singlet oxygen is found to occur by energy transfer both from the singlet and triplet excited states of the sensitizer, although the reaction from the triplet state is highly preferred because singlet–triplet interaction is of very low probability (Stevens et al., 1981). Because of its longer lifetime (milliseconds to several seconds), the excited triplet state may diffuse a significant distance in fluid media and has a much higher probability of interaction with other molecules. Those compounds with long-lived triplet states are therefore more likely to be photosen-sitizers. The lifetime of a molecule in the excited singlet or triplet states is related directly to the fluorescence and phosphorescence lifetimes, respectively.

Information about the possible photoreactivity of the drug substance and its degradation products or metabolites can therefore be obtained from quantum yield measurements and lifetime studies. Additionally, the singlet and triplet energies of the compound are of considerable interest in study of the mechanism of the photo-reaction, especially with respect to energy transfer. The photosensitizing potential of the 4-aminiquinolines is emphasized by the results obtained from the fluorescence and phosphorescence measurements (Nord et al., 1994). These compounds have long-lived triplet states; triplet state energy transfer to form singlet oxygen is likely to occur, and they are good candidates for potential photosensitizing activity *in vivo*.

10.2.7 Quantum Yield

The quantum yield provides information about the effectiveness of a certain photo-induced process. The quantum yield of loss of starting material or product formation will provide valuable information about structure–activity relationships. Fluores-cence and phosphorescence quantum yields will indicate the fraction of molecules likely to be found in the excited singlet and triplet state. The quantum yield is a useful parameter to predict the importance of a certain reaction; for example, an isolated degradation product can have a long phosphorescence lifetime and should therefore be considered as a possible sensitizer. If the quantum yield of formation of this product is very low, however, it is less likely to be formed in a biologically active concentration and may therefore play a minor role in phototoxicity reactions.

To determine quantum yield, the change in drug concentration should not exceed 15% during the irradiation so that the photodecomposition remains linear with irradiation time (Moore, 1987). By comparing the reaction quantum yield for the photodecomposition of mefloquine with its fluorescence quantum yield at physio-logical pH, it is demonstrated that deactivation of the excited state by formation of degradation products is strongly favored compared to radiative deactivation (Tønne-sen and Moore, 1991). This may be important *in vitro* and *in vivo* because the main degradation product is shown to be photoreactive.

10.2.8 Action Spectrum

The action spectrum of a drug substance or drug product consists of wavelengths that elicit degradation *in vitro* or a certain clinical reaction *in vivo*. An action spectrum is obtained by measuring the radiation dose required to evoke the same degree of degradation or the same biological response at different wavelengths. The action spectrum for a pure compound would normally overlap the absorption spectrum of the compound. It should be noted that, in formulations in which sensitized reactions are possible or absorption of radiation by inactive chromophores and radiation scat-tering is considerable, there is not always a direct relationship between the absorption spectrum of the formulation and the action spectrum for the photochemical degrada-tion of the active compound. *In vivo*, the action spectrum will not parallel the absorption spectrum *in vitro* unless the action mechanism and the quantum yield are the same at all wavelengths and the absorption spectra *in vitro* and *in vivo* are identical.

10.2.9 Identification of Intermediates

Free radicals and singlet oxygen generated by the photoexcited drug molecule appear to be the principal intermediate species in the photochemical reaction pathway. These intermediates can react with other components in a drug preparation (e.g., excipients or drug molecules in the ground state), leading to a change in product stability.

It is also widely accepted that the toxicity of many endogenous compounds arises from free radical intermediates. Generally, reactions that lead to more permanent biological damage involve free radical intermediates (Kensler and Taffe, 1986). Because of the oxygen content of blood and tissue, adverse reactions might be ascribed to photosensitized oxidation reactions. There are two mechanisms of photosensitized oxidation. The type I mechanism is a free radical process and the type II involves excited molecular oxygen. For drugs that produce free radicals as well as singlet molecular oxygen, both mechanisms may be observed in photo-oxidation (Spikes, 1977).

A number of techniques have been developed to enable the detection of free radical intermediates and singlet oxygen in photochemical reactions. In most cases, the photo-oxidation potential would be determined by a combination of methods (Chapter 12). Methods like electron spin resonance (ESR) or direct monitoring of singlet oxygen phosphorescence at 1268 nm must be applied for the exact identification of intermediates. Application of these methods in the study of the antimalarial compounds has been demonstrated (Motton et al., 1999). The formation and reactivity of free radicals in mefloquine have been further investigated by pulse radiolysis and flash photolysis (Navaratnam et al., 2000). However, useful information can also be obtained from photo-oxidation experiments by using less advanced and indirect methods like an oxygen electrode and oxidizable substrates in the presence or absence of specific quenchers.

Histidine and 2,5-dimethyl furane are substrates for singlet oxygen. The latter can be used independently of pH, while histidine is suitable at pH above 7.5. L-tryptophane is a substrate for superoxide; suitable quenchers are DABCO and azide (singlet oxygen); mannitol (hydroxyl radical); and 2-mercapto ethylamine and glutathione (other radicals). The specificity or lack of specificity of the various substrates and quenchers should be taken into account. In order to compare the rates of photo-oxidation of various compounds, correction must be made for the differences in absorption spectra between the compounds, and between different ionization states of the same compound. Thus, the rates should be normalized to a constant amount of irradiation absorbed and calculated from the area under the absorption curve (i.e., the overlap integral between the absorption spectrum of the sample and the emission spectrum of the photon source). The photooxidizing potential of several antimalarial compounds has been investigated by using this experimental approach (Moore and Hemmens, 1982; Tønnesen and Moore, 1991).

Evaluation of self-sensitizing properties, i.e., the ability to induce formation of radicals and react with the same radicals in a secondary reaction, is of interest from a formulation viewpoint. The stability of primaquine in solution is strongly influenced by self-sensitized reactions involving oxygen radicals (Kristensen et al., 1993,

1998). A stabilizing effect could be obtained by adding appropriate radical quenchers to the samples.

10.3 *IN VITRO* PHOTOSENSITIZATION STUDIES

The components of the *in vitro* assay discussed thus far are important for a better understanding of the overall photoreactivity of a drug substance or preparation. To assess the potential of a compound or preparation to provoke adverse effects, a number of other aspects must be taken into account. An extended *in vitro* screening will also in most cases be needed. The photosensitization studies should be planned according to the knowledge obtained so far from the *in vitro* assay and from the possible targets for photoinduced adverse reactions for the particular drug.

10.3.1 Examples of Test Parameters

The possible site of action for the antimalarials as potential photosensitizers will be various parts of the eye and the skin. Photohemolytical properties; photosensitized polymerization of lens proteins; and interaction with melanin were therefore selected as *in vitro* test parameters in the further screening of these compounds. According to their photohemolytical capability, quinacrine, quinine, and primaquine need to be considered as potential *in vivo* photosensitizers (Kristensen et al., 1994a).

It is widely accepted that photoinduced lipid peroxidation has detrimental effects on cell membranes and therefore plays an important role in skin phototoxicity. The observed photohemolysis induced by these antimalarials may be an indication of extensive photoperoxidation of the membrane lipids. Photo-oxidation experiments with linoleic acid should be performed supplementary to the photohemolysis studies.

Photosensitized polymerization of lens proteins can be selected as a measure of eye phototoxicity (Chapter 11). Proteins isolated from calf lens are suitable to simulate the conditions in the human eye. The reaction mechanisms can be evaluated by adding various quenchers to the reaction medium during irradiation. Chloroquine, hydroxychloroquine, mefloquine, and quinacrine induced polymerization under the given experimental conditions (Kristensen et al., 1995). These compounds have a large apparent distribution volume and a long elimination half-life and must therefore be considered as potential photosensitizers in the eye. Primaquine and quinine were also shown to induce polymerization of lens proteins *in vitro* but are less likely to reach the eye *in vivo* due to fast elimination from the body.

Some of the antimalarial compounds accumulate in melanin-rich areas of the body (skin, eye, hair). The turnover of melanin in the body is very low, except for epidermal melanin. Compounds with high melanin affinity can be retained in melanin-containing tissues for years (Lindquist, 1986). Information about the interaction between drugs and melanin is therefore of great importance in evaluating drug phototoxicity. The total binding of the drug substance to melanin is of main concern. The surface characteristics of the melanin selected for the experiments must be taken into account because the chemical composition of melanin isolated from different tissues will vary. Binding to melanin was demonstrated for all the antimalarial compounds investigated (Kristensen et al., 1994b).

Chloroquine and hydroxychloroquine possess a large distribution volume and an extremely long half-life in addition to a strong interaction with melanin *in vitro*. Both compounds are therefore distinguished as particularly good candidates for melanin binding *in vivo*. It should be kept in mind, however, that the conditions *in vivo* are quite different from the simple *in vitro* system used in the assay. Interactions between the drug molecule and biological macromolecules like proteins can occur, leading to a change in solubility and absorption characteristics of the drug.

10.3.2 Aspects of Main Importance

10.3.2.1 Pharmacokinetic Parameters

Pharmacokinetic aspects must be taken into account in cases in which the *in vitro* screening indicates that the compound is likely to cause phototoxic reactions *in vivo*. This includes information about *in vivo* metabolites and degradation products (if available). Phototoxicity is generally dose dependent, i.e., dependent on the concentration of the drug sensitizer and the intensity of the incident radiation at the site of action. Prior to a phototoxic reaction, the sensitizer must be distributed to tissues that are exposed to irradiation and further absorb the light that penetrates these tissues. If no pharmacokinetic data are available, the probability of being retained in a lipophilic medium can be predicted to some extent from the pK_a and/or the log P value of the compound.

For the antimalarial drugs, however, essential data are available in the literature. Many of the antimalarials investigated have a large apparent distribution volume (V_d) and a long elimination half-life ($t_{1/2}$) (Table 10.1). This reflects accumulation of the drugs in body tissues. Most of the adverse effects associated with the use of antimalarials are related to the eye and skin. The retina is richly supplied with blood vessels, but the blood–retinal barrier is normally very tight and therefore restricts

Table 10.1 Pharmacokinetics and Absorption Characteristics[a] of Antimalarial Drugs

Compound	Absorption Maximum (nm)	V_d (l/kg)[b]	$t_{1/2}$[c]
Amodiaquine	341	Widely	5 hours
Chloroquine	329, 343	200	25–60 days
Hydroxy-chloroquine	329, 343	Widely	40 days
Mefloquine	284	13–29	19.5 days
Primaquine	352	3–4	7 hours
Proguanil	253	33	16–20 hours
Quinacrine	423, 443	500	5 days
Quinine	331	2	9 hours

[a] Phosphate buffer, pH = 7.4.
[b] V_d: apparent volume of distribution.
[c] $t_{1/2}$: elimination half-life.

movement of substances from the capillaries. The lens has no blood supply and drugs cannot be distributed directly from the systemic circulation to this tissue. On certain occasions, permeability of the blood–retinal barrier can be altered. Drugs will then enter the retinal pigmented epithelium (RPE) and retina to a larger extent. Drugs accumulated in the RPE can further pass through the vitreous cavity and reach the lens.

Quinacrine, which is tricyclic and highly lipophilic, will easily penetrate cell membranes but hardly diffuse through the hydrophilic vitreous cavity. It also forms a stable complex with melanin and will therefore be retained in the RPE. Quinacrine must be considered a potent photosensitizer in the retina due to the absorption maximum of this drug in the visible region of the spectrum (Table 10.1). The more hydrophilic compounds chloroquine and hydroxychloroquine will more easily be transported to the lens. They are known to accumulate in the eye and induce toxic reactions. Primaquine is not likely to be distributed to the eye to any extent due to low distribution volume and fast elimination.

Several of the antimalarial compounds seem likely to accumulate in the lipophilic or melanin-containing tissues of the skin. Adverse dermal effects observed after medication with antimalarials are probably a result of phototoxic reactions.

10.3.2.2 Normalization of Results

One essential aspect in the study of photochemical and photobiological responses is the number of molecules available for light absorption (Megaw and Drake, 1986). Thus, comparison of biological effects induced by the various drugs foresees the drug concentration given in moles/liter. It is, however, a considerable oversimplification to assume that the sensitizer producing a given effect at the lowest molar concentration is the most effective compound. The biological effects should always be normalized to the number of photons absorbed by the sensitizer in the actual medium (Pooler and Valenzeno, 1982). This is often neglected in phototoxicity studies.

The effect of normalization is clearly demonstrated in Figure 10.5 through Figure 10.7. Figure 10.5 shows the photohemolytic effect of the antimalarial drugs quinine, quinacrine, and primaquine at the same molar concentration (without normalization). The medium contains red blood cells (RBC) and absorbs irradiation in the UV-Vis region of the spectrum. A corrected absorption spectrum for each drug compound can be made from the amount of irradiation available for the drug substances in the medium (Figure 10.6) and the photohemolytic effect can be estimated from the photohemolysis percentage and the relative number of photons absorbed by the drug (Kristensen et al., 1994a).

When the antimalarial drugs were evaluated without normalization of the results, quinacrine seemed to be the most potent compound (Figure 10.5). However, compared on a normalized scale, quinacrine turned out to be in the same range as primaquine, while quinine was the most powerful photohemolytic agent (Figure 10.7). This emphasizes the importance of normalizing biological effects to the relative number of photons absorbed by the drug in order to evaluate their relative photosensitizing potencies.

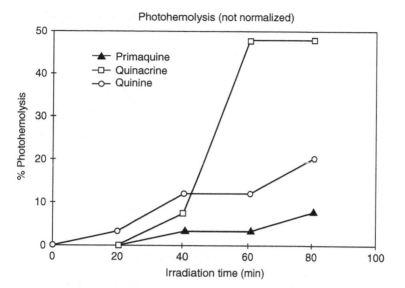

Figure 10.5 Photohemolytic effect of primaquine, quinacrine, and quinine at the same molar concentration (data not normalized).

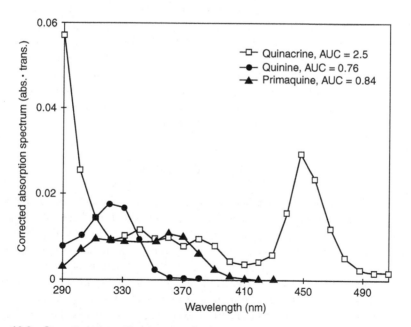

Figure 10.6 Corrected absorption spectra of primaquine, quinacrine, and quinine.

10.3.2.3 *Influence of Degradation Products*

The photochemical degradation of drugs in the medium will run parallel to the phototoxicity reaction studied. The extent of degradation in the actual test medium

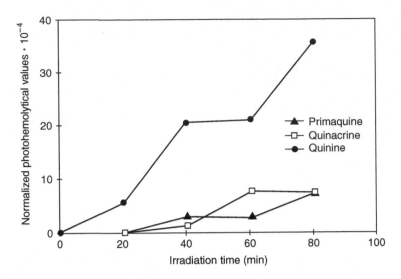

Figure 10.7 Photohemolytic effect of primaquine, quinacrine, and quinine normalized to the amount of light absorbed.

cannot always be estimated from the degradation rate in a pure solution due to filter effect from compounds present in the medium (e.g., red blood cells). The concentration of the parent compound should be maintained at a level of >85% of the initial concentration to minimize the effect of degradation products formed during the study. This requires methods for quantitation of the drug in biological media and is not always a simple task. In most cases, the phototoxic effects observed must be considered as a result of the parent drug and the photochemical (and thermal) degradation products formed during the experiment.

10.4 CONCLUSION

This chapter describes the application of an *in vitro* photoassay as a method to estimate the photoreactivity of drugs. The fundamental photochemical behavior of a series of antimalarial compounds has been examined. All the compounds were demonstrated to photodecompose under the given experimental conditions and to form reactive intermediates during the process. The degradation rate and mechanisms were strongly dependent on the reaction medium. In the solid state, some of the substances demonstrated a change in crystal modification and/or a change in color. This information is of great relevance to the formulation process. The results from the screening were further applied to find a basis for the adverse photobiological effects associated with the clinical use of these drugs.

The *in vitro* experiments, combined with the pharmacokinetic data available from the literature; the absorption characteristics of the drugs; and the optical properties of the tissues, resulted in an estimate of the capability of each compound to act as an *in vivo* photosensitizer. A thorough discussion of the photoreactivity of the antimalarial compounds is presented by Kristensen (1997) and the results are summarized in Table 10.1. In the case of antimalarial compounds, the drug substance can reach the skin surface and, to a various extent, the eye after administration to the body. The drugs interact with melanin *in vitro* and are therefore likely to accumulate in melanin-rich tissues, thus impairing their clearance from skin and eye. Exposed to direct sunlight, the drugs decompose to form certain stable photoproducts. Some of the main degradation products are identical to the *in vivo* metabolites and appear to be photoreactive. Free radical intermediates and reactive oxygen species are generated during the process, which may lead to cell membrane damage and protein polymerization.

No photosensitizer yields only one type of reactive species. The reactive pathways that the excited molecule follows depend on the nature of the sensitizer and its environment. This emphasizes the need for using more than one experimental model in the evaluation of drug photoreactivity and to take into account pharmacokinetic parameters and method specificity in the discussion of the normalized results. Observation of drug photoreactivity *in vitro* does not necessarily mean that the same combination of reactions will take place *in vivo*. A certain *in vitro–in vivo* correlation should be obtained, however, if the *in vitro* photoassay is carefully designed. The assay will then result in a reduction in the number of animal experiments required in development of new drug substances and drug formulations.

REFERENCES

Bachem, A. (1956) Opthalmic ultraviolet action spectra, *Am. J. Opthalmol.*, 41, 969–975.

Baldursdottir, S.G., Kjøniksen, A.-L., Karlsen, J., Nystrøm, B., Roots, J., and Tønnesen, H.H. (2003a) Riboflavin-photosensitized changes in aqueous solutions of alginate. Rheological studies, *Biomacromol.*, 4, 429–436.

Baldursdottir, S.G., Kjøniksen, A.-L., Roots, J., Tønnesen, H.H., and Nystrøm, B. (2003b) Influence of concentration and molecular weight on the photosensitized degradation of alginate in aqueous solutions, *Polym. Bull.*, 50, 373–380.

Bosanquet, A.G. (1986) Stability of solutions of antineoplastic agents during preparation and storage for *in vitro* assays. II. Assay methods, adriamycin and the other antitumor antibiotics, *Cancer Chemother. Pharmacol.*, 17, 1–10.

Boscá, F., Canudas, N., Marín, M.L., and Miranda, M.A. (2002) A photophysical and photochemical study of 6-methoxy-2-naphthylacetic acid, the major metabolite of the phototoxic non-steroidal antiinflammatory drug nabumetone, *Photochem. Photobiol.*, 71, 173–177.

Ehrenfeld, M., Nesher, R., and Merin, S. (1986) Delayed-onset chloroquine retinopathy, *Br. J. Ophtalmol.*, 70, 281–283.

Fraunfelder, F.T. and Meyer, S.M. (Eds.) (1989) *Drug-Induced Ocular Side Effects and Drug Interactions,* 3rd ed., Philadelphia: Lea and Febiger, pp. 58–62.

Kensler, T.W. and Taffe, B.G. (1986) Free radicals in tumor promotion, *Adv. Free Radical Biol. Med.*, 2, 347–387.

Kirk, B. (1987) The evaluation of a light protective giving set. The photosensitivity of intravenous dacarbazine solutions, *Intensive Ther. Clin. Monit.*, 8, 78–86.

Kochevar, I.E. and Lamola, A.A. (1979) Chlorpromazine and protriptyline phototoxicity: photosensitized oxygen-independent red cell hemolysis, *Photochem. Photobiol.*, 29, 1177–1197.

Kristensen, S., Grislingaas, A.-L., Greenhill, J.V., Skjetne, T., Karlsen, J., and Tønnesen, H.H. (1993) Photochemical stability of biologically active compounds. V. Photochemical degradation of primaquine in an aqueous medium, *Int. J. Pharm.*, 100, 15–23.

Kristensen, S., Karlsen, J., and Tønnesen, H.H. (1994a) Photoreactivity of biologically active compounds. VI. Photohemolytical properties of antimalarials *in vitro*, *Pharm. Sci. Commun.*, 4, 183–191.

Kristensen, S., Orsteen, A.-L., Sande, S.A., and Tønnesen, H.H. (1994b) Photoreactivity of biologically active compounds. VII. Interaction of antimalarial drugs with melanin *in vitro* as part of phototoxicity screening, *J. Photochem. Photobiol. B: Biol.*, 26, 87–95.

Kristensen, S., Wang, R.-H., Tønnesen, H.H., Dillon, J., and Roberts, J.E. (1995) Photoreactivity of biologically active compounds. VIII. Photosensitized polymerization of lens proteins by antimalarial drugs *in vitro*, *Photochem. Photobiol.*, 61, 124–130.

Kristensen, S., Grinberg, L., and Tønnesen, H.H. (1997) Photoreactivity of biologically active compounds. XI. Primaquine and its metabolites as radical inducers, *Eur. J. Pharm. Sci.*, 5, 139–146.

Kristensen, S. (1997) Photoreactivity of Antimalarial Drugs, thesis, University of Oslo, Norway.

Kristensen, S., Nord, K., Orsteen, A.-L., and Tønnesen, H.H. (1998) Photoreactivity of biologically active compuonds. XIV. Influence of oxygen on light induced reactions of primaquine, *Pharmazie*, 53, 98–103.

Laurie, W.A., McHale, D., and Saag, K. (1986) Photoreactions of quinine in aqueous citric acid solution, *Tetrahedron*, 42, 3711–3714.

Laurie, W.A., McHale, D., Saag, K., and Sheridan, J.B. (1988) Photoreactions of quinine in aqueous citric acid solution. II. Some end-products, *Tetrahedron*, 44, 5905–5910.

Lindquist, N.G. (1986) Melanin affinity of xenobiotics, *Upsala J. Med. Sci.*, 91, 283–288.

McHale, D., Laurie, W.A., Saag, K., and Sheridan, J.B. (1989) Photoreactions of quinine in aqueous citric acid solution. III. Products formed in aqueous 2-hydroxy-2-methyl-propionic acid, *Tetrahedron*, 45, 2127–2130.

Megaw, J.M. and Drake, L.A. (1986) Photobiology: an overview, in Jackson, E.M. (Ed.) *Photobiology of the Skin and Eye*, New York: Marcel Dekker, pp. 1–31.

Moore, D.E. and Tamat, S.R. (1980) Photosensitization by drugs: photolysis of some chlorine-containing drugs, *J. Pharm. Pharmacol.*, 32, 172–177.

Moore, D.E. and Burt, C.D. (1981) Photosensitization by drugs in surfactant solutions, *Photochem. Photobiol.*, 34, 431–439.

Moore, D.E. and Hemmens, V.J. (1982) Photosensitization by antimalarial drugs, *Photochem. Photobiol.*, 36, 71–77.

Moore, D.E. (1987) Principles and practice of drug decomposition studies, *J. Pharm. Biomed. Anal.*, 5, 441–453.

Moore, D.E. and Chappuis, P.P. (1988) A comparative study of the photochemistry of the non-steroidal anti-inflammatory drugs, naproxen, benoxaprofen and indomethacin, *Photochem. Photobiol.*, 47, 173–180.

Moore, D.E., Roberts–Thomson, S., Zhen, D., and Duke, C.C. (1990) Photochemical studies on the anti-inflammatory drug diclofenac, *Photochem. Photobiol.*, 52, 685–690.

Moore, D.E. (1998) Mechanisms of photosensitization by phototoxic drugs, *Mutat. Res.*, 422, 165–173.

Moore, D.E., Ghebremeskel, K.A., Chen, B.B., and Wong, E.Y. (1998) Electron transfer process in the reactivity of non-steroidal anti-inflammatory drugs in the ground and excited states, *Photochem. Photobiol.*, 68, 685–691.

Moore, D.E. and Wang, J. (1998) Electron-transfer mechanisms in photottsensitization by the anti-inflammatory drug benzydamine, *J. Photochem. Photobiol. B: Biol.*, 43, 175–180.

Motton, A.G., Martinez, L.J., Holt, N., Sik, R.H., Reszka, K., Chignell, C.F., Tønnesen, H.H., and Roberts, J. (1999) Photophysical studies of antimalarial drugs, *Photochem. Photobiol.,* 69, 282–287.

Navaratnam, S., Hamblett, I., and Tønnesen, H.H. (2000) Photoreactivity of biologically active compounds. XVI. Formation and reactivity of free radicals in mefloquine, *J. Photochem. Photobiol. B: Biol.,* 56, 25–38.

Newton, D.W., Fung, E.Y.Y., and Williams, D.A. (1981) Stability of five catecholamines and terbutaline sulphate in 5% dextrose injection in the absence and presence of aminopylline, *Am. J. Hosp. Pharm.,* 38, 1314–1319.

Nord, K., Karlsen, J., and Tønnesen, H.H. (1991) Photochemical stability of biological active compounds. IV. Photochemical degradation of chloroquine, *Int. J. Pharm.,* 72, 11–18.

Nord. K., Karlsen, J., and Tønnesen, H.H. (1994) Photochemical stability of biologically active compounds. IX. Characterization of the spectroscopic properties of the 4-aminoquinolines, chloroquine and hydroxyquinoline, and of selected metabolites by absorption, fluorescence and phosphorescence measurements, *Photochem. Photobiol.,* 60, 427–431.

Pacakova, V., Konas, M., and Kotvalova, V. (1985) Reaction gas chromatography: study of the photodecomposition of selected substances, *Chromatographia,* 20, 164–172.

Pooler, J.P. and Valenzeno, D.P. (1982) A method to quantify the potency of photosensitizers that modify cell membranes, *J. Natl. Cancer Inst.,* 69, 211–215.

Roberts, J.E. (1988) Ocular phototoxicity, in Moreno, G., Pottier, R.H., and Truscott, T.G. (Eds.) *Photosensitization. Molcular, Cellular and Medical Aspects.* Berlin: Springer, pp. 325–330.

Rodighiero, G. and Dal'Acqua, F. (1976) Biochemical and medical aspects of psoralens, *Photochem. Photobiol.,* 24, 647–653.

Sande, S.A. and Karlsen, J. (1993) Evaluation of reaction order. Software in pharmaceutics. III, *Int. J. Pharm.,* 98, 209–218.

Spikes, J.D. (1977) Photosensitization, in Smith, K.C. (Ed.) *The Science of Photobiology,* New York: Plenum Press, pp. 87–112.

Stevens, T.J., Marsh, K.L., and Barltrop, J.A. (1981) Photoperoxidation of unsaturated organic molecules. XXI. Sensitizer yields of O_2 $^1\Delta_g$, *J. Phys. Chem.,* 85, 3079–3082.

Tanenbaum, L. and Tuffanelli, D.L. (1980) Antimalarial agents; chloroquine, hydroxychloroquine and quinacrine, *Arch. Dermatol.,* 116, 587–591.

Taylor, R.B., Moody, R.R., Ochekpe, N.A., Low, A.S., and Harper, M.I.A. (1990) A chemical stability study of proguanil hydrochloride, *Int. J. Pharm.,* 60, 185–190.

Tønnesen, H.H., Grislingaas, A.-L., Woo, S.O., and Karlsen. J. (1988) Photochemical stability of antimalarials. I. Hydroxychloroquine, *Int. J. Pharm.,* 43, 215–219.

Tønnesen, H.H. and Grislingaas, A.-L. (1990) Photochemical stability of biologically active compounds. II. Photochemical decomposition of mefloquine in water, *Int. J. Pharm.,* 60, 157–162.

Tønnesen, H.H. and Moore, D.E. (1991) Photochemical stability of biologically active compounds. III. Mefloquine as a photosensitizer, *Int. J. Pharm.,* 70, 95–101.

Tønnesen, H.H. and Moore, D.E. (1993) Photochemical degradation of components in drug formulations, *Pharm. Technol.,* 5, 27–33.

Tønnesen, H.H. and Karlsen, J. (1995) Photochemical degradation of compounds in drug formulations. III. A discussion of experimental conditions, *Pharmeuropa,* 7, 137–141.

Tønnesen, H.H., Skrede, G., and Martinsen, B.K. (1997) Photoreactivity of biologically active compounds. XIII. Photostability of mefloquine in the solid state, *Drug Stab.,* 1, 249–253.

Webster, L.T. (1990) Drugs used in chemotherapy of protozoal infections, in Goodman, L.S., Gilman, A., Rall, T.W., Nies, A.S., and Taylor, P. (Eds.) *Goodman and Gilman's, The Pharmacological Basis of Therapeutics,* 8th ed., New York: Pergamon Press, pp. 978–1007.

CHAPTER 11

Screening Dyes, Drugs, and Dietary Supplements for Ocular Phototoxicity

Joan E. Roberts

CONTENTS

11.1 INTRODUCTION

The eye is constantly subjected to and interacts with ambient radiation. This light serves to direct vision and circadian rhythm (Roberts, 2000) and therefore, under normal circumstances, must be benign. However, chronic exposure to intense light- and/or age-related changes can lead to light-induced damage to the eye (Sliney, 1999, 2002; Glickman, 2002; Roberts, 2001; Zigman, 1993; Dillon, 1991; Andley, 1987). This danger is enhanced with increased exposure to intense light because of high altitudes (Hu et al., 1989), outdoor employment (Longstreth, 1998), use of sun beds, or during phototherapy for seasonal depression (Roberts et al., 1992a; Terman, 1990).

Many drugs, dietary supplements, and diagnostic dyes absorb in the UV or visible range and can be excited by wavelengths of light transmitted to the lens and retina. This absorption can lead to dramatically enhanced ocular damage through phototoxic side effects of those dyes and drugs (Fraunfelder, 1982; Roberts, 1996). The extent to which a particular photosensitizing drug will affect the human lens or retina *in vivo* depends upon

- Residence time in the lens or retina, which is determined by its structure
- Photoefficiency of the particular sensitizer, due to its ability to absorb the appropriate wavelengths of light transmitted to the lenticular or retinal substrates
- Mechanism by which it causes that damage, including possible binding of the dye to ocular constituents (binding can alter the photochemical mechanism and also increase the retention time of the sensitizer in the eye)
- Presence of endogenous quenchers or free radical scavengers that could stop or retard these photochemical reactions

All of these parameters play a role in the photosensitized oxidation of human ocular tissue and its importance as an underlying mechanism in cataractogenesis and retinal damage.

Presented in this chapter are several screens that can be used to determine potential of drugs to cause light damage to the eye. These include a simple screen that takes into account the optical properties of the eye and the structure and absorption spectra of the various drugs that can be used to eliminate various drugs as potential photooxidants in the eye. In addition, a second, more detailed screen is presented that can be used to determine (1) quantitative potential for a drug to cause photooxidative damage in the eye; (2) mechanism by which it occurs; and (3) *in vivo* verification of phototoxicity.

11.1.1 Structure of the Eye

The human eye is composed of several layers (Figure 11.1). The outermost layer contains the sclera, whose function is to protect the eyeball and the cornea, which

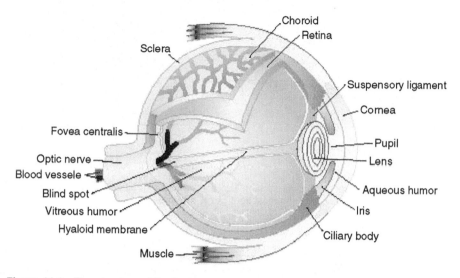

Figure 11.1 The structure of the human eye.

focuses incoming light onto the lens. Beneath this layer is the choroid containing the iris, which is known as the uvea. This region encompasses melanocytes containing the pigment melanin, whose function is to prevent light scattering. The opening in the iris, the pupil, expands and contracts to control the amount of incoming light. The iris and the lens are bathed in the aqueous humor, a fluid that maintains intraocular pressure; this fluid also contains various antioxidants and supports transport to the lens. The lens is positioned behind the iris; its function is to focus light onto the retina.

Behind the lens is the vitreous humor, a fluid that supports the lens and retina and that also contains antioxidants. The retina is composed of the photoreceptor cells (rods and cones) that receive light and the neural portion (ganglion, amacrine, horizontal, and bipolar cells) transducing light signals through the retina to the optic nerve. Behind the photoreceptor cells are the retinal pigment epithelial cells, Bruchs' membrane, and the posterior choroid. The photoreceptor cells are avascular and their nutrient support (ions, fluid, and metabolites) is provided by the retinal pigment epithelial cells. Transport to the retinal pigment epithelial cells is across the Bruch's membrane by the choriocapillaris.

11.1.2 Light Absorption by the Human Eye

For light to damage the eye, it must be absorbed. Each wavelength of light will affect different areas of the eye. Ambient radiation from the sun can contain varying amounts of ultraviolet (UV)-C (100 to 280 nm); UV-B (280 to 320 nm); UV-A (320 to 400 nm); and visible light (400 to 760 nm). Most UV-C and some UV-B are filtered by the ozone layer. Ultraviolet light contains shorter wavelengths of light than visible; the shorter the wavelength, the greater the energy and the greater the potential for biological damage.

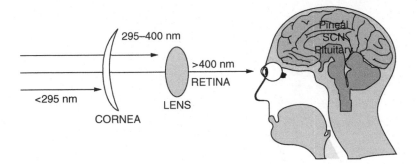

Figure 11.2 The eye as an optical system. The cornea cuts out all light below ca. 295 nm, while the lens absorbs most light below 400 nm.

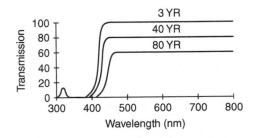

Figure 11.3 The transmission characteristics of the human lens with age showing an increase as a result of an age-related yellowing. The young human lens has a window of light at 320 nm that is transmitted to the retina. The adult human lens transmits only visible light (above 400 nm) to the retina. The aged human lens filters out most of the blue light (400 to 450 nm) from the retina.

The primate/human eye has unique filtering characteristics (Bachem, 1956). The human cornea cuts off all light below 295 nm, so all UV-C and some UV-B are filtered from reaching the lens. The adult human lens absorbs the remaining UV-B and all UV-A (295 to 400 nm); therefore, only visible light reaches the adult human retina (Figure 11.2). However, the young human lens transmits a small window of UV-B light (320 nm) to the retina and the elderly lens filters out much of the short blue visible light (Barker et al., 1991). Aphakia (removal of the lens) and certain forms of blindness may also change the wavelength characteristics of light reaching the retina (Figure 11.3).

11.1.3 Direct Light Damage in the Eye

Short ultraviolet light exposure to the cornea leads to an inflammation reaction (Offord et al., 1999; Pitts et al., 1976); very painful and similar to sunburn, this is known as keratitis. However, these corneal wounds usually heal and do not cause permanent damage. On the other hand, UV damage to the lens and visible light damage to the retina are painless, accumulative, and permanent, possibly leading to formation of cataracts (clouding of the lens) (Dillon, 1991; Balasubramanian, 2000), which can only be corrected by surgery, and to macular and retinal degeneration

(Ham et al., 1982), leading to permanent blindness for which no treatment exists. A number of factors suppress or enhance light damage to the eye, including oxygen, antioxidants, repair mechanisms, and biophysical mechanisms.

- *Oxygen tension in the eye.* The cornea is highly oxygenated. The retina is supplied by the blood so it has varying but high oxygen content in different portions of the retinal tissues. The aqueous humor and the lens have low oxygen content, but it is sufficient for photo-oxidation to occur (McLaren et al., 1998; Roberts et al., 1992a; Kwan et al., 1971). The higher the oxygen content of the tissue, the easier it is for light damage to occur.
- *Defense systems.* Because the eye is constantly subjected to ambient radiation, each portion of the eye contains very efficient defense systems. Antioxidant enzymes (superoxide dismutase (SOD) and catalase) and antioxidants (e.g., vitamin E, vitamin C, lutein, zeaxanthin, lycopene, glutathione, and melanin) serve to protect against oxidative and photoinduced damage (Handelman and Dratz, 1986; Giblin, 2000; Seth and Kharb, 1999; Edge et al., 1998; Khachik et al., 1997). Unfortunately, most of these antioxidants and protective enzymes decrease beginning at 40 years of age (Sarna, 1992; Khachik et al., 1997; Samiec et al., 1998; Sethna et al., 1983). With the protective systems diminished with age, there is loss of protection against all light-induced damage to the eye and the induction of age-related blinding disorders.
- *Repair.* Even if the eye is damaged, the damage does not need to be permanent. The cornea and retina have very efficient repair systems. However, damage to the lens is cumulative and cannot be repaired (Andley, 1987).
- *Biophysical mechanisms.* Ocular damage from light can occur through an inflammatory response or a photo-oxidation reaction. In an inflammatory response, an initial insult to the tissue provokes a cascade of events that eventually results in wider damage to the tissue (Busch et al., 1999; Wang et al., 1999). In photo-oxidation reactions, a sensitizing compound in the eye absorbs light, is excited to a singlet, then a triplet state, and from the triplet produces free radicals and reactive oxygen species, which in turn damage the ocular tissues (Straight and Spikes, 1985) (Figure 11.4).

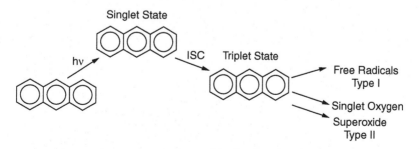

Figure 11.4 The molecular mechanism involved in the phototoxic damage induced in the eye is through a photosensitized oxidation reaction depicted here. This involves the absorption of light by the sensitizer, which excites the compound to the singlet state and then, through intersystem crossing (ISC), goes to the triplet state. The excited triplet state of the drug/dye then proceeds to form free radicals (type I) or singlet oxygen or superoxide (type II) that cause the eventual biological damage.

11.1.4 Drug-Induced Ocular Phototoxicity

Most of the damage to the eye caused by direct irradiation from the sun or artificial sources is from ultraviolet radiation. However, in the presence of a light-activated (photosensitized) diagnostic dye or drug, patients are in danger of enhanced ocular injury from ultraviolet and visible light. The extent to which a particular photosensitizer will affect the human cornea, lens, and/or retina *in vivo* depends upon:

- Residence time in the eye, which is determined by its structure and binding of the dye to ocular constituents; binding can alter the photochemical mechanism as well as increase the retention time of the sensitizer in the eye
- Photoefficiency of the particular sensitizer (i.e., its ability to absorb the appropriate wavelengths of light transmitted to the lenticular or retinal substrates); these transmission characteristics change with age
- Oxygen content of the ocular tissue
- Presence of endogenous quenchers or free radical scavengers and enzyme detoxification systems that could stop or retard these photochemical reactions

11.2 PREDICTING OCULAR PHOTOTOXICITY

Based on the theoretical considerations stated previously, it is relatively easy to predict that a drug cannot cause ocular damage through a photoinduced event. The short screen given in Table 11.1 will dramatically reduce the number of potential substances that need to be considered for ocular phototoxicity.

11.2.1 Short Screen for Potential Ocular Phototoxicity

11.2.1.1 *Measure the Absorbance Spectrum*

In order for a chemical compound (diagnostic dye, drug, endogenous sensitizer) to induce a phototoxic response in any biological tissue, it must first absorb light. This absorption is limited by the filtering characteristics of the biological tissues involved. A comparison of the transmission characteristics of the eye with the absorbance spectra of the drug may be used as a quick screen for phototoxicity. To have the potential to damage the aqueous or the lens, a drug needs a UV spectrum that consists of absorption wavelengths longer than 295 nm (Bachem, 1956). This includes drugs

Table 11.1   Short Screen for Potential Phototoxicity in the Lens/Retina

1. Chemical structure
   Heterocyclic, tricyclic, porphyrin
   Amphiphilic/lipophilic
2. Absorbance spectra
   Longer than 295 nm (lens); 400 nm (retina)
   Binds to DNA, lens protein, melanin
3. Skin phototoxicity

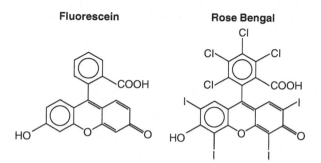

Figure 11.5 The structures of flourescein and Rose Bengal. They have hetercyclic ring systems.

such as chlorpromazine (max 310 nm), tetracycline (max, 365 nm), and the porphyrins (392 nm; 500 to 650 nm) (Roberts, 1984).

In the older human, drugs/dietary supplements, dyes (i.e., hypericin) and diagnostic dyes (photodynamic therapy porphyrins) that absorb in the 400- to 600-nm region could produce phototoxic damage to the lens (Schey et al., 2000) and the retina. In very young humans, all drugs/dyes absorbing at wavelengths longer than 295 nm are potential photosensitizers in the retina as well as the lens. Drugs that do not absorb in these regions cannot cause photodamage.

11.2.1.2 Examine the Chemical Structure

Most drugs must have a tricyclic, heterocyclic, or porphyrin ring to fulfill the energy requirements to produce a stable, long-lived (triplet) reactive molecule that will go on to produce free radicals and reactive oxygen species. The addition of a halide group can enhance the amount of triplet produced (Turro, 1978). As seen in Figure 11.5, fluorescein and Rose Bengal have similar structures except for the attached iodine groups. Fluorescein is highly fluorescent (singlet state) but is not very efficient at reaching the triplet state (poor quantum yield for the triplet: 0.03). The singlet state of Rose Bengal would easily go through intersystem crossing to the triplet (good quantum yield for the triplet: 0.60). By simple inspection of the structures of these two diagnostic dyes, it would be concluded that Rose Bengal has a much greater potential to produce phototoxic damage to the eye than fluorescein.

11.2.1.3 Test Solubility Properties

Dyes or drugs to be examined should be tested for their partitioning in protic and aprotic solvents. Their hydrophobicity will indicate potential for crossing blood ocular barriers. More hydrophilic substances are less likely to cross blood ocular barriers. Compounds that are amphiphilic or lipophilic cross all blood–retinal and/or blood–lenticular barriers. The probable site of damage may also be determined by the hydrophobicity (membranes) or hydrophilicity (cytosol) of the photosensitizing dye or drug.

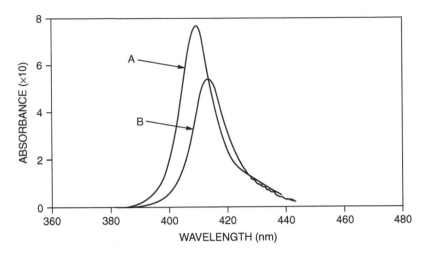

Figure 11.6 The Soret band of a porphyrin (A) and with the addition of lens protein (B). The decrease in the absorbance with the shift to the red seen here is indicative of binding of a drug to the macromolecule.

11.2.1.4 Measure Binding of the Drug to Ocular Tissues

Binding of a drug to ocular tissues (DNA-cornea, lens; proteins-lens; melanin-retina) (Roberts and Dillon, 1987; Dayhaw–Barker and Barker, 1986; Steiner and Buhring, 1990; Sarna, 1992) would increase its retention time in the eye. Furthermore, binding a photosensitizing substance to macromolecules increases the lifetime of its triplet state. It is the triplet state of the dye or drug that leads to further oxidative and free radical reactions. Therefore, drugs/dyes that bind to ocular tissues are very likely to induce phototoxic damage (Roberts et al., 1991a) in that tissue. Binding is determined by measuring the absorption spectra of the drug in the presence and absence of lens proteins, DNA, and/or melanin (Roberts and Dillon, 1987; Roberts et al., 1990; Steiner and Buhring, 1990). A red shift in the absorbance spectra of the drug in the presence of any of these biomolecules is an indication of binding of the sensitizer to the biomolecule (Figure 11.6).

11.2.1.5 Note Reports of Skin Phototoxicity

Finally, any reports of skin phototoxicity for a particular drug should provide a clear warning of potential ocular phototoxicity. Skin phototoxicity is more readily apparent than ocular phototoxicity, although it is induced by compounds with similar chemical features (Oppenlander, 1988).

11.2.2 Detailed Screen for Predicting Ocular Phototoxicity

The short screen can determine a drug's potential for phototoxicity. Once it has been determined that a drug/dye is a possible photosensitizing agent, the more detailed screen given next can determine the potential site of damage (*in situ* fluorescence techniques); predict the type and efficiency of damage (*in vitro* assays); and determine the mechanism of damage (photophysical studies of the short-lived excited state intermediates). Once the excited state intermediates produced by a particular photosensitizing agent have been determined, the introduction of specific quenching agents may stop unwanted reactions (Roberts et al., 1991a).

11.2.2.1 Determine Potential Sites for Ocular Damage

The site of damage is determined by drug penetration and transmission of the appropriate wavelengths of light to that site. At these sites, numerous substrates for phototoxic damage are in the eye.

- *Cornea.* Corneal epithelial and endothelial cells may be easily damaged, leading to keratitis (Cullen, 2002; Pitts et al., 1976; Hull et al., 1983). However, these cells have a very efficient repair mechanism and the damage is rarely permanent.
- *Uvea.* Uveal cells are highly melanized and are ordinarily protected against light-induced damage. However, melanogenesis may be modified with phototoxic reactions leading to a greater risk from UV radiation (Hu et al., 2000, 2002).
- *Lens.* Epithelial cells of the lens have direct contact with the aqueous humor. Their function is to control transport to the lens. They are most vulnerable to phototoxic damage. Damage to these cells would readily compromise the viability of the lens (Roberts et al., 1994). The lens fiber membrane can be photochemically damaged through damage to the lipids and/or the main intrinsic membrane protein (Roberts et al., 1985). This will result in a change in the refractive index causing an opacification. Phototoxic reactions can lead to a modification of certain amino acids (histidine, tryptophan, cysteine) (Roberts, 1984; McDermott et al., 1991) and/or a covalent attachment of sensitizer to cytosol lens proteins. In either case, this changes the physical properties of the protein, leading to aggregation and finally opacification (cataractogenesis). The covalently bound chromophore may now act as an endogenous sensitizer producing prolonged sensitivity to light. Because of little turnover of lens proteins, this damage is cumulative.
- *Retina.* Phototoxic damage can occur in retinal pigment epithelial tissues, the choroid, and the rod outer segments, which contain the photoreceptors. If the damage is not extensive, repair mechanisms are present to allow for recovery of retinal tissues. However, extensive phototoxic damage to the retina can lead to permanent blindness (Glickman, 2002; Dayhaw–Barker, 2002; Dayhaw–Barker and Barker, 1986; Ham et al., 1982).

11.2.2.2 Determine Location/Uptake of Dye/Drug

- *Radiolabeling.* The traditional method of determining uptake into ocular tissues is *in vivo* radiolabeling. This method is time consuming and expensive, although it is effective in determining which ocular tissues have accumulated the drug in question.
- *Fluorescence spectroscopy.* An alternative method to determine uptake of a drug into ocular tissues is ocular fluorometry. After a dye or drug has absorbed light and is excited to the singlet state, it can decay to the ground state and is accompanied by the emission of light; this is known as fluorescence. Because most photosensitizers are fluorescent, transmitted or reflective fluorescence provides an

accurate means of measuring uptake of a sensitizer into ocular tissue that is simpler, less expensive, and less arduous than using radiolabeled materials. This technique may also be used noninvasively, *in vivo* — for instance, using a slit lamp to detect uptake of sensitizers into the human, or scanning or reflective fluorometry to determine the presence of endogenous and exogenous fluorescent materials in the retina (Docchio, 1989; Docchio et al., 1991; Cubeddu et al., 1999; Sgarbossa et al., 2001).

11.2.2.3 Determine the Phototoxic Efficiency

The targets of photooxidative reactions may be proteins, lipids, DNA, RNA, and/or cell membranes (Straight and Spikes, 1985). *In vitro* tests can be designed to determine the specific sites of damage to the various ocular compartments (i.e., lens and retinal epithelial cells and photoreceptor cells) and the products of those reactions. Table 11.2 presents a summary of additional biochemical and photophysical techniques that can be performed to predict the potential for and extent of *in vivo* phototoxicity more accurately.

11.2.2.4 In Situ Assay Focal Length Variability

The Scan Tox™ System, which measures focal length variability, is a method for monitoring lens optical quality in culture conditions that mimic conditions inside the eye (Dovrat et al., 1986, 1993; Dovrat and Weinreb, 1995, 1999; Weinreb and Dovrat, 1996–1997; Sivak et al., 1986a, b, 1987, 1990). The ocular lens is an ideal organ for long-term culture experiments because it has no direct blood supply and no connection to the nervous system. The Scan Tox™ system makes it possible to keep lenses for long-term studies of up to a few weeks. The use of cultured lenses, mainly bovine, replaces the need for testing the effects of potentially damaging agents on live animals. The optical monitoring apparatus uses a computer-operated scanning laser beam, a video camera system, and a video frame analyzer to record the focal length and transmittance of the cultured lens.

The scanner is designed to measure the focal length at points across the diameter of the lens. The lens container permits the lens to be exposed to a vertical laser beam from below. The laser source projects its light onto a plain mirror, which is mounted at 45° on a carriage assembly. The mirror reflects the laser beam directly up through the test lens. The mirror carriage is connected to a positioning motor, which moves the laser beam across the lens. The camera sees the cross section of the beams and, by examining the image at each position of the mirror, Scan Tox™ software is able to measure the quality of the lens by calculating the back vertex distance for each beam position. The cultured lenses continue to maintain their original refractive function. When foreign substances are introduced to a cultured lens, the Scan Tox™ system measures the resulting optical response. This provides a very sensitive means to follow early damage to the eye lens. It has recently been used to define the ocular phototoxicity of hypericin, a phototoxic component of St. John's Wort (Wahlman et al., 2003)

Table 11.2   Detailed Screen for Ocular Phototoxicity

In Vitro/In Situ Studies	Technique
a) Focal length variability	Whole lens damage
b) Cell culture	DNA, RNA, protein
Enzyme assays	Antioxidant enzymes
Histology	Endothelial, epithelial, photoreceptor cell damage
	Protein change
c) Gel electrophoresis	
Amino acid analysis	
d) Mass spectrometry	Lipid changes, peptide maps
e) TLC	Lipid oxidation
f) HPLC	Lipid peroxides, DNA adducts, protein modification
g) Normalization for photons	
Biophysical Techniques	
Fluorescence	Uptake, binding, quantum yields
Laser flash photolysis	Triplet detection, binding/lifetime, quantum yields
Luminescence	Singlet oxygen, oxygen tension
Pulse radiolysis	Radical and oxyradical intermediates
ESR	Radical and oxyradical intermediates

11.2.2.5 In Vitro *Assays*

- *Cell culture/whole tissue.* The first reported assay for phototoxicity in human ocular cells (Roberts, 1981) measured changes in macromolecular synthesis in the presence and absence of a light-activated drug. Other studies have assessed damage to corneal and uveal melanocytes, lenticular and retinal cells by measuring pump function and enzyme activities *in vitro* and *in situ* (Hu et al., 2002; Andley et al., 2004; Roberts et al., 1994; Organisiak and Winkler, 1994; Rao and Zigler, 1992; Dayhaw–Barker, 1987).
- *Gel electrophoresis and amino acid analysis.* Gel electrophoresis has been used to monitor polymerization of ocular proteins (Kristensen et al., 1995; Roberts et al., 1992a, Roberts, 1984; Zigler et al., 1982). Photopolymerization is one of the most apparent changes in ocular protein induced by photosensitizing dyes and drugs. Quantitative changes can be measured by scanning the gel and determining relative reaction rates. Specific amino acid modifications can be determined using amino acid analysis (Roberts, 1996, 1984). Zhu and Crouch (1992) have illustrated the wide variety of classical protein analysis techniques (gel electrophoresis, amino acid analysis, sequencing, isoelectric point determination, western blot, ELISA) that can be used to investigate phototoxic damage induced by dyes and drugs.
- *Mass spectometry.* Recent innovations in the field of mass spectrometry (liquid secondary ion mass spectrometry (LSIMS) and electrospray ionization (ESI) have allowed for identification of specific amino acid modifications within large proteins through molecular weight mapping. These techniques have been applied to determine specific sites of photooxidative damage in corneal and lenticular proteins (Schey et al., 2000; Roberts et al., 2001). These studies can serve as a model for defining damage from any potential phototoxic agent in the eye.
- *Thin layer chromatography.* This technique is particularly effective at separating triacyclycerol, free fatty acid, and phospholipids from lens (Fleschner, 1995) and

retinal (Organisiak et al., 1992) membranes. Thin layer chromotography/gas chromotography/mass spectrometry (TLC/GC/Mass Spec) may be used to measure lenticular or retinal lipid modifications (Handelman and Dratz, 1986). Specific lipids may be modified in the presence of photosensitizing agents and separated on TLC plates. The plates can then be scanned for quantitative analysis of these specific changes.

- *High-pressure liquid chromatography.* HPLC is particularly effective at separating and identifying lipid peroxides from the retina (Akasaka et al., 1993). It has also been used to identify adducts formed between DNA nucleotides and phototoxic agents (Oroskar et al., 1994). HPLC has also been used to assess the rates of photo-oxidation of lens proteins in the presence of a sensitizer. Using this technique, it is possible to determine induced amino acid modification within the protein and its location, and to detect possible binding of sensitizing drugs to specific lens crystallins (McDermott et al., 1991).

- *Normalization for photons absorbed.* Whatever the target tissue or extent of damage, the toxic effects of these dyes and drugs are the result of photochemical reactions. As such, their rate of efficiency is dependent on the number of photons absorbed by the sensitizer in the biological tissue. Therefore, in order to get an accurate comparison of the photosensitizing potency of various dyes and drugs with different structures and absorptive characteristics, it is essential to normalize for the number of photons absorbed by each drug in a particular system.

  This can be done with a simple computer-generated mathematical formula (Roberts, 1996; Kristensen et al., 1995), which takes into account the absorption spectrum of the drug; the output of the lamp source used in the experiments; and the optical properties of the eye. The total relative number of photons absorbed by a drug under particular experimental conditions is the area under the product curve:

$$\text{Photons absorbed} = I \times AB \times \lambda$$

  where I = the intensity of the lamp at various wavelengths adjusted for the transmission characteristics of the cornea or lens; AB is the absorbance of the dye/drug; and λ is the number of photons at those wavelengths. The rates of each photooxidative event are then adjusted accordingly for each sensitizer. This in turn can be corrected for the actual transmission characteristics of the cornea and/or lens and the output of the sun to predict *in vivo* effects.

11.2.2.6 Summary

In vitro techniques determine potential damage done to an ocular substrate, which gives information about the photoefficiency of a drug if it is taken up into the various compartments of the eye. Additional information about the site of potential damage can be predicted based on which ocular substrate (DNA, RNA, protein, lipid) is affected.

11.2.2.7 Photophysical Studies: Determination of Precise Excited State Intermediates Causing Phototoxic Damage

- *Quantum yields.* In predicting the phototoxicity of a dye or drug, it is important to determine the proportion of photons leading to a benign (fluorescence, singlet state) event and the proportion leading to a potentially destructive event (phosphorescence, triplet state). The efficiency of a photoinduced process may be expressed as its quantum yield (Q).

$$Q = \frac{\text{number of photons used to produce an event}}{\text{number of photons originally absorbed}}$$

This gives a measure of how likely a photochemical event is to occur. The quantum yield is often expressed as a percentage. For instance, the quantum yield for fluorescence of fluorescein is 0.92 and that of Rose Bengal is 0.08 (Figure 11.2). This means that most of the light absorbed by fluorescein is given off in the form of fluorescence energy, making it a relative safe diagnostic dye for the eye. On the other hand, the triplet quantum yield for fluorescein is 0.03 and for Rose Bengal is 0.60. This indicates that very little of the singlet fluorescein energy is transformed into a triplet, while most of the energy of Rose Bengal will be available for intersystem crossing and will reach the triplet state from where it can produce ocular damage. Therefore, flourescein is appropriate, but Rose Bengal would be an inappropriate diagnostic dye for the eye.

- *Laser flash photolysis.* This method uses a pulse of monochromatic light to promote a specific dye or drug to an excited state (Rodgers, 1985).

  Triplet detection. Time-resolved techniques (absorption spectroscopy or diffuse reflectance) allow for detection of the triplet state of the excited chromophore even in intact tissues. This technique has been used to determine the absence (Dillon and Atherton, 1990) of triplet formation by the endogenous 3-hydroxy-kynurenine, as well as the presence of triplets from sensitizing drugs in intact lenses (Roberts et al., 1991b).

  Lifetime/binding. All ocular damage from photo-oxidation reactions occurs through the triplet state of the drug. The longer the lifetime is, the greater the potential for damage. The lifetimes of the triplets of sensitizing drugs were found to be greater when bound to macromolecules or in an intact organ than when free in solution (Roberts et al., 1991b). The presence of a triplet and an increase in lifetime when bound to intact ocular tissue are predictive of a drug that is causing photooxidative damage to the eye *in vivo* (Roberts et al., 1991a).

- *Luminescence (singlet oxygen).* The presence and lifetime of singlet oxygen can be determined using time-resolved infrared luminescence measurements at 1270 nm (Rodgers and Snowden, 1982). Using this technique, it can be determined whether and how efficiently a dye or drug can produce singlet oxygen (Roberts et al., 2000; Motten et al., 1999). Because singlet oxygen is the most powerful oxidant in a photo-oxidation reaction, a dye or drug that is an efficient producer of singlet oxygen could be predicted to induce phototoxic damage if present in the eye.

- *Pulse radiolysis (hydroxyl, peroxy radicals, and superoxide).* Pulse radiolysis consists of the delivery of a very short, intense pulse of ionizing radiation to a sample, the resultant changes in light absorption of the sample being followed by a very fast spectrophotometer (Land, 1985). The technique may be used to detect the formation of short-lived radical species of a dye or drug (Roberts et al., 1998, 2000). In addition, the interaction of a dye or drug with excited oxygen intermediates (hydroxyl radical, superoxide, peroxy radicals) (Land et al., 1983), which are cleanly generated in this system, allow for an understanding of a possible mechanism of *in vivo* photooxidative ocular damage.

- *Electron spin resonance (hydroxyl, peroxy, and carbon-centered radicals and superoxide).* Electron spin resonance (ESR or EPR) spectroscopy detects and characterizes species containing an odd number of electrons, namely, free radicals and paramagnetic metal ions. The photo-oxidation reactions responsible for the phototoxic responses in the eye involve free radicals formed via electron transfer (electron exchange) between the sensitizing drug in an excited state and a substrate from the ocular tissues. Although these radicals are very short lived, they can be observed with ESR *in situ* during their photogeneration. For instance, illumination of Rose Bengal (RB), an ophthalmic diagnostic dye, in the presence of an electron donor such as NADH affords a radical anion of the dye RB⁻ that can be directly measured using ESR (Sarna et al., 1991).

  Radicals that are too reactive to accumulate in detectable quantities can frequently be detected by ESR using spin trapping techniques. In this approach, an agent called a spin trap reacts with a short-lived radical R· to give a spin adduct R-T·, which has a much longer lifetime. The original radical R· is identified by the characteristic ESR spectrum of the R-T·. Carbon-, nitrogen-, and sulfur-centered radicals, as well as all of the important oxygen-centered radicals (hydroxyl, superoxide, alkoxyl, and peroxyl), can be identified using ESR directly or in combination with the spin trapping technique.

Using these techniques, the photosensitized generation of superoxide in protic (Reszka et al., 1992) and aprotic media (Reszka, et al., 1993) can be monitored. These are model systems for the hydrophilic (aqueous) and hydrophobic (ROS membrane) portions of the eye. ESR has recently been used to define the photochemical mechanisms involved in the light activation of endogenous pigments in the lens (Reszka et al., 1996) and the retina (Reszka et al., 1995). With these systems defined, ESR can be used to predict potential phototoxic events induced by exogenous photosensitizing dyes and drugs and natural pigments (Roberts et al., 2000; Motten et al., 1999).

11.2.2.8 Summary

The molecular mechanism involved in the phototoxic damage induced in the eye is photosensitized oxidation reactions. This mechanism begins with the absorption of light by the sensitizing compound (endogenous pigment, dye, drug, or dietary supplement), which promotes the compound to an excited singlet state (short lived) and then, through intersystem crossing, goes to the triplet state (long lived). The excited triplet state of the substance then proceeds via a type I (free radical) or type II (singlet oxygen) mechanism, causing the eventual biological damage (Straight and Spikes, 1985) (Figure 11.4).

Therefore, information about the efficiency and excited state intermediates for a phototoxic reaction in the eye obtained by using photophysical techniques (fluorescence, flash photolysis, pulse radiolysis, ESR) can be predictive of phototoxicity *in vivo*. It has been confirmed that photophysical studies collate well with *in vivo* data (Roberts et al., 1991a). For instance, tetrasulphonatophenylporphyrin (TPPS), which binds to lens proteins, shows a long-lived triplet in the intact calf and human lens, produces singlet oxygen efficiently, and causes photooxidative damage *in vivo* in pigmented mouse eyes, whereas uroporphyrin (URO), which produces an efficient triplet but does not bind to ocular tissues, does not cause photooxidative damage *in vivo*.

11.3 *IN VIVO* TESTING

The use of the short or more detailed (Table 11.1 and Table 11.2) screens for ocular phototoxicity will not totally eliminate the need for accurate *in vivo* experiments. The function of these studies is to limit the need for *in vivo* testing for ocular phototoxicity of large numbers of drugs. Those drugs found in screening to be highly likely to produce phototoxic side effects in the eye should be tested further with animal studies to determine the exact site and extent of damage to be expected in humans (Roberts et al., 1991a).

Prolonged use of a phototoxic drug is most probably of greater long-term risk to the eye than short-term dosage because of cumulative damage. Because the cornea has an active repair system, little or no long-term side effects of phototoxicity should occur there. However, because no turnover occurs in the lens constituents, any modification in that tissue will tend to stay and accumulate with age. Thus, cataractogenesis may not develop until much later than the initial insult. In addition, phototoxic damage to the lens may not only cause direct damage to cell viability, but also may undermine its defense system so that gross morphological effects may appear much later than the original insult.

The susceptibility of the lens and retina to light-induced damage increases with the age-related changes in chromophores (Roberts et al., 2000, 2001, 2002) with concurrent decrease in antioxidant status (Giblin, 2000; Samiec et al., 1998; Sarna, 1992). Therefore, the elderly may be particularly sensitive to drugs and other agents that induce phototoxic side effects. Sight may be regained after cataract surgery; however, damage to the retina that is not repaired leads to permanent blindness. The environmental lighting, particularly the constant presence of intense ambient light, must always be taken into account when assessing the potential *in vivo* ocular toxicity of a drug.

11.4 PROTECTION

Even if a drug has the potential to produce phototoxic side effects in the eye, no damage will be done if the specific wavelengths of optical radiation absorbed by the drug are blocked from transmittance to the eye. This can be easily done with wrap-around eyeglasses (Gallas and Eisner, 2001; Sliney, 1999; Merriam, 1996) that incorporate specific filters. Furthermore, nontoxic quenchers and scavengers could be given in conjunction with the phototoxic drug to negate its ocular side effects while allowing for the primary effect of the drug (Roberts, 1981; Roberts et al., 1991a; Roberts and Mathews–Roth, 1993).

11.5 CONCLUSION

With simple, inexpensive *in vitro* testing, compounds can be monitored at their developmental stage for potential ocular phototoxicity. It may be that a portion of the molecule can be modified to reduce phototoxicity while leaving the primary drug effect intact. This may reduce the necessity of later, more costly, drug recalls. In the future, more effective use of ocular fluorometry will allow for a more accurate assessment of the uptake and location of exogenous photosensitizing dyes, drugs, and herbal supplements in the eye with the potential to harm the eye. Also, for those drugs that must be continued in spite of their phototoxicity (i.e., antimalarial drugs; Motten et al., 1999), appropriate protection against light (sunglasses) and specific supplementary antioxidants may be prescribed to retard or eliminate the most severe blinding side effects (Roberts and Mathews–Roth, 1993).

ACKNOWLEDGMENTS

The author wishes to thank the Hugoton Foundation for its financial support and Dr. Ann Motten, NIEHS, North Carolina, for help in preparing this manuscript.

REFERENCES

Akasaka, K., Ohrui, H., and Meguro, H. (1993) Normal-phase high-performance liquid chromatography with a fluorimetric postcolumn detection system for lipid hydroperoxides, *J. Chromatogr.*, 628, 31–35.

Andley, U.P., Hebert, J.S., Marrison, A., Reddan, J.R., and Pentland, A.P. (1994) Modulation of lens epithelial cell proliferation by enhanced prostaglandin synthesis after UVB exposure, *Invest. Ophthalmol. Vis. Sci.*, 35, 374–381.

Andley, U.P. (1987) Photodamage to the eye, *Photochem. Photobiol.*, 46, 1057–1060.

Bachem, A. (1956) Ophthalmic action spectra, *Am. J. Ophthal.*, 41, 969–975.

Balasubramanian, D. (2000) Ultraviolet radiation and cataract, *J. Ocul. Pharmacol. Ther.*, 16, 285–297.

Barker, F.M., Brainard, G.C., and Dayhaw–Barker, P. (1991) Transmission of the human lens as a function of age, *Invest. Ophthalmol. Vis. Sci.*, 32S, 1083.

Busch, E.M., Gorgels, T.G., Roberts, J.E., and van Norren, D. (1999) The effects of two stereoisomers of *N*-acetylcysteine on photochemical damage by UVA and blue light in rat retina, *Photochem. Photobiol.*, 70, 353–358.

Cubeddu, R., Taroni, P., Hu, D.N., Sakai, N., Nakanishi, K., and Roberts, J.E. (1999) Photophysical studies of A2-E, putative precursor of lipofuscin, in human retinal pigment epithelial cells, *Photochem. Photobiol.*, 70, 172–175.

Cullen, A.P. (2002) Photokeratitis and other phototoxic effects on the cornea and conjunctiva, *Int. J. Tox.*, 21, 455–464.

Dayhaw–Barker, P. (1987) Ocular photosensitization, *Photochem. Photobiol.*, 46, 1051–1056.

Dayhaw–Barker, P. (2002) Retinal pigment epithelium melanin and ocular toxicity, *Int. J. Tox.*, 21, 451–454.

Dayhaw–Barker, P. and Barker, F.M. (1986) Photoeffects on the eye, in Jackson, E.M. (Ed.) *Photobiology of the Skin and the Eye*, New York: Marcel Dekker, 117–147.

Dillon, J. (1991) Photophysics and photobiology of the eye, *J. Photochem. Photobiol. B Biol.*, 10, 23–40.

Dillon, J. and Atherton, S.J. 1990 Time resolved spectroscopic studies on the intact human lens, *Photochem. Photobiol.*, 51, 465–468.

Docchio, F. (1989) Ocular fluorometry: principles, fluorophores, instrumentation and clinical applications, *Lasers Surg. Med.*, 9, 515–532.

Docchio, F., Boulton, M., Cubeddu, R., Ramponi, R., and Dayhaw–Barker, P. (1991) Age-related changes in the flourescence of melanin and lipofuscin granules of the retinal pigment epithelium: a time-resolved fluorescence spectroscopy study, *Photochem. Photobiol.*, 54, 247–253.

Dovrat, A., Sivak, J.G., and Gershon, D. (1986). Novel approach to monitoring lens function during organ culture, *Lens Res.*, 3, 207–215.

Dovrat, A., Horwitz, J., Sivak, J.G., Weinreb, O., Scharf, J., and Silbermann, M. (1993) DL-propranolol inhibits lens hexokinase activity and affects lens optics, *Exp. Eye. Res.*, 57, 747–751.

Dovrat, A. and Weinreb, O. (1995). Recovery of lens optics and epithelial enzymes after ultraviolet A radiation, *Invest. Ophthalmol. Vis. Sci.*, 36, 2417–2424.

Dovrat, A. and Weinreb, O. (1999) Effects of UV-A radiation on lens epithelial Na,K,ATPase in organ culture, *Invest. Ophthalmol. Vis. Sci.*, 40, 1616–1620.

Edge, R., Land, E.J., McGarvey, D., Mulroy, L., and Truscott, T.G. (1998) Relative one-electron reduction potentials of carotenoid radical cations and the interactions of carotenoids with the vitamin E radical cation, *J. Am. Chem. Soc.*, 120, 4087–4090.

Fleschner, C.R. (1995) Fatty acid composition of triacylglycerols, free fatty acid and phospholipids from bovine lens membrane fractions, *Invest. Ophthal. Vis. Sci.*, 36, 261–264.

Fraunfelder, F.T. (1982) *Drug-Induced Ocular Side Effects and Drug Interactions*, 2nd ed., Philadelphia: Lea & Febiger.

Gallas, J. and Eisner, M. (2001) Eye protection from sunlight damage, in Giacomoni, P.U. (Ed.) *Sunlight Protection in Man*, Amsterdam: Elsevier, 437–455.

Glickman, R.D. (2002) Retinal light damage mechanisms, *Int. J. Tox.*, 21, 473–490.

Giblin, F.J. (2000) Glutathione: a vital lens antioxidant, *J. Ocul. Pharmacol. Ther.*, 16, 121–135.

Ham, W.T., Mueller, H.A., Ruffolo, J.J., Guerry, D., and Guerry, R.K. (1982) Action spectrum for retinal injury from near ultraviolet radiation in the aphakic monkey, *Am. J. Ophthal.*, 93, 299–305.

Handelman, G.J. and Dratz, E.A. (1986) The role of antioxidants in the retina and retinal pigment epithelium and the nature of prooxidant induced damage, *Adv. Free Rad. Biol. Med.*, 2, 1–89.

Hu, D.N. (2000) Regulation of growth and melanogenesis of uveal melanocytes, *Pigment Cell Res.*, 13, 81–86.

Hu, D.-N., Savage, H.E., and Roberts. J.E. (2002) Uveal melanocytes, ocular pigment epithelium and muller cells in culture: *in vitro* toxicology, *Int. J. Tox.*, 21, 465–472.

Hu, T.-S., Zhen, Q., Sperduto, R.D., Zhao, J.-L., Milton, R.C., Nakajima, A., and the Tibet Eye Study Group (1989) Age-related cataract in the Tibet Eye Study, *Arch. Phthalmol.*, 107, 666–671.

Hull, D.S., Csukas, S., and Green, K. (1983) Trifluoperazine: corneal endothelial phototoxicity, *Photochem. Photobiol.*, 38, 425–428.

Khachik, F., Bernstein P.S., and Garland, D.L. (1997) Identification of lutein and zeaxanthin oxidation products in human and monkey retinas, *Invest. Ophthal. Vis. Sci.*, 38, 1802–1811.

Kwan, M., Niinikoske, J., and Hunt, T.K. (1971) Oxygen tension in the aqueous and the lens, *Invest. Ophthal.*, 11, 108–111.

Kristensen, S., Wang, R.H., Tønnesen, H., Dillon, J., and Roberts, J.E. (1995) Photoreactivity of biologically active compounds. VII. Photosensitized polymerization of lens proteins by antimalarial drugs *in vitro*, *Photochem. Photobiol.*, 61, 124–130.

Land, E.J. (1985) Pulse radiolysis, in Bensasson, R.V., Jori, G., Land, E.J., and Truscott, T.J. (Eds.) *Primary Photoprocesses in Biology and Medicine*, Amsterdam: Elsevier, 35–44.

Land, E.J., Mukherjee, T., Swallow, A.J., and Bruce, J.M. (1983) One-electron reduction of adriamycin: properties of the semiquinone, *Arch. Biochem. Biophys.*, 225, 116–121.

Longstreth, J., de Gruijl, F.R., Kripke, M.L., Abseck, S., Arnold, F., Slaper, H.I., Yelders, G., Takizawa, Y., and van der Leun, J.C. (1998) Health risks, *J. Photochem. Photobiol. B*, 46, 20–39.

McLaren, J.W., Dinslage, S., Dillon, J.P., Roberts, J.E., and Brubaker, R.F. (1998) Measuring oxygen tension in the anterior chamber of rabbits, *Invest. Ophthalmol. Vis. Sci.*, 39, 1899–1909.

McDermott, M., Chiesa, R., Roberts, J.E., and Dillon, J. (1991) Photo-oxidation of specific residues in a-crystallin polypeptides, *Biochemistry*, 30, 8653–8660.

Merriam, J.C. (1996) The concentration of light in the human lens, *Trans. Am. Ophthalmol. Soc.*, 94, 803–918.

Motten, A.G., Martinez, L.J., Holt, N., Sik, R.H., Reszka, K., Chignell, C.F., Tønnesen, H.H., and Roberts, J.E. (1999) Photophysical studies on antimalarial drugs, *Photochem. Photobiol.*, 69, 282–287.

Offord, W.A., Sharif, N.A., Mace, K., Tromvoukis, Y., Spillare, E.A., Avanti, O., Howe, W.E., and Pfeifer, A.M.A. (1999) Immortalized human corneal epithelial cells for ocular toxicity and inflammation studies, *Invest. Ophthalmol. Vis. Sci.*, 40, 1091–1101.

Oppenlander, T. (1988) A comprehensive photochemical and photophysical assay exploring the photoreactivity of drugs, *Chimia*, 42, 331–342.

Organisiak, D.T. and Winkler, B.S. (1994) Retinal light damage: practical and theoretical considerations, *Prog. Retinal Res.*, 13, 1–29.

Organisciak, D.T., Darrow, M.A., Jiang, Y.-L., Marak, G.E., and Blanks, J.C. (1992) Protection by dimethylthiourea against retinal light damage in rats, *Invest. Ophthalmol. Vis. Sci.*, 33, 1187–1192.

Oroskar, A., Olack, G., Peak, M.J., and Gasparro, F.P. (1994) 4'-Aminomethyl-4,5",8-trimethylpsoralen photochemistry: the effect of concentration and UVA fluence on photoadduct formation in poly(dA-dT) and calf thymus DNA, *Photochem. Photobiol.*, 60, 567–573.

Pitts, D.G., Cullen, A.P., and Parr, W.H. (1976). Ocular ultraviolet effects in the rabbit eye, DHEW (NIOSH) Publication, 77, 130–138.

Rao, D.M. and Zigler, J.S., Jr. (1992) Levels of reduced pyridine nucleotides and lens photodamage, *Photochem. Photobiol.*, 56, 523–528.

Reszka, K., Bilski, P., Chignell, C.F., and Dillon, J. (1996). Photosensitization by the human lens component kynurenine: an EPR and spin trapping investigation, *Free Rad. Biol. Med.*, 20, 23–34.

Reszka, K., Eldred, G., Wang, R.H., Chignell, C.F., and Dillon, J. (1995) Photochemistry of human lipofuscin as studied by EPR, *Photochem. Photobiol.*, 62, 1005–1008.

Reszka, K., Bilski, P., Sik, R.H., and Chignell, C.F. (1993) Photosensitized generation of superoxide radical in aprotic solvents: an EPR and spin trapping study, *Free Rad. Res. Comm.*, 19, 33–44.

Reszka, K., Lown, J.W., and Chignell, C.F. (1992) Photosensitization by anicancer agents. X. *ortho*-Semiquinone and superoxide radicals produced during anthapyrazole-sensitized oxidation of catechols, *Photochem. Photobiol.*, 55, 359–366.

Roberts, J.E. (1981) The effects of photo-oxidation by proflavin in HeLa cells. I. The molecular mechanisms, *Photochem. Photobiol.,* 33, 55–60.

Roberts, J.E. (1984) The photodynamic effect of chlorpromazine, promazine and hematoporphyrin on lens protein, *Invest. Ophthalmol. Vis. Sci.*, 25, 746–750.

Roberts, J.E. (1996) Ocular phototoxicity, in Marzulli, F. and Maiback, M. (Eds.) *Dermatotoxicology*, 5th ed., Washington, D.C.: Taylor & Francis, 307–313.

Roberts, J.E. (2000) Light and immunomodulation, *Ann. N.Y. Acad. Sci.*, 917, 435–445.

Roberts, J.E. (2001) Ocular phototoxicity, *J. Photochem. Photobiol. B Biol.*, 64, 136–143.

Roberts, J.E. and Dillon, J. (1987) *In vitro* studies on the photosensitized oxidation of lens proteins by porphyrins, *Photochem. Photobiol.*, 46, 683–688.

Roberts, J.E. and Mathews–Roth, M. (1993) Cysteine ameliorates photosensitivity in Erythropoietic Protoporphyria, *Arch. Derm.*, 129, 1350–1351.

Roberts, J.E., Roy, D., and Dillon, J. (1985) The photosensitized oxidation of the calf lens main intrinsic protein (MP26) with hematoporphyrin, *Curr. Eye Res.*, 4, 181–185.

Roberts, J.E., Atherton, S.J., and Dillon, J. (1990) Photophysical studies on the binding of tetrasulphonatophenylporphyrin (TPPS) to lens proteins, *Photochem. Photobiol.*, 52, 845–848.

Roberts, J.E., Kinley, J., Young, A., Jenkins, G., Atherton, S.J., and Dillon, J. (1991a) *In vivo* and photophysical studies on photooxidative damage to lens proteins and their protection by radioprotectors, *Photochem. Photobiol.*, 53, 33–38.

Roberts, J.E., Atherton, S.J., and Dillon, J. (1991b) Detection of porphyrin excited states in the intact bovine lens, *Photochem. Photobiol.*, 54, 855–857.

Roberts, J.E., Harriman, A., Atherton, S.J., and Dillon, J. (1992a) A non-invasive method to detect oxygen tensions and other environmental factors in the lens, *Int. Soc. Ocular Phototox,* Sedona, AR, Abstract 64.

Roberts, J.E., Reme, C., Terman, M., and Dillon, J. (1992b) Exposure to bright light and the concurrent use of photosensitizing drugs, *N. Engl. J. Med.*, 326, 1500–1501.

Roberts, J.E., Schieb, S., Garner, W.H., and Lou, M. (1994) Development of a new photooxidative induced cataract model using TPPS in an intact lens, *Invest. Ophthal. Vis. Sci.*, 35, 2137.

Roberts, J.E., Hu, D.N., and Wishart, J.F. (1998) Pulse radiolysis studies of melatonin and chloromelatonin, *J. Photochem. Photobiol. B Biol.*, 42, 125–132.

Roberts, J.E., Wishart, J.F., Martinez, L., and Chignell, C.F. (2000) Photochemical studies on xanthurenic acid, *Photochem. Photobiol.*, 72, 467–471.

Roberts J.E., Finley, E.L., Patat, S.A., and Schey, K.L. (2001) Photo-oxidation of lens proteins with xanthurenic acid: a putative chromophore for cataractogenesis, *Photochem. Photobiol.*, 74, 740–744.

Roberts, J.E., Kukielczak, B.M., Hu, D.-N., Miller, D.S., Bilski, P., Sik, R.H., Motten, A.G., and Chignell, C.F. (2002) The role of A2E in prevention or enhancement of light damage in human retinal pigment epithelial cells, *Photochem. Photobiol.*, 75, 184–190.

Rodgers, M.A.J. (1985) Instrumentation for the generation and detection of transient species, in Bensasson, R.V., Jori, G., Land, E.J., and Truscott, T.J. (Eds.) *Primary Photoprocesses in Biology and Medicine,* Amsterdam: Elsevier/North-Holland, 1–24.

Rodgers, M.A.J. and Snowden, P.T. (1982) Lifetime of O_2 (singlet delta) in liquid water as determined by time resolved infra-red luminescence measurements, *J. Am. Chem. Soc.*, 104, 5541–5543.

Samiec, P.S., Drews–Botsch, C., Flagge, E.W., Kurtz, J.C., Sternberg, P., Reed, R.L., and Jones, D.P. (1998) Glutathione in human plasma decline in association with aging, age-related macular degeneration and diabetes, *Free Rad. Biol. Med.*, 24, 699–704.

Sarna, T. (1992) Ocular melanins, *J. Photochem. Photobiol.*, 12, 215–258.

Sarna, T., Zajac, J., Bowman, M.K., and Truscott, T.G. (1991) Photoinduced electron transfer reactions of Rose Bengal and selected electron donors, *J. Photochem. Photobiol. A: Chem.*, 60, 295–310.

Schey, K.L., Patat, S., Chignell, C.F., Datillo, M., Wang, R.H., and Roberts, J.E. (2000) Photo-oxidation of lens proteins by hypericin (active ingredient in St. John's Wort), *Photochem. Photobiol.*, 72, 200–207.

Seth, R.K. and Kharb, S. (1999) Protective function of alpha-tocopherol against process of cataractogenesis in humans, *Ann. Nutr. Metab.*, 43, 286–289.

Sethna, S.S., Holleschau, A.M., and Rathbun, W.B. (1983) Activity of glutathione synthesis enzymes in human lens related to age, *Curr. Eye Res.*, 2, 735–742.

Sgarbossa, A., Angelini, N., Gioffre, D., Youssef, T., Lenci, F., and Roberts, J.E. (2001) The uptake, location and fluorescence of hypericin in bovine intact lens, *Curr. Eye Res.*, 21, 597–601.

Sivak, J.G., Gershon, D., Dovrat, A., and Weerheim, J. (1986a) Scanning lens for *in vitro* toxicology, in Goldberg A.M. (Ed.) *In Vitro Toxicology, Approaches to Validation, Alternative Methods in Toxicology*, New York: M.A. Liebert, Inc., 181–188.

Sivak, J.G., Gershon, D., Dovrat, A., and Weerheim, J. (1986b) Computer assisted scanning laser monitor of optical quality of the excised crystalline lens, *Vis. Res.*, 26, 1873–1879.

Sivak, J.G., Gershon, D., Dovrat, A., and Weerheim, J. (1987) Ocular lens organ culture system for *in vitro* toxicology, in Goldberg, A.M. (Ed.) *In Vitro Toxicology, Approaches to Validation, Alternative Methods in Toxicology*, New York: M.A. Liebert, Inc., 499–506.

Sivak, J.G., Yoshimura, M., Weerheim, J., and Dovrat, A. (1990) Effect of hydrogen peroxide, DL-propranolol and prednisone on bovine lens optical function in culture, *Invest. Ophthalmol. Vis. Sci.*, 31, 954–963.

Sliney, D.H. (1999) Geometrical assessment of ocular exposure to environmental UV radiation — implications for ophthalmic epidemiology, *J. Epidemiol.*, 9, 22–32.

Sliney, D.H. (2002) How light reaches the eye, *Int. J. Tox.*, 21, 501–509.

Steiner, K. and Buhring, K.U. (1990) The melanin binding of bisoprolol and its toxicological relevance, *Lens Eye Toxicity Res.*, 7, 319–333.

Straight, R. and Spikes, J.D. (1985) Photosensitized oxidation of biomolecules, in Frimer, A. (Ed.) *Singlet O, Vol. IV. Polymers and Biopolymers*, Boca Raton, FL: CRC Press, 91–143.

Terman, M., Amira, L., Terman, J.S., and Ross, D.C. (1996) Predictors of response and nonresponse to light treatment for winter depression, *Am. J. Psychiatry*, 153, 1423–1429.

Turro, N. (1978) *Modern Molecular Photochemistry.* Menlo Park, CA: Benjamin/Cummings Publishing Co.

Wahlman, J., Hirst, M., Roberts, J.E., and Trevithick, J.R. (2003) Focal length variability and protein leakage as tools for measuring photooxidative damage to the lens, *Photochem. Photobiol.*, 78, 88–92.

Wang, X.C., Jobin, C., Allen, J.B., Roberts, W., and Jaffe, G.J. (1999) Suppression of NF-kappaB-dependent proinflammatory gene expression in human RPE cells by a proteasome inhibitor, *Invest. Ophthalmol. Vis. Sci.*, 40, 477–486.

Weinreb, O. and Dovrat, A. (1996–1997) Long-term organ culture system to study the effects of UV-A irradiation on transglutaminase, *Metabolic Pediatr. Syst. Ophthalmol.*, 19/20, 23–26.

Zhu, L. and Crouch, R.K. (1992) Albumin in the cornea is oxidized by hydrogen peroxide, *Cornea*, 11, 567–572.

Zigler, J.S., Jr., Jernigan, H.M., Jr., Perlmutter, N.S., and Kinoshita, J.H. (1982) Photodynamic cross-linking of polypeptides in intact rat lens, *Exp. Eye Res.*, 35, 239–249.

Zigman, S. (1993) Ocular light damage, *Photochem. Photobiol.*, 57, 1060–1068.

CHAPTER 12

Photochemical and Photophysical Methods Used in Study of Drug Photoreactivity

Suppiah Navaratnam

CONTENTS

0-415-30323-0/04/$0.00+$1.50

12.1 INTRODUCTION

The first law of photochemistry, established by Grotthus and Draper in 1818, states that for a photoreaction to occur, a molecule must absorb photons of an appropriate energy equal to the difference in energy between two electronic levels. A general sequential scheme for the photoactivation and photophysical deactivation processes is shown as a modified Jablonski diagram in Figure 12.1. The thick horizontal lines represent the electronic energy levels of a molecule and the thin horizontal lines are vibrational levels associated with each electronic energy level. The boxes at the side of each energy level represent the molecular orbitals. The bottom of the three small boxes is the highest occupied molecular orbital of the ground state (S_o). The two boxes above it represent the lowest and second lowest unoccupied molecular orbitals of the ground state. The arrows in the boxes represent the position and spin of the electrons in the molecular orbitals for each electronic energy level. S_1 and S_2 are the singlet excited states and T_1, T_2, and T_3 are the excited triplet states.

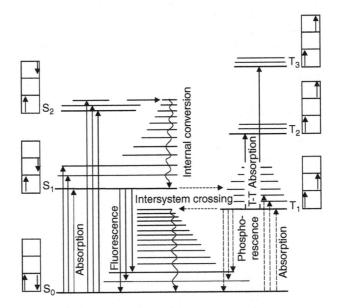

Figure 12.1   Modified Jablonski diagram showing photophysical processes that can occur following absorption of a photon of light by a drug molecule. S_o is the ground state; S_1 and S_2 are excited singlet states and T_1, T_2, and T_3 are excited triplet states.

12.1.1 Absorption

Upon absorption of this energy, the valence electron is raised to an upper excited state, conserving the spin. Because most organic compounds have singlet ground state, the corresponding upper excited state will also be a singlet. This is supposed to take place in the femtosecond time scale. If the energy absorbed is equal to the difference between the ground state and second or higher excited state, it is generally believed that the valence electron first gets excited to the corresponding upper excited state but finds its way into the first excited state via internal conversion.

Among other information, the nature of the transition can be acquired from the absorption spectrum of a drug. For example, the lowest singlet excited state is of n,π^* in nature if the molar absorption coefficient of the longest wavelength absorption is small and the band shows a blue shift (i.e., toward shorter wavelength) in going from a nonpolar to a polar solvent. Elisei et al. (1996) investigated the solvent effect on a series of methylated angelicins. They found that the longest wavelength shoulder (around 330 nm) showed a blue shift in going from cyclohexane (nonpolar solvent) to ethanol (polar solvent), whereas the band centered around 300 nm showed a red shift under those conditions. They assigned the absorption centered around 330 nm to π,π^* and that around 300 nm to n,π^* transitions, which they confirmed by molecular orbital calculations.

Many drugs show spectral shifts or changes in absorption intensity with pH. By plotting the absorption changes against pH one can obtain a sigmoid curve from which the ground state protonation constant (pKa) can be determined. A pKa of around 6 was obtained, by absorption measurements, for a number of fluoroquinolones and was assigned to the dissociation of the carboxylic group (Bilski et al., 1996; Sortino et al., 1998, 1999; Navaratnam and Claridge, 2000).

12.1.2 Fluorescence

Fluorescence is the emission of light from an excited singlet state as it deactivates into the ground state. Even though organic molecules have many electronically excited states, fluorescence emission is observed almost exclusively from the lowest vibrational level of a thermally stabilized low-lying singlet state (Kasha's rule). Azulene, in which fluorescence is known to occur from the second excited singlet level (S_2), is a known exception to this rule. The energy of the emitted photon is always less than that of the absorbed photon. Generally, fluorescence spectra are a mirror image of their absorption spectra, but are shifted to longer wavelengths due to loss of vibrational energy (known as Stoke's shift).

Fluorescence measurements give useful information about the excited singlet state of a drug. Because the emission takes place from the lowest vibrational level of the lowest excited singlet state, the emission maximum is a measure of the energy level of the excited singlet state. The effect of solvent on the fluorescence emission can be used to assign the nature of the singlet state transitions. A red shift in the emission band with increase in solvent polarity would indicate a π,π^* transition, whereas an n,π^* transition would result in a blue shift.

Table 12.1   Effect of Solvent Polarity on the Fluorescence Emission Maximum (λ_{max}/nm) of Some Angelicin Derivatives

Compound/Solvent	Cyclohexane	Benzene	Dioxane	Acetonitrile	Ethanol
3,4′-dimethylangelicin	—	—	414	420	436
4′,5′-dimethylangelicin	408	434	436	473	487
4,4′,5′-trimethylangelicin	420	422	427	445	465
6,4′,5′-trimethyl angelicin	390	440	442	470	482
4,6,4′,5′-tetramethylangelicin	395	425	428	445	462
4,6,4′-trimethylangelicin	386	395	395	418	422
4,6,4′-trimethylthioangelicin	405	412	414	422	426
4,6,5-trimethylthioangelicin	409	413	415	422	425
4,4′,5′-trimethylthioangelicin	412	415	417	432	437
4,6,4′,5′-tetramethylthioangelicin	410	420	422	434	441

Source: Adapted from Elisei, F. et al., *Photochem. Photobiol.*, 64, 67–74, 1996, and Elisei, F. et al., *Photochem. Photobiol.*, 68, 164–172, 1998.

The effect of solvent polarity on the emission properties of a series of methyl-substituted angelicins (Elisei et al., 1996) and thioangelicins (Elisei et al., 1998) was investigated. Elisei et al. found the emission band progressively shifted to longer wavelengths, for all the compounds studied, upon changing the solvent polarity from nonpolar to polar as shown in Table 12.1. They concluded that the lowest excited singlet is a π,π^* transition. The confirmation for this came from their molecular orbital calculations. The acidity constant of the excited singlet state is determined in a way similar to the determination of the ground state pKa by absorption measurements.

12.1.3   Internal Conversion

Internal conversion is the radiationless process in which a molecule in the S_1 electronic state is converted to the isoenergetic state of S_0, which then cascades to the ground state.

12.1.4   Intersystem Crossing

The transition between two excited states with similar energy but of different multiplicity is called intersystem crossing. It is a spin-forbidden process; thus the rate is slower than that of internal conversion. The rate varies from 10^{11} to 10^7 s^{-1}.

12.1.5   Phosphorescence

Phosphorescence is the emission of light from the lowest vibrational level of the lowest triplet state to the ground state. As a spin-forbidden process, its radiative lifetime is long and, normally, of the order of 10^{-4} to 10^{-2} s. Because of the long lifetime of phosphorescence, collisional deactivation of the triplet competes effectively with emission. Phosphorescence emission is observed with many drugs at 77 K. Any standard spectrofluorimeter has a phosphorescence accessory attachment. The drug under investigation is dissolved in a glass matrix. A mixture of ether–isopentane–ethanol (3:2:1 by volume) will form a transparent glass at 77 K and is widely used in phosphorescence measurements. The triplet energy level of the drug is estimated from the emission maximum. The triplet lifetime at 77 K can be also determined. Emitting triplet states could be characterized as π,π^* or n,π^* on the basis of phosphorescence lifetime measurements. The (π,π^*) triplet lifetimes are much greater than those of (n,π^*) triplet states. Photosensitizing abilities of 4-amino-quilnolones were predicted by determining their emission properties (Nord et al., 1994).

12.2 PHOTOCHEMICAL REACTIONS

The preceding deactivation processes are generally considered to be photophysical processes. In addition to these, deactivation can take place via photochemical processes such as photoionization, decarboxylation, dechlorination, and permanent product formation. All these processes can be studied under steady-state or time-resolved domains.

12.2.1 Steady-State Photolysis

The simple steady-state photolysis apparatus shown in Figure 12.2 consists of a light source such as a high- or medium-pressure mercury vapor lamp, temperature-controlled water bath fitted with optically flat quartz window, shutter, and light filter. Analysis of the reaction products may be carried out using one or more of the standard methods, such as optical absorption, emission, ESR, etc.

12.2.1.1 Optical Absorption Spectroscopy

UV-Vis absorption spectroscopy is the most widely used technique in the study of drug photochemistry. The instrument is inexpensive and easy to use. Concentrations of the drug and its product can be easily monitored by this method because they normally have distinct spectra. Not only the rate of photodegradation but also the quantum yield for the process is determined in some cases. Moreover, the excited state responsible for the reaction can be ascertained by carrying out photolysis in

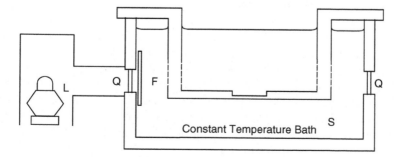

Figure 12.2 A simple steady state photolysis apparatus. L = lamp; Q = optically flat quartz window; F = filter; S = stand.

the presence of well-known scavengers. Davies et al. (1976) irradiated an oxygen-free isopropanol solution of chlorpromazine (ClP), a major tranquilizer, and found that within 15 min the absorption spectrum changed to that of promazine. However, no spectral changes were observed in an oxygenated solution even after 2 h of irradiation, from which they concluded that the dechlorination of ClP occurs from its excited triplet state.

However, in many cases the spectra of the drug and its photoproducts overlap, making it difficult for accurate kinetic analysis or quantum yield determination. Under these conditions, separation techniques such as TLC and HPLC are used to separate the components and then spectral analysis of the components is carried out (Thoma and Holzmann, 1998; Andrisano et al., 1999, 2000, 2001a, b; De Filippis et al., 2002; Lopez et al., 2003). Recently, first-derivative and dual-wavelength spectrophotometric methods have been employed to overcome the problem of overlap absorption of the parent drug and its products (Ulvi, 1998). Using this methodology, Ulvi has determined the quantum yield for the photodecomposition of three diuretics — chlorothiazide, hydrochlorothiazide, and trichloromethazide — in ethanolic solutions.

12.2.1.2 Optical Emission Spectroscopy

Many drugs and/or their products emit fluorescence when excited with light of appropriate wavelength, so the technique of fluorescence spectroscopy is used in drug photoreactivity studies. This is a highly sensitive and greatly specific method with an excellent instrumentation and thus widespread usage. Spectrofluorometry was used to study the long-term stability of norfloxacin; it was found that norfloxacin fluorescence increased by 160% when exposed to fluorescent light over a period of 15 months, while using HPLC resulted in a loss of 5% of the drug (Cordoba–Diaz et al., 1998). Photodegradation of alprazolam was observed at different pHs by measuring the fluorescence emission from the photoproducts (Nudelman and Cabrera, 2002). This shows that spectrofluorometry is a very sensitive technique in the study of the photodegradation of drugs. The application of fluorescence and phosphorescence in photoreactivity of antimalarials is discussed in more detail in Chapter 10.

12.2.1.3 Separation Techniques

When photolysis results in a product that has no distinct absorption or emission spectra or in a mixture of products, then a separation technique such as HPLC, TLC, GC, or GC/MS is used to analyze the products. HPLC is the most common tool in this context; reasonably priced, versatile commercial instruments are a standard fixture in many laboratories. Decarboxylation of benoxaprofen and its analogues were determined using HPLC with a fluorescence detector (Navaratnam et al., 1985, 1993). Photoproducts of chloroquine in isopropanol and primaquine in phosphate buffer have been isolated by TLC and identified using MS and NMR (Nord et al., 1991; Kristensen et al., 1993). This work is discussed in detail in Chapter 10. Davies et al. (1976) used TLC in combination with NMR and MS to identify photoproducts of ClP.

GC has limited application because most photoproducts are less volatile. However, volatile products could easily be detected by GC. For example, acetone was detected and quantified when an oxygen-free isopropanolic solution of ClP was irradiated (Davies et al., 1976)

12.2.1.4 Electron Spin Resonance Spectroscopy

Free radicals as well as triplets formed on photolysis can be detected and characterized by ESR. This is based on the measurement of paramagnetism due to unpaired electrons when a molecule is in a magnetic field; because triplets and free radicals have unpaired electrons, they can be monitored by ESR. An ESR spectrum is obtained by measuring the microwave absorption as the magnetic field is modified. The spectrum is usually displayed as a first derivative of the imaginary part of the magnetic susceptibility. This is suitable to observe radicals under steady state conditions, for example, by continuous photolysis. An ESR spectrum of the cation radical of chlorpromazine, produced under continuous illumination, in acidic solution showed an overall splitting of 25 gauss, which is typical of a π radical on a single aromatic ring.

Motten et al. (1983) irradiated musk ambrette in basic oxygen-free methanol and observed the ESR spectra due to two distinct nitro anion radicals. One radical was centered on the nitro group in the plane of the aromatic ring, while the other was centered on a nitro group twisted out of the plane of the ring. Moreover, they found the two radicals interconvert and maintain an equilibrium concentration ratio. They found two closely related, but nonphotosensitizing compounds, musk xylene and musk ketone, to show ESR spectra on photolysis. The ESR signal they observed at 77 K, when they irradiated an oxygenated basic methanolic solution of musk ambrette and froze it in liquid nitrogen, resembled that of superoxide anion ($O_2^{\bullet-}$). They were not able to observe the ESR signal due to superoxide anion at room temperature because of the short lifetime of $O_2^{\bullet-}$.

For such short-lived radicals, the technique of spin trapping may be employed (Janzen, 1980). This technique involves the addition of the radical ($R^{\bullet}$) to an organic diamagnetic nitrone or nitroso compound to form a longer lived nitroxide free radical. The structure of the parent free radical may then be determined from the hyperfine coupling of the ESR spectrum of the resultant spin adduct. Nitrones are very reactive and their adducts are stable, even though they do not provide much structural information. On the other hand, nitroso adducts have unique ESR spectra but they are photolytically unstable.

Chignell and co-workers have used these techniques (spin trapping and ESR) extensively in the study of free radicals formed on photolysis of drugs such as sulphanilamide and its analogues (Chignell et al., 1980, 1981; Motten and Chignell, 1983) and azathioprine and its metabolites (Moore et al., 1994). They have also detected the formation of active oxygen species sensitized by amiodarone (Li and Chignell, 1987a), adriamycin (Li and Chignell, 1987b), chlorpromazine and its metabolites, and promazine (Motten et al., 1985; Li and Chignell, 1987c; Hall et al., 1991), tetracycline (Li et al., 1987a, b), sulfoxides of ClP and promazine (Buettner et al., 1986), 2-mecaptopyridines (Reszka and Chignell, 1994a, b, 1995), and fluoroquinolones (Martinez et al., 1998).

Some of the widely used spin traps include 5,5′-dimethyl-1-oxide (DMPO), di-t-butylnitroxide (DTBN), 2-methyl-2-nitosopropane (MNP), and α-(4-pyridyl-1-oxide)-N-t-butylnitrone (POBN), and nitromethane (NM). Singlet oxygen sensitized by the drug may also be detected by ESR by photolyzing an aerated solution of the drug in the presence of a hindered amine, for example 2,2,6,6,-tetramethylpiperidine, which reacts with $^1O_2^*$ to form a stable nitroxide free radical TEMPO. The ESR signal would be reduced in the presence of singlet oxygen quenchers such as DABCO.

12.2.1.5 Singlet Oxygen ($^1O_2^*$) Measurements

One of the modes of deactivation of an excited triplet is by energy transfer to the molecular oxygen. The reaction may be written as:

$$^3D^* + O_2 \rightarrow D + {}^1O_2^* \qquad (12.1)$$

where $^3D^*$ is excited drug triplet and $^1O_2^*$ is excited singlet oxygen. This energy transfer reaction is feasible as long as the triplet energy level of the drug is above 96 kJ mol^{-1}.

Detection and quantification of the formation of excited singlet oxygen sensitized by drug molecules by the time-resolved method is discussed in Section 12.2.2.5. In steady state photolysis, in addition to the spin trapping method (Section 12.2.1.4), formation and reactivity of singlet oxygen can be studied by measuring the rate of consumption of oxygen during photolysis in the presence and absence of quenchers (Davies et al., 1976; Moore, 1977). The rate of oxygen uptake is normally measured using an oxygen electrode.

In organic solvents, quenchers such as 2,5-dimethyl furan (DMF), 1,4-diazabicyclo[2.2.2]octane (DABCO), β-carotene, and diphenylisobenzofuran (DPBF) are used. In aqueous solutions, histidine, DABCO, and sodium azide have been the choice of many groups. Chlorpromazine was shown to sensitize the formation of excited singlet oxygen in isopropanol; it was shown that no oxygen uptake was observed when ClP was irradiated in the absence of DABCO, while the rate of uptake increased linearly with DABCO concentration (Davies et al., 1976). Quantum yield for the formation of $^1O_2^*$ sensitized by benoxaprofen was determined to be 0.18 using histidine as the quencher (Navaratnam et al., 1984, 1985). In addition to measuring the rate of oxygen uptake, product separation and quantification are also employed; for example, cholesterol chemically quenches singlet oxygen to yield 5α-hydroperoxide. The oxygen consumption method can be effectively used in determining the quantum yield for the formation of $^1O_2^*$ provided (1) neither substrate nor $^1O_2^*$ reacts with the excited sensitizer, (2) products do not react with molecular oxygen, and (3) the sensitizer does not undergo peroxygenation.

Kraljic and El Mohsni (1978) developed a method based on bleaching of N,N-dimethyl-4-nitrosoaniline (RNO) by $^1O_2^*$ in the presence of imidazole. $^1O_2^*$ generated during photolysis reacts with imidazole to form a transannular peroxide that bleaches RNO; the bleaching can be monitored spectrophotometrically at 440 nm. Hall and

Table 12.2   Quantum Yields for Singlet Oxygen Formation, Φ_Δ, of Antimalarial Drugs in D_2O/Phosphate at pD = 7[a]

Antimalarial Drug	Φ_Δ
Amodiaquine dichloride	0.011
Chloroquine diphosphate	<0.0005
Hydroxychloroquine sulphate	<0.0005
Mefloquine hydrochloride	0.38
Primaquine diphosphate	<0.0005
Quinine sulphate	0.36
Quinacrine dichloride	0.013

[a] Measured by steady-state phosphorescence emission.

Source: Motten, A.G. et al., Photochem. Photobiol., 69, 282–287, 1999.

Chignell (1987) developed a near-infrared detective system, based on the phosphorescence emission of the excited singlet oxygen, under steady state conditions. Using this technique, Motten et al. (1999) have determined the quantum yields of singlet oxygen formation for a series of antimalarial drugs in D_2O as shown in Table 12.2.

12.2.2   Time-Resolved Studies

Upon absorption of light, any drug molecule gets excited to a higher energy state from which a number of photochemical and photophysical reactions may occur. The products of these reactions may be permanent, in which case they can be analyzed and quantified by the methods discussed in Section 12.2.1. However, these permanent products can arrive via short-lived free radicals. In order to have a comprehensive understanding of the photoreactions, it is necessary to study the characteristics and the reactivities of these short-lived free radicals and the excited states. Such studies can be conveniently carried out using flash photolysis or pulse radiolysis.

12.2.2.1   Laser Flash Photolysis

Norrish and Porter developed the technique of flash photolysis in 1949 (Norrish and Porter, 1949; Porter, 1950). This technique, in which a flash lamp was used as an excitation source, had limitations in intensity and duration of the flash. Short pulses and high intensities of flash lamps are mutually exclusive. However, with the advent of pulsed lasers, flash photolysis equipment operating even in the femtosecond time scale is available. Table 12.3 lists some commonly used pulsed lasers.

   The main workhorse in drug photochemistry studies is the nanosecond laser flash photolysis equipment. Many laboratories, including the author's, have used nanosecond laser flash photolysis to study and characterize triplet states and free radicals formed on photoexcitation of drugs. The system used in this laboratory (see Figure 12.3 for a block diagram of the set-up) (Navaratnam et al., 1985; Navaratnam and Phillips, 1991) essentially consists of a Q-switched Nd:YAG laser (JK Lasers 2000

Table 12.3   Representative Sample of Types of Pulsed Lasers Used in Laser Flash Photolysis

Laser	Pulsewidth (ns)	Wavelength (nm)
Q-switched Nd:YAG	8–15	1064, 532, 355, 266
Mode-locked Nd:YAG	0.025	1064, 532, 355, 266
Q-switched ruby	10–30	694, 347
Mode locked ruby	0.025	694, 347
Super-radiant N_2-gas	2–10	337
ArF excimer	8–10	193
KrF excimer	12–16	249
XeCl excimer	8–10	308
XeF excimer	8–12	350
Flash-lamp pumped dye	200	Tunable
Laser-pumped dye	Depends on pump laser	Tunable

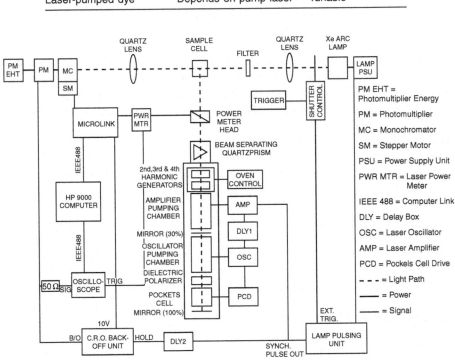

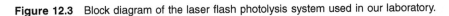

Figure 12.3   Block diagram of the laser flash photolysis system used in our laboratory.

series) capable of delivering up to 1 J of energy at 1064 nm in pulses of 12-ns duration as the excitation source.

The fundamental beam can be frequency doubled, tripled, and quadrupled to deliver light of 532, 355, and 266 nm, respectively. The harmonics are separated using a prism and/or beam splitter. The sample under investigation is contained in a quartz cell and changed after each laser pulse, using a remote-control flow system. The analyzing beam generated by a Xenon lamp, which can be pulsed to increase the analyzing light up to 200 times in the UV region, passes through the sample at right angles to the laser beam. A shutter that opens a few microseconds before the laser pulse and closes after the event under examination is completed, and appropriate filters are placed between the analyzing lamp and the sample in order to minimize the photolysis of the sample by the monitoring light. The monitoring light is then dispersed in wavelength by a monochromator and then on to a photodetector. The transmittance of light at this wavelength is detected by this photodetector before, during and after the laser pulse. The detector is also connected to an automatic back-off box that enables changes in transmittance to be observed by feeding back a signal equal and opposite to the detector anode current prior to the laser pulse, thus maintaining the anode current close to zero.

The signal from the detector is captured and stored using a programmable digital oscilloscope. Back-off and energy meter readings are digitized using an analogue to digital converter (ADC) attached to a DI-AN data acquisition system. Data acquisition and processing are carried out using a PC. Time-resolved spectra are obtained by repeating the procedure at successive wavelengths.

12.2.2.1.1 Triplet–Triplet Absorption Spectrum

In the author's laboratory, triplet–triplet absorption spectra have been determined for a number of drugs, such as:

- Chlorpromazine (Davies et al., 1976; Navaratnam et al., 1978)
- Suphacetamide (Land et al., 1982)
- Furocoumarins (Beaumont et al., 1979, 1980, 1983)
- Amiloride (Hamoudi et al., 1984)
- Benoxaprofen and derivatives (Navaratnam et al., 1984, 1985, 1993)
- Propranolol (Navaratnam et al., 1987)
- Fenbufen (Navaratnam and Jones, 2000)
- Ofloxacine (Navaratnam and Claridge, 2000)
- Mefloquine (Navaratnam et al., 2000)

The spectra were obtained in deoxygenated solutions. As a first step, the assignment of triplet was based on the reactivity with oxygen where the decay of the absorption at its maximum wavelength was monitored in the presence and absence of oxygen. In all cases, the rate of oxygen quenching was found to be diffusion controlled. The assignment was confirmed by energy transfer experiments. For example, Navaratnam and Jones (2000) laser flash photolyzed (λ_{ex} = 266 nm) a nitrogen-saturated solution of fenbufen in acetonitrile and obtained a transient absorption spectrum with a maximum at 420 nm. This transient was assigned to the triplet state on the basis that it was quenched by oxygen. To confirm this assignment, a sensitization experiment was carried out using benzophenone as the sensitizer. A nitrogen-saturated, acetonitrile solution of benzophenone (2×10^{-3} mol^{-1} dm^{-3}) containing 4×10^{-5} mol^{-1} dm^{-3} fenbufen was subjected to a 355-nm laser pulse. The initial spectrum showed maxima at 520 and 310 nm and was in agreement with the triplet–triplet absorption spectrum of benzophenone in acetonitrile.

The transient spectra at subsequent times showed a decrease in absorption of these bands with a concomitant increase of a band centered at 420 nm with isosbestic points at 360 and 460 nm. The spectrum after 9 μs, when benzophenone triplet has

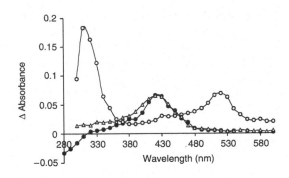

Figure 12.4   Transient absorption spectra obtained 120 ns (–●–) after subjecting a nitrogen-saturated acetonitrile solution of 2×10^{-5} M fenbufen to 5 mJ of 266 nm laser pulse and 80 ns (–○–) and 9 μs (–△–) after subjecting a nitrogen-saturated acetonitrile solution of 2×10^{-3} M benzophenone and 4×10^{-5} M fenbufen to 7 mJ of 355 nm laser pulse.

completely decayed, showed a maximum at 420 nm and was identical to the spectrum obtained when fenbufen solution was subjected to a 266-nm laser pulse (Figure 12.4). Furthermore, the decay at 520 nm matched the build-up at 420 nm.

From these activities, the triplet–triplet absorption spectrum of fenbufen in acetonitrile was confirmed and the molar absorption coefficient of the absorption at 420 nm was determined, as discussed in the next section. Xanthone was used as the sensitizer to confirm the triplet–triplet absorption spectrum of benoxaprofen (Navaratnam et al., 1984, 1985) and mefloquinone in water (Navaratnam et al., 2000); benzophenone as the sensitizer for benoxaprofen and derivatives in acetonotrile (Navaratnam et al., 1993); and acetonaphthone as a quencher for ofloxacine in water (Navaratnam and Claridge, 2000). Selection of a sensitizer depends on a number of factors; for example, the sensitizer must be such that

- Its triplet energy is higher than that of the acceptor.
- It must be excitable at a wavelength different from the acceptor.
- The triplet–triplet spectrum has an absorption band distinct from that of the acceptor.
- The triplet state has a reasonable (at least a microsecond) lifetime.

Table 12.4 lists some common sensitizers/quenchers.

12.2.2.1.2   Determination of Molar Absorption Coefficient

The energy transfer experiments described previously can be utilized to determine the molar absorption coefficient. Under ideal conditions, a given concentration of donor triplet would produce an equal amount of acceptor triplet, i.e.,

$$[^3D^*] = [^3A^*] \tag{12.2}$$

$$OD_D/\varepsilon_D = OD_A/\varepsilon_A \tag{12.3}$$

Table 12.4 Common Sensitizers and Quenchers Used in Energy Transfer Studies

Sensitizer/Quencher	Solvent	E_t kJ/mol	Φ_t	λ_{max}	ε_t	$\tau_t/\mu s$
Acenaphthene	Nonpolar	250	0.46	422		
	Polar	248	0.58	430	6,000	3300
Benzophenone	Benzene	287	1	530	7,220	6.9
	Polar	289	1	525	6,250	50
Biphenyl	Benzene	65.4	0.84	367	27,100	130
	Polar	65.5		361	42,800	
Beta-carotene	Benzene	88	0.001	520		70
	Polar	85		515	187,000	9
Naphthalene	Nonpolar	253	0.75	425	13,200	175
	Polar	255	0.8	415	24,500	1800
Xanthone	Nonpolar	310		610	5,300	0.02
	Polar	310		605	6,480	17.9

Source: Adapted from Murov, S.L. et al., *Handbook of Photochemistry*, New York, Marcel Dekker, Inc., 1993.

Thus,

$$\varepsilon_A = OD_A \times \varepsilon_D / OD_D \tag{12.4}$$

where OD_D and OD_A are the maximum absorbance of the donor triplet in the absence of the acceptor and the maximum absorbance of the acceptor triplet in the presence of the acceptor and the donor, respectively, and ε_D and ε_A are the molar absorption coefficients of donor and acceptor triplet at their respective absorption maxima.

Using xanthone as the sensitizer, Navaratnam et al. (1985) calculated the molar absorption coefficient of benoxaprofen triplet–triplet absorption in aqueous solution at 410 nm as 17,500 dm³ mol⁻¹ cm⁻¹. In this case this simple treatment was possible because

- The sensitizer (xanthone in water) triplet is reasonably long lived (20 μs) and its intrinsic decay is negligible in the presence of benoxaprofen (acceptor).
- Benoxaprofen is readily soluble in water with no absorption at the excitation wavelength of the sensitizer (355 nm), which means that it was possible to have a high enough concentration of acceptor (benoxaprofen) to quench all donor triplets at the same time as only the donor is excited.
- Benoxaprofen triplet decay is practically zero during the time that energy transfer takes place and can be neglected.

Other cases may not be this simple and thus a number of corrections must be made. For example, if the decay of the donor, by any means other than that by energy transfer, is not negligible, then the real absorbance, OD_A (real), is given by

$$OD_A \text{ (real)} = OD_A \text{ (maximum observed)} \times \left(1 + k_1/k_2[A]\right) \tag{12.5}$$

where OD_A (maximum observed) is the maximum absorption measured, k_1 is the rate of decay in the absence of acceptor, and k_2 is the rate in the presence of the acceptor with a concentration of [A].

If the rate of decay of the acceptor triplet (k_3) is not negligible compared to $(k_1 + k_2[A])$, then a further correction must be applied. Capellos and Bielski (1972) showed that the corrected absorbance, OD_A (real), is then given by

$$OD_A \text{ (real)} =$$
$$OD_A \text{ (maximum observed) } \exp\left(k_3 \ln\left(k_1 + k_2[A]\right)/\left(k_3 - k_2[A] - k_1\right)\right) \quad (12.6)$$

Further corrections would be needed if, for example, direct excitation of the acceptor occurs (Amouyal et al., 1974).

This method could also be used with quenchers instead of sensitizers. In this case, the triplet state of the drug molecule sensitizes the formation of the quencher triplet whose molar absorption coefficient is established. Beta-carotene and its water-soluble analogue, crocetin, have been widely used for this purpose (Craw and Lambert, 1983). Navaratnam and Claridge (2000) have used 2-acetonaphthone as the quencher to determine the molar absorption coefficient of ofloxacine triplet.

In addition to the energy transfer method, singlet state depletion and complete conversion methods have been used to determine the triplet–triplet molar absorption coefficient. However, these methods assume that the loss in ground state is equivalent to the gain in triplet state; thus, its application is restricted to drugs that do not take part in any other primary photoreactions from the excited singlet state, e.g., photo-ionization, photodecarboxylation, etc. Furthermore, the singlet depletion method involves measuring the depletion of the ground state at a wavelength where, it is assumed, the triplet does not absorb; however, this is questionable and so this method is subject to uncertainties. However, Ferguson (1998) obtained a value of 38,000 dm^3 mol^{-1} cm^{-1} at 500 nm for tetra-polyethoxy-phthalocyanine in toluene by this method, which is in agreement with that of 37,000 in benzene obtained using the energy transfer method. Using a modified singlet state depletion method, Craw et al. (1983) have determined a value of 16,500 dm^3 mol^{-1} cm^{-1} at 370 nm for 8-MOP in water.

Bonnet et al. (1983) and Reddi et al. (1983) have used the complete conversion method to determine values for the molar absorption coefficients of hematoporphy-rins. The method is suitable for drugs with a very high ground state molar absorption coefficient.

12.2.2.1.3 Determination of Quantum Yields

This is a comparative method introduced by Richards and Thomas (1970). The method involves comparing the triplet formed when two solutions with low and identical absorbance at the excitation wavelength are subjected to two identical laser pulses successively. Because the ground state absorption is low and identical at the excitation wavelength, equal numbers of photons are absorbed by the two solutions, provided the laser energy remains unchanged. Rather than relying on the reproduc-ibility of laser energy from pulse to pulse, the experiment is carried out over a range of laser energies (keeping them as low as possible) and a graph is plotted for absorbance at λ_{max} against laser intensity. These plots would be linear if the energy absorbed is low enough to convert only a small percentage of the ground state into the excited state. The slopes give the triplet absorbance per energy of the standard (OD_s) and the unknown (OD_x). The quantum yield (Φ_x) of the drug under investigation is given by

$$\Phi_x = \Phi_s \times OD_x \times \varepsilon_s / (OD_s \times \varepsilon_x) \qquad (12.7)$$

where (Φ_s) is the quantum yield of the standard; ε_s and ε_x are the molar absorption coefficients of the standard and the drug at the respective wavelengths of measurement. As discussed in the previous section, ε_x has already been determined and ε_s is taken from the literature. A representative sample of drugs whose triplet state properties have been determined using this technique is given in Table 12.5.

12.2.2.1.4 Characterization of Radicals

Photoionization is one of the modes of deactivation of an excited molecule. A number of drugs are shown to photoionize when excited with the appropriate light, e.g.,

- Chlorpromazine (Navaratnam et al., 1978; Buettner et al., 1989)
- Amiloride (Hamoudi et al., 1984)
- Sulfacetamide (Land et al., 1982)
- Ephedrine (Navaratnam et al., 1983)
- Propranolol (Navaratnam et al., 1987)
- 4-Hydroxybenzothiazole (Chedekel et al., 1980)
- Hydrochlorothiazide (Tamat and Moore, 1983)
- Naproxen (Moore and Chappuis, 1988)
- Tolmetin (Sortino and Sciano, 1999)
- Mefloquine (Navaratnam et al., 2000)

The mechanism of photoionization may be monophotonic or biphotonic. Monophotonic reactions are generally believed to occur from the excited singlet state or an upper vibrational level of the singlet state; however, in the biphotonic process either of the excited states (i.e., singlet or triplet) can absorb the second photon. Whatever the mechanism may be, the products are the solvated electron and the corresponding cation radical. In aqueous solutions, the hydrated electron has a strong absorption band centered at 720 nm (Fielden and Hart, 1967). Because of its high molar absorption coefficient, it is very easy to observe this absorption even if the yield is very small. To confirm the absorption is due to a hydrated electron, scavenging experiments are performed. For example, in the presence of nitrous oxide, the absorption due to the hydrated electron should be removed within a few nanoseconds in accordance with the reaction:

$$e_{aq}^- + H_2O + N_2O \rightarrow N_2 + {}^\bullet OH + {}^-OH \qquad (12.8)$$

Transient absorption spectra were obtained at various times after subjecting an argon-saturated aqueous solution of ofloxacine to a 355-nm laser pulse (Navaratnam

Table 12.5   Triplet State Properties of Some Drugs Studied Using Laser Flash Photolysis Technique

Drug	Solvent	λ_{max} (nm)	e_t	Φ_t	Reference
Benoxaprofen (Bp)	Water	410	17,500	0.19	Navaratnam et al., 1985
Bp	Acetonitrile	410	26,000	0.38	Navaratnam et al., 1993
Dichloro-Bp	Acetonitrile	420	175,000	0.36	Navaratnam et al., 1993
Difluoro-Bp	Acetonitrile	400	18,000	0.36	Navaratnam et al., 1993
Chloro-Bp	Acetonitrile	410	17,500	0.36	Navaratnam et al., 1993
Carprofen	Ethanol	490	14,200	0.37	Bosca et al., 1997
Ketoprofen	Water	526	9,600	1.0	Bosca et al., 1997
Tiaprofenic acid	Water	600	2,300	0.9	Encinas et al., 1998
Suprofen	Water	360, 600		1.0	Bosca et al., 2001
Naproxen	Acetonitrile	440	9,600	0.28	Martinez and Scaiano, 1998
Chlorpromazine	Water	460	21,900	0.08	Navaratnam et al., 1978
Flumaquine	Water	575	14,000	0.9	Bazin et al., 2000
Propranolol	Water	430	5,200	0.82	Navaratnam et al., 1987
Amiloride	Water	400	6,300	0.023	Hamoudi et al., 1984
Fenbufen	Acetonitrile	420	30,000	1.0	Navaratnam and Jones, 2000
Ofloxacin	Water	610	11,000	0.33	Navaratnam and Claridge, 2000
Rufloxacin	Water	640	3,900	0.7	Sortino et al., 1999
Mefloquine	Water	430	3,600	1.0	Navaratnam et al., 2000
Psoralen	Benzene	450	8,100	0.034	Bensasson et al., 1978
Psoralen	Methanol	440	30,100	0.06	Sa e Melo et al., 1979
Angelicin	Benzene	450	4,700	0.009	Bensasson et al., 1978
5-Methoxypsoralen	Benzene	450	10,200	0.067	Bensasson et al., 1978
8-Methoxypsoralen	Benzene	480	10,000	0.011	Bensasson et al., 1978
8-Methoxypsoralen	Ethanol	370	24,000	0.03	Sa e Melo et al., 1979
Trimethylpsoralen	Methanol	470	33,000	0.093	Beaumont et al., 1979
4'-Aminomethyl-4,5',8'-trimethylpsoralen	Methanol	460	24,200	0.2	Salet et al., 1980
3'-Carboxypsoralen	Benzene	410	11,500	0.3	Ronhard–Haret et al., 1987
3'-Carboxypsoralen	Water	580	8,300	≥0.33	Ronhard–Haret et al., 1987
7H-Thiopsoralen	Benzene	510	13,100	0.37	Aloisi et al., 2000

Table 12.5 (continued)   Triplet State Properties of Some Drugs Studied Using Laser Flash Photolysis Technique

Drug	Solvent	λ_{max} (nm)	e_t	Φ_t	Reference
7H-Thiopyranpsoralen	Trifluoroethanol	300, 560, 680	7,500	0.6	Aloisi et al., 2000
2H-Selenolpsoralen	Benzene	420, 560	16,600	0.42	Aloisi et al., 2000
2H-Selenolpsoralen	Ethanol			0.6	Aloisi et al., 2000
2H-Selenolpsoralen	Trifluoroethanol			1.0	Aloisi et al., 2000
2H-1-Benzothiopyran	Trifluoroethanol	380, 690	5,200	0.97	Aloisi et al., 2000
7H-Selenin-benzopsoralen	Trifluoroethanol	380, 550, 640	5,200	1	Aloisi et al., 2000
7H-Selenolpsoralen	Trifluoroethanol	300, 560, 640	4,000	0.09	Aloisi et al., 2000
Tetra-polyethoxy-phthalocyanine	Toluene	500	37,500	0.1	Ferguson, 1998
Tetra-polyethoxy-zinc-phthalocyanine	Acetonitrile	490	22,000	0.4	Ferguson, 1998
Nabumetone	Acetonitrile	440	11,700	0.29	Martinez and Scaiano, 1998
4,6,4'-Trimethyl angelicin	Ethanol	450	4,700	0.5	Elsei et al., 1998
4,6,4'-Trimethyl angelicin	Ethanol	500	7,200	0.87	Elsei et al., 1998
4,6,5-Trimethyl angelicin	Ethanol	500	8,700	0.89	Elsei et al., 1998
4,4',5'-Trimethyl angelicin	Ethanol	500	7,000	0.83	Elsei et al., 1998
4,6,4',5'-Tetramethyl angelicin	Ethanol	510	5,800	0.86	Elsei et al., 1998

and Claridge, 2000). It was found that the absorption in the long wavelength region decayed by a biphasic process — one with a lifetime of 700 ns and the other greater than 50 µs. It was concluded that the short-lived species is the hydrated electron by its reaction with nitrous oxide. Other fluoroquinolones, such as rufloxacin, enoxacin, norfloxacin, and lomefloxacin, are also shown to photoionize (Monti and Sortino, 2002).

In very rare cases in which photoionization is the only photoprocess, the absorption observed in a nitrous oxide-saturated solution will be that of the cation radical. However, generally, other processes such as intersystem crossing also occur in parallel. In the presence of oxygen, the hydrated electron and the triplet will be scavenged at diffusion-controlled rates, and the absorption observed in the oxygenated solution will be due to the cation radical. Under these conditions, the cation radical spectrum is easily determined. The molar absorption coefficient of the cation radical can also be calculated using the hydrated electron as an internal standard. The molar absorption coefficient for sulphacetamide cation radical was determined in this manner (Land et al., 1982) and later confirmed by pulse radiolysis (see Section 12.2.2.6).

12.2.2.1.5   Monophotonic or Biphotonic

To determine whether the photoionization is monophotonic, the yield of electron is measured as a function of laser intensity. If the yield varies linearly as a function of laser intensity, it is monophotonic; in biphotonic processes the yield increases linearly with the square of laser intensity. Chloropromazine (Navaratnam et al., 1978), azaproppazone (Jones et al., 1985, 1986), and ofloxacine (Navaratnam and Claridge, 2000) have been shown to photoionize by a monophotonic process. A biphotonic process was observed with ephedrine (Navaratnam et al., 1983), propranolol (Navaratnam et al., 1987), sulphacetamide (Land et al., 1982), mefloquine (Navaratnam et al., 2000), enoxacin, norfloxacin, and lomefloxacin (Monti and Sortino, 2002).

12.2.2.2 Time-Resolved Raman Resonance Spectroscopy (TR³)

The optical absorption method discussed in the previous section is the analytical method most widely used in conjunction with laser flash photolysis and pulse radiolysis. However, the transient absorption spectra are usually broad and structureless. With development of the technique of TR³ spectroscopy, it is now possible to determine the structure of the excited molecules and free radicals by measuring their vibrational spectra. The method is based on the inelastic scattering (Raman scattering) of light by molecular vibrations. For Raman and resonance Raman scattering, the wavenumber of a vibrational band is characteristic of a particular group of atoms vibrating. The vibrational spectrum provides detailed information on the structure and bonding of the molecules. Resonance Raman also has increased sensitivity and selectivity.

Sarata et al. (2000) used nanosecond time-resolved resonance Raman to investigate the transients formed on photolyzing aqueous solutions of chlorpromazine and related phenothiazine derivatives. They obtained four transient species absorbing at 380, 480, 525, and 580 nm, respectively. They identified two of them as triplet (absorption peak at 480 nm) and cation radical (λ_{max} at 525 nm). Furthermore, they concluded that the triplet is an n,π* state on the basis of resonance Raman results. Bisby and Parker (2001) obtained the TR³ spectrum of 4-hydroxycinnamate radical by one-electron oxidation of 4-hydroxycinnamate using duroquinone triplet. The spectrum showed prominent bands ascribable to C–O and ring C–C stretching vibrations. They interpreted the spectrum as indicating strong delocalization of the radical site to the double bond in conjunction with the aromatic ring in 4-hydroxycinnamate. They concluded that this contributes to the low reduction potential of the radical and the antioxidant properties of 4-hydoxycinnamate.

12.2.2.3 Time-Resolved Photoacoustic Method

Photoacoustic spectroscopy is based on measurement of the heat generated by the radiationless processes for the deactivation of excited molecules (Rosencwaig, 1980). The population of the excited species and the heat emission will be modulated with the same frequency as the exciting light source. At the appropriate frequencies, the excited species will emit their excess heat phase-shifted with respect to the heat emitted by the fast relaxation processes. A gas-coupled microphone and a phase-sensitive detector in combination with an intensity-modulated light source can be used for the measurements. The method is limited to species with lifetimes longer than 1 ms.

Patel and Tam (1981) suggested that a better time resolution would be possible if the heat dissipated after absorption of a laser pulse is measured. Braslavsky et al. (1983) developed laser-induced optoacoustic spectroscopy (LIOAS) to study the radiationless processes of biliverdin dimethyl ester in ethanol. Here, the heat generated by the radiationless processes within the volume excited by a laser pulse generates a pressure wave that, in turn, initiates an acoustics wave traveling to a transducer sensitive to longitudinal displacement waves. The mechanism for the photodecarboxylation of ketoprofen was determined by Borsarelli et al. (2000), using LIOAS. A subset of this method is time-resolved thermal lensing (TRTL). This method is applied to measure singlet oxygen (Section 12.2.2.5).

12.2.2.4 Time-Resolved Conductivity

This is a very useful method for investigating fast chemical processes involving the creation and destruction of charged species (Asmus, 1973; Asmus and Janata, 1982). The method is based on the principle that, when a voltage is applied to a solution (via a pair of electrodes), the change in conductance induced by irradiation can be monitored by measuring the change in current flow between the electrodes. The current flow can be monitored as a voltage in a load resistor. Thus, change in conductance, which is proportional to the change in ion concentration, is measured as a change in voltage ($\Delta V_L(t)$) in a load resistor (R_L) and is given by

$$\Delta V_L(t) = (V_B \times R_L)/(k_z \times 10^3) \sum \Delta c_i z_i l_i \qquad (12.9)$$

where V_B is the voltage applied, Δc_i the change in concentration of ith ion, z_i its charge, l_i specific conductance of the species involved, and k_z the cell constant. The method is most useful in the investigation of acid base equilibria. The time resolution is limited to about 10 ns for aqueous systems but subnanosecond time resolution is feasible for nonpolar solvents. The sensitivity is high and concentrations of less than 10^{-7} M can be detected by this method.

12.2.2.5 Time-Resolved Measurement of Excited Singlet Oxygen ($^1O_2^*$)

Among other methods (see Section 12.2.1.5), excited singlet oxygen can be detected by its phosphorescence emission in the infrared region. Khan (1984, 1985) showed the λ_{max} of the emission to be at 1268 nm in the gas phase; Bromberg and Foote (1989) found that the spectrum showed a small red shift to 1279 in benzene. The transition ($^1\Delta_g \rightarrow {}^3\Sigma_g^-$) is strongly forbidden, so the radiative lifetime is very long (up to 1 h in the gas phase at low pressure) in the absence of any collisional deactivation. However, in solution, nonradiative deactivation predominates and the decay rate constant (k_{nr}) is much higher than the radiative rate constant (k_r). As a consequence, the quantum yield in solution is very low. The quantum yield and the overall rate constant for deactivation of excited singlet oxygen are solvent dependent (Rodgers, 1983).

In spite of low quantum efficiency and short lifetimes, it is now routine to use fast infrared photodiodes to measure singlet oxygen in time-resolved and steady-state modes. The first time-resolved detection system for singlet oxygen measurement was developed in Rodgers' laboratory (Rodgers and Snowdon, 1982). The system in the author's laboratory is essentially similar to it and consists of a liquid nitrogen cooled North Coast EO-817P Germanium photodiode/amplifier, closely coupled to the laser photolysis cell in right angle geometry. A 1-mm-thick, 20-mm-diameter piece of AR-coated silicon (II-IV Inc) is placed between the diode and the cell to act as a cut-off filter for light below 1100 nm.

The quantum yield of excited singlet oxygen (Φ_Δ) can be determined by the comparison method discussed for the triplet state. Compounds such as phenalenone (Schmidt et al., 1984) and benzophenone (Gorman et al., 1987) are generally used as standards for this method. Various research groups have used this method to determine the quantum yield for a number of drugs. A representative sample is given in Table 12.6.

As mentioned in Section 12.2.2.3, TRTL may be used to detect and quantify singlet oxygen (Fuke et al., 1983; Rossbroich et al., 1984). As seen earlier, one mode of deactivation of an excited molecule is by radiationless decay, in which the excess energy is lost as heat. The heat evolved causes a temperature gradient along the laser beam. This leads to changes in density and refractive index, which causes the system to act like a diverging lens. The time-resolved thermal lensing due to release of energy by decaying excited states can be used to measure the lifetimes of $^1O_2^*$ in the range of 0.1 to 100 μs. Usually, a pulsed laser is used as pump and the probe beam is provided by a continuous wave (cw) laser. The probe beam is dispersed by the thermal lens, reducing the light reaching the photodiode through a pinhole. The photodiode gives the relative magnitudes of heat contribution for fast and slow nonradiative processes relative to the acoustic transit time. The signal is resolved into amplitudes due to fast (U1) and slow (ΔU) heat evolution, and Φ_Δ can be evaluated using

$$\Delta U / U_{tot} = \Phi_\Delta E_\Delta / N_A h \left(\upsilon_1 - \upsilon_f \Phi_f \right) \qquad (12.10)$$

where N_A is Avogadro's number; h Planck's constant; υ_1 the pump frequency; Φ_f and υ_f the quantum yield and integrated average frequency of sensitizer fluorescence, respectively; $U_{tot} = \Delta U + U1$; Φ_Δ is the quantum yield of singlet oxygen; and E_Δ is the molar energy content of $^1O_2^*$.

Redmond and Braslavsky (1988) have measured Φ_Δ for a number of sensitizers in benzene by TRTL and TRPD and found a very good agreement between them.

12.2.2.6 Pulse Radiolysis

This technique is very similar to flash photolysis in every aspect except for the initiating step in the generation of excited species. In flash photolysis, the light is

Table 12.6 Singlet Oxygen Quantum Yields, Φ_Δ, for Some Drugs, Measured by Time-Resolved Phosphorescence Emission

Drug	Φ_Δ	Solvent
Benoxaprofen (Bp)	0.19	D_2O
Bp	0.38	Acetonitrile
Dichloro-Bp	0.44	Acetonitrile
Difluoro-Bp	0.4	Acetonitrile
Chloro-Bp	0.44	Acetonitrile
Dichloro-Bp	0.16	D_2O
Difluoro-Bp	0.15	D_2O
Chloro-Bp	0.22	D_2O
Nabumetone	0.19	Acetonitrile
Chlorpromazine	0.21	Acetonitrile
Promazine	0.23	Acetonitrile
Pericyanine	0.2	Acetonitrile
Thioproperazine	0.23	Acetonitrile
Methotrimeperazine	0.19	Acetonitrile
Prochlorperazine	0.25	Acetonitrile
Propiomazine	0.12	Acetonitrile
Fenbufen	1.0	Acetonitrile
Mefloquine	1.0	D_2O
Ofloxacin	0.13	D_2O
Enoxacin	0.2	D_2O
Norfloxacin	0.16	D_2O
Lomofloxacin	0.08	D_2O
7H-Thiopyranpsoralen	0.33	Benzene
7H-Thiopyranpsoralen	0.58	Trifluoroethanol
2H-Selenolpsoralen	0.33	Benzene
2H-Selenolpsoralen	0.46	Ethanol
2H-Selenolpsoralen	0.96	Trifluoroethanol
2H-1-Benzothiopyran	0.96	Trifluoroethanol
7H-Selenin-benzopsoralen	0.77	Trifluoroethanol
7H-Selenolpsoralen	0.74	Trifluoroethanol
Tetra-polyethoxy-phthalocyanine	0.1	Toluene
Tetra-polyethoxy-zinc-phthalocyanine	0.43	Acetonitrile
3,4′-Dimethylangelicin	0.2	Ethanol
4,5′-Dimethylthioangelicin	0.14	Ethanol
4,4′,5′-Trimethylangelicin	0.15	Ethanol
6,4′,5′-Trimethylangelicin	0.29	Ethanol
4,6,4′,5′-Tetraimethylangelicin	0.23	Ethanol
3,4′-Dimethylangelicin	0.1	Acetonitrile
4,5′-Dimethylthioangelicin	0.16	Acetonitrile
4,4′,5′-Trimethylangelicin	0.30	Acetonitrile
6,4′,5′-Trimethylangelicin	0.23	Acetonitrile
4,6,4′,5′-Tetraimethylangelicin	0.27	Acetonitrile
4,6,4′-Trimethylangelicin	0.27	Ethanol
4,6,4′-Trimethylthioangelicin	0.42	Ethanol
4,6,5-Trimethylthioangelicin	0.34	Ethanol
4,4′,5′-Trimethylthioangelicin	0.45	Ethanol
4,6,4′,5′-Tetramethylthioangelicin	0.42	Ethanol
4,6,4′-Trimethylangelicin	0.07	Acetonitrile
4,6,4′-Trimethylthioangelicin	0.26	Acetonitrile
4,6,5-Trimethylthioangelicin	0.29	Acetonitrile

Table 12.6 (continued)   Singlet Oxygen Quantum Yields, Φ_Δ, for Some Drugs, Measured by Time-Resolved Phosphorescence Emission

Drug	Φ_Δ	Solvent
4,4',5'-Trimethylthioangelicin	0.37	Acetonitrile
4,6,4',5'-Tetramethylthioangelicin	0.31	Acetonitrile

absorbed by the compound under investigation, whereas in pulse radiolysis ionizing radiation is absorbed by the solvent to generate transient species (Keene, 1960; Matheson and Dorfman, 1960; McCarthy and MacLachlen, 1960). When a dilute solution is subjected to a short pulse of high energy electrons, the solvent absorbs almost all the radiation and is ionized to give e^- and positive ion. What happens subsequently depends on the polarity of the solvent.

In polar solvents such as water, the electron is solvated within picoseconds and the main reactive species (with their G value, i.e., number of species produced per 100 eV absorbed; shown in parentheses) are e^-_{aq} (2.7), $^\bullet OH$ (2.7), and $H^\bullet$ (0.55). The drug under investigation reacts with one or more of these species to produce oxidized and/or reduced species. It is possible to scavenge the oxidizing species ($^\bullet OH$) or the reducing species (e^-_{aq}, $H^\bullet$) before it reacts with the compound. For example, Navaratnam et al. (2000) pulse radiolyzed a nitrous oxide-saturated aqueous solution of mefloquine (MQ) to produce the cation radical of MQ. Under these conditions, hydrated electrons were rapidly scavenged by nitrous oxide according to Equation 12.11 to leave only hydroxyl radicals as the main reactive radical:

$$e^-_{aq} + H_2O + N_2O \rightarrow N_2 + {}^\bullet OH + {}^-OH \tag{12.11}$$

The hydroxyl radical then reacted with MQ to give the cation radical of MQ according to

$$^\bullet OH + MQ \rightarrow MQ^{\bullet+} + {}^-OH \tag{12.12}$$

However, hydroxyl radicals are very reactive and known to react with aromatic compounds not only by electron abstraction but also by adding to the ring. Well-established, one-electron oxidizing radicals such as $Br_2^{\bullet-}$ and $N_3^\bullet$, formed by pulsing a nitrous oxide saturated solution of potassium bromide or sodium azide, are used to produce cation radicals of the drug molecules. For example, the reactions that take place when a nitrous oxide aqueous solution of 10^{-2} M potassium bromide in the presence of 10^{-4} M chlorpromazine (ClP) is subjected to pulse radiolysis (Asmus et al., 1979; Davies et al., 1979) are given below:

$$^\bullet OH + Br^- \rightarrow Br^\bullet + {}^-OH \tag{12.13}$$

$$Br^\bullet + Br^- \rightarrow Br_2^{\bullet-} \tag{12.14}$$

$$Br_2^{\bullet-} + ClP \rightarrow 2Br^- + ClP^{\bullet+} \tag{12.15}$$

The molar absorption coefficients of the radicals can be easily determined by this method because the primary radical yields, as well as the properties of the one-electron oxidizing species are well known. The molar absorption coefficient of ClP$^{\bullet+}$ was determined to be 1×10^4 M^{-1} cm^{-1} at 525 nm, which is agreement with the literature value obtained by chemical oxidation (Fenner, 1974).

Similarly, the hydroxyl radicals and hydrogen atoms can be scavenged with t-butanol, leaving the hydrated electron to react with the compound to produce its anion radical. MQ anion radical was produced by pulsing a nitrogen-saturated aqueous solution of MQ in the presence of t-butanol (Navaratnam et al., 2000). Another one-electron reductant, $CO_2^{\bullet-}$, is formed by pulsing a nitrous oxide-saturated solution of sodium formate:

$$^{\bullet}OH + HCO_2^- \rightarrow CO_2^{\bullet-} + H_2O \qquad (12.16)$$

The spectral properties of the semireduced radical of amiloride were determined using the $CO_2^{\bullet-}$ radical (Hamoudi et al., 1984):

$$CO_2^{\bullet-} + AmH^+ \rightarrow AmH^{\bullet} + CO_2 \qquad (12.17)$$

Because electron transfer from $CO_2^{\bullet-}$ to oxygen is very rapid, oxygenated aqueous solutions of sodium formate are used to study the superoxide anion radical and its reactions:

$$CO_2^{\bullet-} + O_2 \rightarrow O_2^{\bullet-} + CO_2 \qquad (12.18)$$

Land et al. (1982) showed that sufacetamide photoionizes when subjected to a 266-nm laser pulse. When the experiment was repeated with an oxygenated solution, they found an increase in absorption, in the visible region, over 800 ns. An absorption spectrum with maximum at 480 nm was obtained when the end of the pulse spectrum was subtracted from that after 800 ns. This was tentatively ascribed to the reaction product of superoxide anion formed in accordance with:

$$e_{aq}^- + O_2 \rightarrow O_2^{\bullet-} \qquad (12.19)$$

$$SA + O_2^{\bullet-} \rightarrow \text{product} \qquad (12.20)$$

They confirmed this assignment by pulse radiolysis; a similar spectrum was obtained when they pulsed an oxygen-saturated aqueous solution of sulphacetamide in the presence of 0.1 M sodium formate. Under these conditions, superoxide anion is formed in accordance with Equation 12.18, which then reacted with SA in accordance with Equation 12.20 to form the same product as in the laser flash photolysis experiment, thus illustrating that the two techniques are complementary.

In nonpolar solvents, such as benzene (B), electrons are not easily solvated, but tend to recombine with the parent ion to form the excited state of the solvent molecule, as shown by the following reactions:

$$B \rightarrow B^{\bullet+} + e^- \qquad (12.21)$$

$$e^- + B \rightarrow B^{\bullet-} \qquad (12.22)$$

$$B^{\bullet-} + B^{\bullet+} \rightarrow {}^1B^* + {}^3B^* \qquad (12.23)$$

Because the lifetimes of benzene-excited states are very short, reasonably high concentrations of the solute are needed to scavenge most of the excited states. In practice, solutes like biphenyl (Bp), naphthalene, etc., which have longer triplet lifetimes, are used at concentrations in the range of 10 to 100 mM:

$$Bp + {}^1B^* \rightarrow {}^1Bp^* + B \qquad (12.24)$$

$${}^1Bp^* \rightarrow {}^3Bp^* \qquad (12.25)$$

$$Bp + {}^3B^* \rightarrow {}^3Bp^* + B \qquad (12.26)$$

In the presence of lower concentrations (usually of the order of 10^{-4} M) of the drug molecule, the triplet state of the drug is formed by energy transfer from biphenyl triplet:

$$Drug + {}^3Bp^* \rightarrow {}^3drug^* + Bp \qquad (12.27)$$

Thus, another method of producing the triplet excited state of the drug exists, and its properties and reactivity can then be studied in the way discussed for laser flash photolysis. Triplet state properties of two new phthalocyanine derivatives (tetra-polyethoxy-phthalocyanine and poly-ethoxy-zinc-phthalocyanine) in benzene were studied using laser flash photolysis and pulse radiolysis techniques and found to be in good agreement (Ferguson, 1998). Pulse radiolysis is a very attractive tool with which to study the triplet state properties of drugs with very low intersystem crossing efficiencies — carotenoids, for example.

REFERENCES

Aloisi, G.G., Elisei, F., Moro, S., Miolo, G., and Dall'Acqua, F. (2000) Photophysical properties of the lowest excited singlet and triplet states of thio- and seleno-psoralens, *Photochem. Photobiol.*, 71, 506–513.

Amouyal, E., Bensasson, R.V., and Land, E.J. (1974) Triplet states of ubiquinone analogs studied by ultra violet and electron nanosecond irradiation, *Photochem. Photobiol.*, 20, 415–422.

Andrisano,V., Gotti, R., Leoni, A., and Cavrini, V. (1999) Photodegradation studies on atenolol by liquid chromatography, *J. Pharm. Biomed. Anal.*, 21, 851–857.

Andrisano, V., Bertucci, C., Battaglia, A., and Cavrini, V. (2000) Photostability of drugs: photodegradation of melatonin and its determination in commercial formulations, *J. Pharm. Biomed. Anal.*, 23, 15–23.

Andrisano, V., Ballardini, R., Hrelia, P., Cameli, N., Tosti, A., Gotti, R., and Cavrini, V. (2001a) Studies on the photostability and *in vitro* phototoxicity of labetalol, *Eur. J. Pharm. Sci.*, 12, 495–504.

Andrisano, V., Hrelia, P., Gotti, R., Leoni, A., and Cavrini, V. (2001b) Photostability and phototoxicity studies on diltiazem, *J. Pharm. Biomed. Anal.*, 25, 589–597.

Asmus, K.-D. (1973) Application of conductivity techniques in pulse radiolysis, in Adams, G.M., Fielden, E.M., and Michael, B.D. (Eds.), *Fast Processes in Radiation Chemistry and Biology*, London, The Institute of Physics and John Wiley & Sons, pp. 40–59.

Asmus, K.-D. and Janata, E. (1982) Conductivity monitoring techniques, in Baxendale, J.H. and Busi, F. (Eds.) *The Study of Fast Processes and Transient Species by Electron Pulse Radiolysis*, New York, D. Reidel, pp. 91–113.

Asmus, K.-D., Bahnemann, D., Monig, J., Searle, A., and Wilson, R.L. (1979) Electrophilic free radicals and phenothiazine: a pulse radiolysis study, in Edwards, H.E., Navaratnam, S., Parsons, B.J., and Phillips, G.O. (Eds.), *Radiation Biology and Chemistry: Research and Development*, Amsterdam, Elservier, pp. 39–47.

Bazin, M., Bosca, F., Marin, M.L., Miranda, M.A., Patterson, L.K., and Santus, R.A. (2000) Laser flash photolysis and pulse radiolysis study of primary photochemical processes of flumequine, *Photochem. Photobiol.*, 72, 451–457.

Beaumont, P.C., Parsons, B.J., Phillips, G.O., and Allen, J.C. (1979) A laser flash photolysis study of the triplet states of 8-methoxy psoralen and 4,5',8-trimethoxypsoralen with nucleic acids in solution, *Biochim. Biophys. Acta*, 562, 214–221.

Beaumont, P.C., Parsons, B.J., Navaratnam, S., Phillips, G.O., and Allen, J.C. (1980) The reactivities of furocoumarin excited states with DNA in solution. A laser flash photolysis and fluorescence study, *Biochim. Biophys. Acta*, 608, 259–265.

Beaumont, P.C., Parsons, B.J., Navaratnam, S., and Phillips, G.O. (1983) A laser flash photolysis and fluorescence study of aminomethyl trimethylpsoralen in the presence and absence of DNA, *Photobiochem. Photobiophys.*, 5, 359–364.

Bensasson, R.V., Land, E.J., and Salet, C. (1978) Triplet excited state of furocoumarins: reaction with nucleic acid bases and amino acids, *Photochem. Photobiol.*, 27, 273–280.

Bilski, P., Martinez, L.J., Koker, E.B., and Chignell, C.F. (1996) Photosensitization by norfloxacin is a function of pH, *Photochem. Photobiol.*, 64, 496–500.

Bisby, R.H. and Parker, A.W. (2001) Structure of the radical from one-electron oxidation of 4-hydroxycinnamate, *Free Rad. Res.*, 35, 85–91.

Bonnet, R., Lambert, C., Land, E.J., Scourides, P.A., Siclair, R.S., and Truscott, T.G. (1983) The triplet and radical species of haematoporphyrin and some of its derivatives, *Photochem. Photobiol.*, 38, 1–8.

Borsarelli, C.D., Braslavsky, S.E., Sortino, S., Marconi, G., and Monti, S. (2000) Photodecarboxylation of ketoprofen in aqueous solution. A time-resolved laser-induced optoacoustic study, *Photochem. Photobiol.*, 72, 163–171.

Bosca, F., Encinas, S., Heelis, P.F., and Miranda, M.A. (1997) Photophysical and photochemical characterization of a photosensitizing drug: a combined steady state photolysis and laser flash photolysis study on carprofen, *Chem. Res. Toxicol.*, 10, 820–827.

Bosca, F., Marin, M.L., and Miranda, M.A. (2001) Photoreactivity of the non-steroidal anti-inflammatory 2-arylpropionic acids with photosensitizing side effects, *Photochem. Photobiol.*, 74, 237–255.

Braslavsky, S.E., Ellul, R.M., Weiss, R.G., Al-Ekabi, H., and Schaffner, K. (1983) Pytochrome model 7: photoprocessesin biliverdin dimethyl ester in ethanol studied by laser-induced optoacoustic spectroscopy (LIOAS), *Tetrahedron*, 39, 1909–1913.

Bromberg, A. and Foote, C.S. (1989) Solvent shift of singlet oxygen emission wavelength, *J. Phys. Chem.*, 93, 3968–3969.

Buettner, G.R., Motten, A.G., Hall, R.D., and Chignell, C.F. (1986) Free radical production by chlorpromazine sulfoxide, an ESR spin-trapping and flash photolysis study, *Photochem. Photobiol.*, 44, 5–10.

Buettner, G.R., Hall, R.D., Chignell, C.F., and Motten, A.G. (1989) The stepwise biphotonic photoionization of chlorpromazine as seen by laser flash photolysis, *Photochem. Photobiol.*, 49, 249–256.

Capellos, C. and Bielski, B.H.J. (1972) *Kinetic Systems*, New York, Wiley Interscience.

Chedekel, M.R., Land, E.J., Sinclair, R.S., Tait, D., and Truscott, T.G. (1980) Photochemistry of 4-hydroxybenzathiazole: a model for pheomelanin degradation, *J. Am. Chem. Soc.*, 102, 6587–6590.

Chignell, C.F. and Sik, R.H (1989) Spectroscopic studies of cutaneous photosensitizing agents. XIV. The spin trapping of free radicals formed during the photolysis of halogenated salicylanilide antibacterial agents, *Photochem. Photobiol.*, 50, 287–295.

Chignell, C.F., Kalyanaraman, B., Mason, R.P., and Sik, R.H. (1980) Spectroscopic studies of cutaneous photosensitizing agents. I. Spin trapping of photolysis products from sulfanilamide, 4-aminobenzoic acid and related compounds, *Photochem. Photobiol.*, 32, 563–571.

Chignell, C.F., Kalyanaraman, B., Sik, R.H., and Mason, R.P. (1981) Spectroscopic studies of cutaneous photosensitizing agents. II. Spin trapping of photolysis products from sulfanilamide and 4-aminobenzoic acid using 5,5-dimethyl-1-pyrroline-1-oxide, *Photochem. Photobiol.*, 34, 147–156.

Cordoba–Diaz, M., Cordoba–Borrego, M., and Cordoba–Diaz, D. (1998) The effect of photodegradation on fluorescent properties of norfloxacin, *J. Pharm. Biomed. Anal.*, 18, 865–870.

Craw, M. and Lambert, C. (1983) The characterization of the triplet state of crocetin, a water soluble carotenoid, by nanosecond laser flash photolysis, *Photochem. Photobiol.*, 38, 241–243.

Craw, M., Bensasson, R.V., Ronfard–Haret, J.C., Sa e Melo, M.T., and Truscott, T.G. (1983) Some phtophysical properties of 3-carbethoxypsoralen, 8-methoxypsoralen and 5-methoxypsoralen triplet states, *Photochem. Photobiol.*, 37, 611–615.

Davies, A.K., Navaratnam, S., and Phillips, G.O. (1976) Photochemistry of chlorpromazine in propan-2-ol solution, *J. Chem. Soc. Perkin Trans. II*, 25–29.

Davies, A.K., Land, E.J., Navaratnam, S., Parsons, B.J., and Phillips, G.O. (1979) Pulse radiolysis study of chlorpromazine and promazine free radicals in aqueous solution, *J. Chem. Soc. Faraday Trans. I*, 75, 22–35.

De Filippis, P., Bovina, E., Da Ros, L., Fiori, J., and Cavrini, V., (2002) Photodegradation studies on lacidipine in solution: basic experiments with a *cis–trans* reversible photoequilibrium under UV-A radiation exposure, *J. Pharm. Biomed. Anal.*, 27, 803–812.

Elisei, F., Aloisi, G.G., Lattarini, C., Latterini, L., Dall'Acqua, F., and Guiotto, A. (1996) Photophysical properties of some methyl-substituted angelicins: fluorometric and flash photolytic studies, *Photochem. Photobiol.*, 64, 67–74.

Elisei, F., Aloisi, G.G., Dall'Acqua, F., Latterini, L., Masetti, F., and Rodighiero, P. (1998) Photophysical behavior of angelicins and thioangelicins: semiempirical calculations and experimental study, *Photochem. Photobiol.*, 68, 164–172.

Encinas, S., Miranda, M.A., Marconi, G., and Monti, S. (1998) Transient species in the photochemistry of tiaprofenic acid and its decarboxylated photoproduct, *Photochem. Photobiol.*, 68, 633–639.

Fenner, H. (1974) EPR studies on the mechanism of biotransformation of tricyclic neuroleptics and antidepressants, in Forrest, I.S., Carr, C.J., and Usdin, E. (Eds.), *The Phenothiazine and Structurally Related Drugs*, New York, Raven Press, pp. 5–13.

Fielden, E.M. and Hart, E.J. (1967) Primary radical yields in pule-irradiated alkaline aqueous solution, *Radiat. Res.*, 32, 564–580.

Ferguson, M.W. (1998) Photophysical and photochemical properties of photosensitizers for photodynamic therapy of tumours, Ph.D. thesis, University of Wales.

Fuke, K., Ueda, M., and Itoh, M. (1983) Thermal lensing study of singlet oxygen reactions, *J. Am. Chem. Soc.*, 105, 1091–1096.

Gorman, A.A., Hamblett, I., Lambert, C., Prescott, A.L., Rodgers, M.A.J., and Spence, H.M. (1987) Aromatic ketone naphthalene systems as absolute standards for the triplet sensitized formation of singlet oxygen, $O_2(^1\Delta_g)$, in organic and aqueous-media-A time-resolved luminescence study, *J. Am. Chem. Soc.*, 109, 3091–3097.

Hall, R.D. and Chignell, C.F. (1987) Steady-state near-infrared detection of singlet molecular oxygen: a Stern–Volmer quenching experiment with sodium azide, *Photochem. Photobiol.*, 45, 459–464.

Hall, R.D., Buettner, G.R., and Chignell, C.F. (1991) The biphotonic photoionization of chlorpromazine during conventional flash photolysis: spin trapping results with 5,5-dimethyl-1-pyrroline-*N*-oxide, *Photochem. Photobiol.*, 54, 167–173.

Hamoudi, H.I., Heelis, P.F., Jones, R.A., Navaratnam, S., Parsons, B.J., Phillips, G.O., Vandenburgh, M.J., and Currie, W.J.C. (1984) A laser flash photolysis and pulse radiolysis study of amiloride in aqueous and alcoholic solutions, *Photochem. Photobiol.*, 40, 35–39.

Janzen, E.G. (1980) A critical review of spin trapping in biological systems, in Pryor, W.A. (Ed.) *Free Radicals in Biology*, Vol. IV, New York, Academic Press, pp. 115–154.

Jones, R.A., Parsons, B.J., Navaratnam, S., and Phillips, G.O. (1985.) Photodegradation of azapropazone and its relevance to photodermatology, in Bensasson, R.V., Jori, G., Land, E.J., and Truscott, T.G. (Eds.), *Primary Photoprocesses in Biology and Medicine*, New York, Plenum Press, pp. 61–64.

Jones, R.A., Navaratnam, S., Parsons, B.J., and Phillips, G.O. (1986) Photosensitivity due to anti-inflammatory analgesic drugs: a laser flash photolysis study of azapropazone, in Rainsford, K.D. and Velo, G.P. (Eds.), *Side Effects of Anti-Inflammatory Drugs, Part 2*, Lancaster, MTP Press, pp. 345–354.

Khan, A.U. (1984) Discovery of enzyme generation of $^1\Delta_g$ molecular oxygen: spectra of (0,0) $^1\Delta_g \rightarrow {}^3\Sigma_g$-IR emission, *J. Photochem.*, 25, 327–334.

Khan, A.U. (1985) Singlet molecular oxygen spectroscopy: chemical and photosensitized, in Frimer, A.A. (Ed.), *Singlet O_2*, Boca Raton, FL, CRC Press, pp. 39–79.

Keene, J.P. (1960) Kinetics of radiation-induced chemical reactions, *Nature*, 188, 843–844.

Kraljic, I. and El Moshni, S. (1978) A new method for the detection of singlet oxygen in aqueous solutions, *Photochem. Photobiol.*, 51, 119–121.

Kristensen, S., Grislingas, A., Greenhill, J.V., Skjetne, T., Karlsen, J., and Tønnesen, H.H. (1993) Photochemical stability of biologically active compounds.V. Photochemical degradation of primaquine in an aqueous medium, *Int. J. Pharm.*, 100, 15–23.

Land, E.J., Navaratnam, S., Parsons, B.J., and Phillips, G.O. (1982) Primary processes in the photochemistry of aqueous sulphacetamide: a laser flash photolysis and pulse radiolysis study, *Photochem. Photobiol.*, 35, 637–642.

Li, A.S.W. and Chignell, C.F. (1987a) Spectroscopic studies of cutaneous photosensitizing agents. IX. A spin trapping study of the photolysis of amiodarone and desethylamiodarone, *Photochem. Photobiol.*, 45, 191–197.

Li, A.S.W. and Chignell, C.F. (1987b) Spectroscopic studies of cutaneous photosensitizing agents. X. A spin-trapping and direct electron spin resonance study of the photochemical pathways of daunomycin and adriamycin, *Photochem. Photobiol.*, 45, 565–570.

Li, A.S.W. and Chignell, C.F. (1987c) Spectroscopic studies of cutaneous photosensitizing agents. XI. Photolysis of chlorpromazine metabolites: a spin-trapping study, *Photochem. Photobiol.*, 45, 695–701.

Li, A.S.W., Roethling, H.P., Cummings, K.B., and Chignell, C.F. (1987a) O_2^- photogenerated from aqueous solutions of tetracycline antibiotics (pH 7.3) as evidenced by DMPO spin trapping and cytochrome C reduction, *Biochem. Biophys. Res. Commun.*, 146, 1191–1195.

Li, A.S.W., Chignell, C.F., and Hall, R.D. (1987b) Cutaneous phototoxicity of tetracycline antibiotics: generation of free radicals and singlet oxygen during photolysis as measured by spin-trapping and the phosphorescence of singlet molecular oxygen, *Photochem. Photobiol.*, 46, 379–382.

Lopez, A., Bozzi, A., Mascolo, G., and Kiwi, J. (2003) Kinetic investigation on UV and UV/H_2O_2 degradations of pharmaceutical intermediates in aqueous solution, *J. Photochem. Photobiol. A Chem.*, 156, 121–126.

Martinez, L.J. and Scaiano, J.C. (1998) Characterization of the transient intermediates generated from the photoexcitation of nabumetone: a comparison with naproxen, *Photochem. Photobiol.*, 68, 646–651.

Martinez, L.J., Sik, R.H., and Chignell, C.F. (1998) Fluoroquinolone antimicrobials: singlet oxygen, superoxide and phototoxicity, *Photochem. Photobiol.*, 67, 399–403.

Matheson, M.S. and Dorfman, L.M. (1960) Detection of short lived transients in radiation chemistry, *J. Chem. Phys.*, 32, 1870–1871.

McCarthy, R.L. and McLachlan, A. (1960) Transient benzyl radical reactions produced by high energy radiation, *Trans. Faraday. Soc.*, 56, 1187–1200.

Monti, S. and Sortino, S. (2002) Laser flash photolysis study of photoionization in fluoroquinolones, *Photochem. Photobiol. Sci.*, 1, 877–881.

Moore, D.E. (1977) Photosensitization by drugs, *J. Pharm. Sci.*, 66, 1282–1284.

Moore, D.E. and Chappuis, P.P. (1988) A comparative study of the photochemistry of the non-steroidal anti-inflammatory drugs, naproxen, benoxaprofen and indomethacin, *Photochem. Photobiol.*, 47, 173–181.

Moore, D.E., Sik, R.H., Bilski, P., Chignell, C.F., and Reszka, K.J. (1994) Photochemical sensitization by azathioprine and its metabolites. III. A direct EPR and spin-trapping study of light-induced free radicals from 6-mercaptopurine and its oxidation products, *Photochem. Photobiol.*, 60, 574–581.

Murov, S.L., Carmichael, I., and Hug, G.L. (1993) *Handbook of Photochemistry*, New York, Marcel Dekker, Inc.

Motten, A.G. and Chignell, C.F. (1983) Spectroscopic studies of cutaneous photosensitizing agents. III. Spin trapping of photolysis products from sulfanilamide analogs, *Photochem. Photobiol.*, 37, 17–26.

Motten, A.G., Chignell, C.F., and Mason, R.P. (1983) Spectroscopic studies of cutaneous photosensitizing agents. VI. Identification of the free radicals generated during the photolysis of musk ambrette, musk xylene and musk ketone, *Photochem. Photobiol.*, 38, 671–678.

Motten, A.G., Buettner, G.R., and Chignell, C.F. (1985) Spectroscopic studies of cutaneous photosensitizing agents. VIII. A spin-trapping study of light induced free radicals from chlorpromazine and promazine, *Photochem. Photobiol.*, 42, 9–15.

Motten, A.G., Martinez, L.J., Holt, N., Sik, R.H., Reszka, K.J., Chignell, C.F., Tønnesen, H.H., and Roberts, J.E. (1999) Photophysical studies on antimalarial drugs, *Photochem. Photobiol.*, 69, 282–287.

Navaratnam, S., Parsons, B.J., Phillips, G.O., and Davies, A.K. (1978) Laser flash photolysis study of the photoionization of chlorpromazine and promazine in solution, *J. Chem. Soc. Faraday Trans. I*, 74, 1811–1819.

Navaratnam, S., Land, E.J., Parsons, B.J., Ahmad, I., and Phillips, G.O. (1983) Primary processes in the laser flash photolysis and pulse radiolysis of L-ephedrine, *Photochem. Photobiol.*, 38, 153–159.

Navaratnam, S., Parsons, B.J., and Phillips, G.O. (1984) Laser flash and steady state photolysis of benoxaprofen in solution: relevance to singlet oxygen production, in Bors, W. and Saran, M. (Eds.), *Oxygen Radicals in Chemistry and Biology*, Berlin, Walter de Gruyter, pp. 479–484.

Navaratnam, S., Hughes, J.L.L., Parsons, B.J., and Phillips, G.O. (1985) Laser flash and steady state photolysis of benoxaprofen in aqueous solutions, *Photochem. Photobiol.*, 41, 375–380.

Navaratnam, S., Hughes, J.L.L., Parsons, B.J., and Phillips, G.O. (1987) Laser flash photolysis study of propranolol, a cutaneous photosensitizing drug, in Cronly–Dillon, J., Rosen, E.S., and Marshall, J. (Eds.), *Hazards of Light Myths and Realities Eye and Skin*, Oxford, Pergamon, pp. 89–94.

Navaratnam, S. and Phillips, G.O. (1991) Application of laser flash photolysis to study photoreceptor pigments, in Lenci, F., Ghetti, F., Colombetti, G., Hader, D.-P., and Song, P.-S. (Eds.), *Biophysics of Photoreceptors and Photomovements in Microorganisms*, New York, Plenum Press, pp. 139–148.

Navaratnam, S., Parsons, B.J., and Hughes, J.L.L. (1993) Laser-flash photolysis studies of benoxaprofen and its analogues. I. Yields of triplet states and singlet oxygen in acetonitrile solutions, *J. Photochem. Photobiol. A Chem.*, 73, 97–103.

Navaratnam, S. and Jones, S.A. (2000) Primary process in the photochemistry of fenbufen in acetonitrile, *J. Photchem. Photobiol. A Chem.*, 132, 175–180.

Navaratnam, S. and Claridge, J. (2000) Primary photophysical properties of ofloxacin, *Photochem. Photobiol.*, 72, 283–290.

Navaratnam, S., Hamblett, I., and Tønnesen, H.H. (2000) Photoreactivity of biologically active compounds. XVI. Formation and reactivity of free radicals in mefloquine, *J. Photochem. Photobiol. B Biol.*, 56, 25–38.

Nord, K., Karlsen, J., and Tønnesen, H.H. (1991) Photochemical stability of biologically active compounds. IV. Photochemical degradation of chloroquine, *Int. J. Pharm.*, 72, 11–18.

Nord, K., Karlsen, J., and Tønnesen, H.H. (1994) Photochemical stability of biologically active compounds. IX. Characterization of the spectroscopic properties of the 4-aminoquinolones, chloroquine and hydroxychloroquine, and of selected metabolites by absorption, fluorescence and phosphorescence measurements, *Photochem. Photobiol.*, 60, 427–431.

Norrish, R.G.W. and Porter, G. (1949) Chemical reactions produced by very high light intensities, *Nature*, 164, 658.

Nudelman, N.S. and Cabrera, C.G. (2002) Spectrofluorimetric assay for the photodegradation products of alprazolam, *J. Pharm. Biomed. Anal.*, 30, 887–893.

Patel, C.K.N. and Tam, A.C. (1981) Pulsed optoacoustic spectroscopy of the condensed media, *Rev. Mod. Phys.*, 53, 517–520.

Porter, G. (1950) Flash photolysis and spectroscopy: a new method for the study of free radical reactions, *Proc. R. Soc.*, A200, 284–300.

Reddi, E., Jori, G., Rodgers, M.A.J., and Spikes, J.D. (1983) Flash photolysis studies of hemato- and copro-porphyrins in homogeneous and microheterogeneous aqueous dispersions, *Photochem. Photobiol.*, 38, 639–648.

Redmond, R.W. and Braslavsky, S.E. (1988) Time-resolved thermal lensing and phosphorescence studies on photosensitized molecular oxygen formation. Influence of the electronic configuration of the sensitizer on sensitization efficiency, *Chem. Phys. Lett.*, 148, 523–529.

Reszka, K.J. and Chignell, C.F. (1994a) Photochemistry of 2-mercaptopyridines. I. An EPR and spin-trapping investigation using 5,5-dimethyl-1-pyrroline N-oxide in aqueous and toluene solutions, *Photochem. Photobiol.*, 60, 442–449.

Reszka, K.J. and Chignell, C.F. (1994b) Photochemistry of 2-mercaptopyridines. II. An EPR and spin-trapping investigation using 2-methyl-2-nitrosopropane and aci-nitromethane as spin traps in aqueous solutions, *Photochem. Photobiol.*, 60, 450–454.

Reszka, K.J. and Chignell, C.F. (1995) Photochemistry of 2-mercaptopyridines. III. EPR study of photoproduction of hydroxyl radicals by N-hydroxypyridine-2-thione using 5,5-dimethyl-1-pyrroline N-oxide in aqueous solutions, *Photochem. Photobiol.*, 61, 269–275.

Richards, J.T. and Thomas, J.K. (1970) Laser and flash photolytic studies on the effects of various solvents and solutes on the excited singlet and triplet states of N,N,N′,N′,-tetramethyl paraphenylene diamine (TMPD), *Trans. Faraday Soc.*, 66, 621–632.

Rodgers, M.A.J. (1983) Solvent-induced deactivation of singlet oxygen: additivity relationship in nonaromatic solvents, *J. Am. Chem. Soc.*, 105, 6201–6205.

Rodgers, M.A.J. and Snowden, P.T. (1982) The lifetime of $O_2(^1\Delta_2)$ in liquid water as determined by time resolved infrared luminescence, *J. Am. Chem. Soc.*, 104, 5541–5543.

Ronhard–Haret, J.C., Averbeck, D., Bensasson, R.V., Bisagni, E., and Land, E.J. (1987) Some properties of the triplet excited state of the photosensitizing furocoumarin: 3-carboethoxypsoralen, *Photochem. Photobiol.*, 44, 479–489.

Rosencwaig, A. (1980) *Photoacoustics and Photoacoustic Spectroscopy,* New York, Wiley.

Rossbroich, G., Garcia, N.A., and Braslavsky, S.E. (1985) Lifetime of singlet molecular oxygen determined by time-resolved thermal lensing, in Bensasson, R.V., Jori, G., Land, E.J., and Truscott, T.G. (Eds.), *Primary Photoprocesses in Biology and Medicine*, New York, Plenum Press, pp. 197–199.

Sa e Melo, M.T., Averbeck, D., Bensasson, R.V., Land, E.J., and Salet, C. (1979) Some furocoumarins and analogs: comparison of triplet properties in solution with photobiological activities in yeast, *Photochem. Photobiol.*, 30, 645–651.

Salet, C., Sa e Melo, M.T., Bensasson, R.V., and Land, E.J. (1980) Photophysical properties of aminomethylpsoralen in the presence and absence of DNA, *Biochim. Biophys. Acta*, 607, 379–383.

Sarata, G., Sakai, M., and Takahashi, H. (2000) Nanosecond time-resolved resonance raman and absorption studies of the photochemistry of chlorpromazine and related phenothiazine derivatives, *J. Raman Spectrosc.*, 31, 785–790.

Schmidt, R., Tanielian, C., Dunsbach, R., and Wolff, C. (1984) Phenalenone, a universal reference compound for the determination of quantum yields of singlet oxygen $O_2(^1\Delta_g)$ sensitization, *J. Photochem. Photobiol. A Chem.*, 84, 11–17.

Sortino, S. and Scaiano, J.C. (1999) Laser flash photolysis of tolmetin: a photoadiabatic decarboxylation with a triplet carbanion as the key intermediate in the photodecomposition, *Photochem. Photobiol.*, 69, 167–172.

Sortino, S., De Guidi, G., Giuffrida, S., Monti, S., and Velardita, A. (1998) pH Effects on the spectroscopic and photochemical behaviour of enoxacin: a steady-state and time-resolved study, *Photochem. Photobiol.*, 67, 167–173.

Sortino, S., Marconi, G., Giuffrida, S., De Guidi, G., and Monti, S. (1999) Photophysical properties of rufloxacin in neutral aqueous solution, *Photochem. Photobiol.*, 70, 731–736.

Tamat, S.R. and Moore, D.E. (1983) Photolytic decomposition of hydrochlorothiazide, *J. Pharm. Sci.*, 72, 180–184.

Thoma, K. and Holzmann, C. (1998) Photostability of dithranol, *Eur. J. Pharm. Biopharm.*, 46, 201–208.

Ulvi, V. (1998) Spectrophotometric studies on the photostability of some thiazide diuretics in ethanolic solution, *J. Pharm. Biomed. Anal.*, 17, 77–82.

CHAPTER 13

Addressing the Problem of Light Instability during Formulation Development

D. R. Merrifield, P. L. Carter, D. Clapham, and F. D. Sanderson

CONTENTS

13.1 INTRODUCTION

A number of case histories are presented in this chapter that illustrate the formulation problems raised by light-sensitive drug substances. These range from the study and formulation of parenteral presentations, in which the product is in solution or suspension form and the instability most closely mirrors that of the drug substance; pharmaceutical ointments, in which the drug substance is incorporated as a suspended material in a semisolid base; to the physical effects on a tablet formulation where the active is present as solid component.

It is argued that it is necessary to have a good understanding of the nature and extent of the photoinstability; the mechanism of the light-induced degradation reaction; and the wavelengths causing the instability. The potential problems raised by photounstable formulations are such that it is necessary to develop a strategy to quantify and understand them at an early stage of development. This will permit formulation development to progress in a timely and efficient fashion.

Whilst there is a reasonable measure of agreement between regulatory bodies as to the nature of testing conditions for the stresses of temperature and humidity, there has been significantly more discussion concerning the light exposure conditions to be applied to photosensitive drugs (Anderson et al., 1991a, b). These discussions have culminated in agreed guidelines on light-stability testing, with recommendations as to the extent of light exposure, the way in which this is conducted, how the results are interpreted, and the implications for product labeling. The ICH Q1B guidelines and the recommendations for testing and interpretation arising from them will allow a common approach in drug product registration.

A number of reviews have been produced to aid the practitioner in the practical application of these guidelines (Thatcher et al., 2001a, b; Quattrocchi et al., 2001; Helboe, 1998; Drew, 1998). The guidelines describe a simple screening test for photostability that allows the use of a range of light sources. However, they do not control some test aspects, which can be important in certain instances, such as environmental humidity. As such they do not seek to address all the aspects of the photochemistry of a drug substance or drug product, and they will not provide a complete strategy for assessing the light sensitivity of a drug product, and the restrictions that light sensitivity places on the usage conditions of a drug product.

The conditions that the product will experience in use will clearly vary as a result of presentation type (especially depending on whether the drug is present in the solid or solution phase), the extent to which the dosage form is protected by the packaging, mode of administration, the geographical location of the market, and the local conditions of the clinic. Although the conditions recommended in the ICH (International Conference on Harmonization) guidelines provide a common standard by which light sensitivity may be judged and compared, they can only act as a basic guide to understanding the nature of light instability. The conditions of the ICH tests may be supplemented by other tests specifically designed to help define conditions of use of the formulated product.

It is also worth remembering that formulation may alter the photostability of a drug substance, thus rendering it either more or less sensitive to light. As well as inducing chemical changes in a drug substance or its formulation, light may induce other important changes such as in physical state, coloration or discoloration, or changes in the toxicity profile (EMEA guideline CPMP/SWP/398/01; Spielmann et al., 1998).

Some of these issues and the types of test required to address them are illustrated in the case histories presented in this chapter.

13.2 PHOTOINSTABILITY OF THE DRUG SUBSTANCE

It should be established at an early stage that a development compound is susceptible to light-induced degradation. Depending on the nature of product development within an organization, this will probably be known at an early stage of chemical development and may be established as an issue for a family of chemically similar compounds. Even when this is not the case, photoinstability will become evident very early in analytical method development and this will probably provide the first opportunity to quantify the extent of photoinstability. A commonly experienced problem at this stage relates to the degree of confidence in the purity of the compound, validity of the analytical methods, the degree of knowledge regarding degradation products, availability of the new chemical entity (NCE), and the relevance of testing conditions. Merely knowing that a material is light sensitive, by any given criteria, does not guarantee a clear appreciation of the relevance of that light instability as it relates to product usage.

The ICH guidelines recommend exposure periods under given light intensities that enable a categorization of a material as light sensitive, along with a guide as to the requirement for protective packaging and subsequent labeling. These guidelines also provide a range of options as to the nature of the light source, with the degree of challenge being determined as a cumulative light exposure. Given the intensity of the sources used, the extent of exposure recommended can represent a severe test, likely to considerably exceed the conditions experienced by the drug product in use. It is nevertheless very valuable in setting a common criterion by which a material may be categorized as light sensitive or not.

At this early stage of development, the available information on photostability is likely to be anecdotal in nature, based on laboratory observations made by development chemists. It is important that where light sensitivity is suspected, it is confirmed and quantified at the earliest opportunity, either by testing under the ICH recommended conditions or by an accepted in-house testing protocol (Nema et al., 1995). This is important so that any photoinstability may be taken into account when considering results from early evaluation of the compound in screening tests and that a toxicology program can be designed appropriately. Furthermore, the nature and extent of photoinstability are likely to impact on the objectives and timings of the development program for the NCE and on the aims for its ultimate use as a product.

Recognizing that a candidate drug substance has potential photoinstability will require the development groups to consider several approaches to assessing and solving the problem. First, it will be necessary to understand how to handle the drug substance during product manufacture. In some cases the use of controlled lighting will be needed (Torres Suarez et al., 1994). It will also be necessary to know the intended indication and likely mode of administration of the final formulated product, to establish whether the problem may be largely resolved by the use of suitable packaging or protection. If the problem cannot be resolved in this way, the extent of the instability will need to be determined and decisions made as to whether this may be overcome by modifications to the formulation, or the process by which it is prepared. Both primary (chemical) and secondary (pharmaceutical) processes will need to be reviewed to establish the extent to which the light susceptibility of the compound is likely to compromise the quality of the product during manufacture and use. Special conditions for handling the material at various stages of the process may need to be defined to alleviate this aspect of the problem.

Analytical development work should rapidly identify the extent of the problem caused by the presence of photodegradation products. Identification of the degradation products if they are present at levels above predetermined thresholds is a necessary aspect of elucidating the reaction mechanism and will be required in any submissions to regulatory groups. Their early identification and quantification is important in deciding the appropriate strategy for the toxicological assessment of the drug substance. Here it must be remembered that the photodegradation reaction will need to be sufficiently characterized and understood so that any degradation in the product occurs by the same mechanism, allowing the toxicity assessment of degradation products to apply equally to the final formulated product.

An important part in further elucidation of the degradation reaction relates to the specific wavelengths causing the degradation (Allwood and Plane, 1986; Moore, 1987). Knowledge of this allows the formulation group to select appropriate packaging to protect the product (Bhadresa and Sugden, 1981), so that the packaging components attenuate those wavelengths that cause the most significant degradation. Knowledge of the causative wavelengths also enables the formulator to consider the use of formulation excipients (Lin et al., 2000; Thoma and Klimek, 1991; Thoma and Kübler, 1997), where appropriate, to absorb the light responsible for degradation, thereby reducing the rate at which the degradation occurs to a manageable and acceptable degree.

Establishing the kinetics of the degradation is a further important aspect of characterizing the nature of the photoinstability (Tønnesen, 1991). This will take the form of a series of experiments determining the reaction rate with the objective of establishing its dependence on concentration. The information provided from this study will assist in deciding the development strategy for the formulation and in identifying any limitations that the degradation rate places on the product use.

13.3 PHOTOINSTABILITY OF FORMULATIONS

The approach to dealing with a drug's instability will clearly depend on the nature of the presentation and its mode of use (Tønnesen and Moore, 1993). Four case histories are presented here that exemplify an approach to identifying, quantifying, and resolving the issues raised by a photolabile drug substance or formulation.

13.3.1 Intravenous Infusions

An intravenous infusion presentation offers considerable challenge to the formulation group because the problems associated with light instability will most closely resemble those of the drug substance. The drug is likely to be presented as a dilute solution and used over a number of hours in a clinical situation (Colardyn et al., 1993). Furthermore, it is likely that, given hospital practices, the presentation will be made up by the pharmacy team some time before administration to the patient. It will be necessary that the reconstituted presentation is capable of withstanding a variable light-intensity challenge depending on geographical location of the clinic, the level of ambient light within the pharmacy and ward, and the duration of use. Given the highly variable nature of all these factors, it is important that the directions for use of the product are very clear and that the presentation is capable of withstanding the harshest challenge likely to be met in practice. To evaluate the challenges that the product may face, studies often involve the use of fluorescent light, simulated sunlight, and window light (Vandenbossche et al., 1993).

An additional complication in the formulation of light-sensitive drugs in parenteral infusion presentations arises from the need to demonstrate the usual freedom from contamination by extractives of the giving set and that the drug substance remains compatible with the range of infusion fluids likely to be used in clinical practice. Thus, it is often inappropriate to utilize opaque packaging. It must be considered that any restrictions applying to the mode of use of the infusion fluid, or any contraindicated infusion fluid, will add to complication in use and be perceived as a competitive disadvantage for the product.

Compound A was observed to be light sensitive at an early stage of its development, consistent with its having a highly conjugated functional ring group. The degradation products were identified during early analytical development work and were consistent with a likely and recognized light-degradation mechanism. It was possible to isolate these degradation products following chemical synthesis by column chromatography, or they could be prepared *in situ* in solution by following a standard light exposure regime. A series of photodegradation studies were conducted to establish the extent of the light-sensitivity problem, the implications for the proposed formulation, and the challenges for the manufacturing process.

13.3.1.1 *Light Degradation of the Drug Substance as a Powdered Solid*

An early assessment of the substance was conducted on the dry powder material. It was expected that the drug substance was likely to be prepared as a precipitated crystalline solid or as a lyophile. The stability of the drug substance raw material was determined by reverse-phase HPLC analysis to help judge the feasibility of the chemical and pharmaceutical production processes, and to assess the likely shelf-life prior to reconstitution. This was conducted both as accelerated testing using a xenon light source (Xenotest) and metal halide light source (Dr. Höhnle SOL2 light cabinet), presenting the sample as thin powder sample (about 2 mm thick), and in real time in glass vials using sources of artificial (laboratory) and natural light.

The results of this test are shown in Table 13.1. Apart from the expected problems arising from the variability of natural light, it was evident that degradation was occurring by the same pathway in each case (forming the degradation products A1 and A2), except in the instance of storage in the dark (vials overwrapped in aluminum foil). In the dark, the major degradation product was an alkaline degradation product (A3), with no evidence of formation of the main light-induced degradation product (A1).

Table 13.1   Stability Assessment of Compound A
             as a Powdered Solid

Condition and Exposure Period	Purity	Degradation Products		
		A1	A2	A3
Initial	85.3	nd	nd	0.2
6 weeks sunlight	75.1 (87%)	6.9	4.9	0.2
4 months sunlight	55.2 (65%)	22.2	14.2	1.2
6 months artificial	82.3 (96%)	5.0	2.8	0.4
6 months dark	85.0 (100%)	nd	0.3	0.3

Note: The powder was exposed to the light source as a shallow bed of about 2 mm depth.

Each content is given as %w/w in the dry powder, with the drug substance purity also expressed as % initial given in parentheses.

The content was determined in each case by HPLC analysis.

nd = not detected.

In summary, the solid stability profile was extremely good for the class of compound when it was stored protected from light, but the material degraded rapidly in the presence of any level of natural or artificial light. These findings confirmed that the extent of instability would require it to be protected during storage and that special precautions were likely to be needed during handling, especially during freeze drying of the compound.

13.3.1.2 Light Degradation of the Drug Substance in Solution

The rate of degradation of the drug substance as a function of concentration in solution is a crucial factor when considering the suitability of the NCE for use as an infusion presentation. Although a number of experiments had been conducted attempting to quantify the rate of degradation of solution under in-use conditions, the most valuable assessment of the extent of instability was gained by a series of studies conducted at different concentrations under standard lighting conditions (Table 13.2).

The experiment was conducted in a light cabinet specially constructed in-house to provide a high intensity of artificial light (2× Philips white 35 fluorescent bulbs) at a constant temperature. Constant temperature was achieved by fitting the cabinet with a powerful fan and situating the light cabinet in a constant temperature storage facility. Samples were taken over the course of the reaction and assayed for compound A content by reverse phase HPLC analysis. The reaction was allowed to proceed to a sufficient extent to allow first-order rate constants to be determined.

The results in Table 13.2 demonstrate an interesting paradox for assessing the stability of a pharmaceutical active. It is customary within the industry to assess the proposed product by reference to the time taken to lose a certain proportion of the active constituent, in this case 10%, otherwise referred to as T_{90}. Reference to this figure demonstrates a very unacceptable stability profile for the most dilute solutions (0.05 and 0.025 mg/ml), making handling of such a product impractical. However,

Table 13.2   Degradation Rate of Compound A in Solution

Concentration (mg/ml)	Rate (mol/h)	T_{90}
2.5	3.6×10^{-4}	2.4 h
0.5	2.0×10^{-4}	0.9 h
0.25	1.8×10^{-4}	28 min
0.05	0.9×10^{-4}	11 min
0.025	0.5×10^{-4}	7 min

Note: Determined in clear glass vials at 25°C and 10 klx. (Light-testing cabinet fitted with 2× Philips white 35 fluorescent bulbs.)

T_{90} = time taken for the compound to degrade to 90% of its initial purity value.

Table 13.3   Dependency of Reaction Rate on Light Intensity

Source	Intensity (klx)	Rate (mol mn^{-1})	T_{90} (min)
Artificial[a]	10	1.24×10^{-5}	66
North light	7	0.99×10^{-5}	83
South light	70	8.22×10^{-5}	10
Metal halide[b]	120	4.11×10^{-4}	2

Note: The solutions prepared at 2.5 mg/ml were tested in clear glass injection vials.

T_{90} = time taken for the active to degrade to 90% of initial concentration.

[a] Light testing cabinet, 2× Philips white 35 fluorescent bulbs.
[b] Dr. Höhnle SOL 2 simulated sunlight tester.

assessment of the data in molecular terms clearly shows a faster rate of reaction at the higher concentrations; there is an approximate doubling of rate with a 10-fold increase in concentration. In this instance, the acceptability of the product will need to be judged on the nature of degradation products as well as the rate of disappearance of the active constituent since these will clearly be generated at a higher rate and to a greater extent, with increasing concentration. Further experiments were conducted at a constant concentration of 2.5 mg/ml in clear glass vials to determine the effects of light intensity and light type. The data presented in Table 13.3 and Table 13.4 present the results of these studies.

Taken together it can be seen that when degradation is allowed to proceed to an approximately equal extent (intended to be about 85% of initial purity), the pattern of degradation products formed is similar in each case. Compounds A1 and A2 are the main degradation products in each instance and are formed in the same proportions. No other significant peaks were observed in the chromatogram, and the other known degradation product, due to alkaline decomposition A3, was not detected to any significant extent. Given this common pattern of degradation, consistent with that of the powdered solid and the expected degradation mechanism, it is reasonable to interpret the results shown in Table 13.3 in terms of the effect of light intensity. The expected pattern is observed with the low-intensity north light being a far less severe

Table 13.4   Degradation Product Formation as Function
of Light Type, Conducted on Compound A
Solutions

Source	Potency and (% initial value)	Degradation Products		
		A1	A2	A3
Artificial[a]	0.224	13.6	4.9	0.5
(30 klx.h)	(82%)	(0.7)	(0.25)	(0.02)
South light	0.204	3.9	1.5	nd
(7 klx.h)	(94%)	(0.71)	(0.27)	
North light	0.196	13.3	4.7	0.2
(15 klx.h)	(84%)	(0.72)	(0.25)	(0.01)
Metal halide[b]	0.204	10.2	3.7	0.02
(6 klx.h)	(86%)	(0.71)	(0.26)	(0.001)

Note: These studies were conducted at 2.5 mg/ml initial concentration, stored in clear glass injection vials. It was intended
that each solution should be degraded by 15% of initial
value.

Contents of degradation products are expressed as a percentage of the main peak in the chromatogram (intact
drug) and as a relative proportion of the total degradation
products in the chromatogram, given in parentheses.

The light exposure is given as a cumulative challenge, for
example, in klx h.

nd = not detected.

[a] Light testing cabinet, 2 × Philips white 35 fluorescent bulbs.
[b] Dr. Höhnle SOL 2 simulated sunlight tester.

stress than the metal halide source of the SOL2 sunlight simulator, even though the
relative spectral distribution of these sources would be expected to be similar.

These experiments, along with others, defined the in-use and processing limitations of the product. Given its nature as a clear solution for injection and infusion
use, many of the approaches normally used to overcome light instability in the
formulation were not available in this instance. The approach used for this product
was via protection afforded by secondary packaging. Further experiments enabled
the selection of a suitable protective plastic sleeving that greatly reduced the extent
of degradation, allowing a usable solution shelf-life.

These experiments were conducted using the tunable wavelength source of a
spectrofluorimeter to provide light of known wavelength to a solution held in the
sample cuvette of the instrument. A number of wavelength-specific experiments
were carried out to determine the dependency of degradation rate on wavelength. It
can be seen from Figure 13.1 that the wavelength dependency of the measured
degradation rate closely mirrors the absorption spectrum of the compound. One
drawback of this experiment was that the total irradiance available at narrow wavelength bands from the spectrofluorimeter was extremely low, leading to an extended
experimental time. Higher power tunable systems are now available for conducting
such experiments.

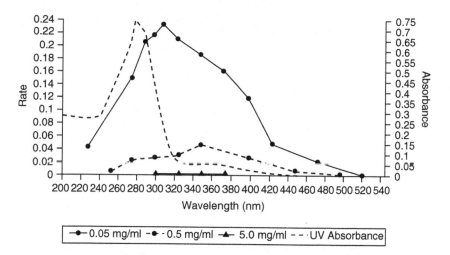

Figure 13.1 Degradation rate of compound A as a function of wavelength and its UV absorbance spectrum.

13.3.2 Topical Ointments

The second case study represents an interesting situation combining the problems of a light-sensitive drug substance incorporated into a presentation that will be exposed to sunlight during patient use. It also represents a situation partway between the solid stability of the drug substance and its photostability in the solution phase.

Compound B is intended for topical administration and is incorporated into an oily base. It was known to be photoinstable from an early stage of development and accordingly a number of salt forms of the drug substance were evaluated in preliminary screening tests. The parent compound (free acid form) was eventually selected as the appropriate form for development, and photostability of the ointment was studied incorporating the drug substance both in the precipitate form and after micronization. Because the clinical dose remained to be established, the drug was incorporated at a range of concentrations to assess whether this has any effect on the photoinstability.

The results in Table 13.5 show a number of interesting aspects. There is a clear dependence of the light stability of the compound on the nature of the salt form used. The effect is particularly striking for the disodium salt when compared with the free acid, given that the former is incorporated as an unmicronized powder.

For the free acid, the micronized material is less stable than the unmicronized form. This is indicative of either a particle size effect, where the finer particle size receives a greater effective exposure to light because of the higher surface area available, or that there is a higher proportion of the drug present actually in solution in this instance. From these limited data, it does not appear that the more finely divided drug substance offers any self-protection by physical obscuration of the light. The results indicate that the extent of loss of the active is reduced as the concentration is increased. However, exactly as in the first case, when this is considered in molar terms it can be seen that the actual quantity of drug lost is greater at the higher concentrations. The complexity of the system must be considered.

Table 13.5   Light Stability of Compound B as Free Acid and Various Salt Forms Incorporated into a White Soft Paraffin Base

	Percent Initial Value Compound B after 1-h Exposure in SOL 2		
Salt/Form	0.1% Conc.	0.5% Conc.	2.0% Conc.
Free acid (micronized)	51.0	80.0	85.9
Free acid (unmicronized)	69.4	nd	nd
Disodium salt (unmicronized)	9.9	3.6	nd
Ethylenediamine salt (unmicronized)	51.9	65.1	nd
Piperazine salt (unmicronized)	79.9	88.2	nd

Note: The samples of ointment were exposed as thin films for 1 h in the Dr. Höhnle SOL 2 simulated sunlight tester. The concentration was determined by HPLC analysis.

nd = not done.

Table 13.6   Light Stability of Compound B Free Acid in Various Brands of Paraffin Ointment

Paraffin Base	Color	Percent of Initial Value after 1-h Exposure in SOL 2
1	White	53.2
2	White	49.6
3	White	50.2
4	White	56.6
5	Yellow	70.0
6	White	73.2
7	Yellow	75.6
8	Yellow	77.4

Note: The samples of ointment were exposed as thin films for 1 h in the Dr. Höhnle SOL 2 simulated sunlight tester. The concentration was determined by HPLC analysis.

The solubility of the drug in the ointment base is approximately 0.05% w/w. It is present both in solution and in suspension in the higher concentration ointments. The relative proportions in the solution and suspension states depend on the ointment strength. Various reactions will be occurring simultaneously, especially if the degradation mechanism differs between the solution and suspension phase.

The drug compound was incorporated at 0.1% concentration, as a micronized powder, into a range of paraffin bases. Each was subjected to a 1-h exposure in the SOL2 light simulation cabinet. The results of this screening test are shown in Table 13.6.

The yellow paraffin bases can be observed to offer more protection to the drug substance than the white brands, with the exception of brand 6. It was known that this brand contained about 10 ppm of the antioxidant α-tocopherol, and it was considered that this constituent may be affording some protection. A further experiment was therefore conducted spiking the control blend, brand 1, with α-tocopherol. This did not produce any noticeable effect on the stability of the drug and neither did the incorporation of varying levels of α-tocopherol in the best white paraffin brand, brand 6. These results are shown in Table 13.7.

Table 13.7   Effect of α-Tocopherol on
Compound B Light Stability

Formulation	Percent Initial Value after 1-h Exposure in SOL 2
Brand 1	53.2
Brand 1 (+10 ppm)	48.3
Brand 6	74.2
Brand 6 (+250 ppm)	73.5
Brand 6 (+500 ppm)	72.5

Note: The samples of ointment were exposed as thln films for 1 h in the Dr. Höhnle SOL 2 simulated sunlight tester. The concentration was determined by HPLC analysis.

UV scans of the various paraffin bases dissolved in chloroform were obtained. This indicated a clear difference for the white paraffin base 6 compared with the other white bases. The UV spectrum of this brand was more similar to those of the yellow bases 5, 7, and 8. It is likely that the light-absorption properties of the paraffins account for their relative abilities to protect the active. Compound B has a UV spectrum with maxima at 215 and 330 nm; it was inferred from the chemical structure that the maximum at 330 nm is most likely to be due to the part of the molecule susceptible to light. White soft paraffins generally have a minimal absorbance in this region of the UV spectrum, 290 nm being the longest wavelength absorbed compared with the yellow grades, which showed a significant absorbance up to 340 nm. It is this broad absorbance that probably confers protection on the active compound B.

These studies demonstrate an important approach to addressing the light instability of a drug substance within a formulation. If it is possible and acceptable to include a material that can block the passage (physical obscuration) (Bechard et al., 1992) or compete with the drug substance for the available light energy (similar light absorbance spectrum) (Thoma and Klimek, 1991; Thoma and Kübler, 1997), then the photoinstability of the active compound should be reduced (Asker and Harris, 1988; Asker and Habib, 1989).

This approach was adopted for this compound, investigating the use of both light blockers and UV absorbers, resulting in an acceptable formulation with minimum photoinstability. Protection of the product by means of formulation is preferable since topical products will often be used on exposed skin. However, where the photoinstability on the skin is not an issue, long-term protection can be provided by the packaging (Takada et al., 1998). The majority of pharmaceutical topical formulations are packed in opaque containers contained within a cardboard outer container.

13.3.3 Solid Oral Dosage Forms

Many compounds of pharmaceutical interest presented as tablets contain drug substances that are photoinstable and a large literature is available dealing with individual drug substances (e.g., Thoma and Kerker, 1992; Thoma and Kübler, 1997; Teraoka et al., 2001). The effect of formulation and processing variables have also been assessed (e.g., Aman and Thoma, 2002).

The third case study deals with a drug, not itself photolabile, incorporated into a solid dosage form and demonstrates a different but equally important aspect of light stability: physical appearance (Jansson et al., 1980; Nyquist, 1984; Augustine et al., 1986). The material in this study is a marketed product, presented as a range of film-coated tablets of different active strengths. They are actually prepared from a single formulation and are tabletted at a range of weights to provide the required dosages. It is known that the active compound itself is not light sensitive even under harsh conditions. In this study, the effect of light on the physical appearance of the formulated product was examined.

The range of tablets, packaged in the intended market pack, was exposed to D65 white light radiation using the SOL2 sunlight simulator light cabinet. The packs (clear PVC/PVdC blister backed with aluminum foil) were exposed clear side uppermost to about 115 klx radiation for 24 h, using foil-wrapped packages as a control. After exposure, the color of the tablets was quantitatively determined using a Tristimulus colorimeter. The tablets were also analyzed by reverse phase HPLC to determine the intact drug content and the two most significant degradation products.

The principle of operation of the Tristimulus colorimeter essentially mimics the red, green, and blue detection of the human eye by measuring the amount of light reflected through wavelength-specific filters or via a spectrophotometric technique. These raw data are transformed mathematically to provide a uniform, reproducible numerical scale that can detect very slight differences in color. The scale is depicted by a three-dimensional model, the L, a, b color space. This represents color on two scales: "a" (red/green) and "b" (yellow/blue), and brightness on a third, "L."

The technique is capable of being refined so that the light spot of the instrument can be tightly focused to a point on the tablet surface. This can allow comparisons between front (exposed) and back (unexposed) faces to be made. The results in Table 13.8 illustrate an increased yellowing of the tablets, shown by an increasing yellow/decreasing blue "b" component. It is also apparent from these data that the effect is more marked in the lower-strength tablets.

It was necessary to demonstrate that this appearance effect was not due to any degradation of the active material. To do this, two additional experiments were conducted; these are shown in Table 13.9 and Table 13.10. These experiments demonstrate that the yellowing is not due to detectable degradation of the active component. The second experiment shows this most clearly; in each case an individual tablet from each of the three classifications — yellowest, palest, and midrange — were analyzed for main peak and degradation products. This was done for tablets in each classification for each strength. Only the results for the 125- and 250-mg strengths are shown in Table 13.10.

Additionally, the yellowest tablet seen was analyzed. In this case the surface yellow layer was scraped off and assayed. Again, there was no observable increase in degradation products and no additional degradation products were observed in the HPLC chromatograms.

In this instance, the nature of the effect of light could be overcome by appropriate packaging. In general this very sensitive technique allows a thorough quantification of the effect of light exposure on physical appearance of solid dosage forms.

Table 13.8   Measurement of Tablet Color Change[a]

Tablet Strength	Control/ Exposed	Mean Color			Standard Deviation		
		L	a	b	L	a	b
125 mg	Control	99.5	2.60	5.32	2.4	0.6	0.6
	Exposed	85.5	1.71	14.58	9.7	1.5	3.2
250 mg	Control	99.0	2.60	4.92	2.7	0.6	0.2
	Exposed	93.0	2.20	11.83	2.5	0.8	2.8
500 mg	Control	95.5	3.02	5.36	3.1	1.0	0.3
	Exposed	90.3	1.48	11.10	2.3	1.1	3.2
750 mg	Control	95.8	2.58	5.18	1.5	0.3	0.2
	Exposed	88.7	2.21	7.39	1.3	0.6	1.9

Note: The Tristimulus color description is based on three axes. This allows any color perceived by the human eye to be expressed in terms of three numbers.

L scale = color brightness; a scale = red/green; b scale = yellow/blue, where a positive increase denotes a yellowing of tablets.

The tablets were exposed to 115 klx intensity in the Dr. Höhnle simulated sunlight tester for 24 h; the tablets were exposed on one side only.

[a]Determined by Tristimulus colorimetry.

Table 13.9   HPLC Analysis of Tablets Subjected to Simulated Sunlight Testing

Tablet Strength	Main Peak	Mean Assay (%)	
		Impurity 1	Impurity 2
125 mg Control	72.6	0.25	nd
125 mg Exposed	73.0	0.23	nd
250 mg Control	71.3	0.31	nd
250 mg Exposed	72.1	0.31	nd
500 mg Control	73.4	0.33	nd
500 mg Exposed	73.2	0.42	nd
750 mg Control	72.6	0.61	nd
750 mg Exposed	73.1	0.54	nd

Note: Contents of degradation products 1 and 2 are given as % w/w.

Exposed tablets were subjected to 115 klx for 24 h (Dr. Höhnle SOL 2), exposed on one side only.

Control tablets were overwrapped with foil.

nd = not detected.

As well as being of aesthetic concern, the appearance of a tablet may be important in determining the correct dose strength. Some drugs are required to be available in a range of dose strengths and these are often distinguished by the color of their core or film coat. It is fairly well recognized that light will cause all dyes to fade; however, other factors such as the relative humidity of the environment are also involved. Quantitative color measurement permits these processes to be studied.

Table 13.10 HPLC Analysis on Selected Individual Tablets

		Mean Assay (%)	
Tablet Strength	Main Peak	Impurity 1	Impurity 2
125 mg yellowest	73.4	0.26	nd
125 mg mid-range	73.0	0.21	nd
125 mg palest	72.0	0.22	nd
250 mg yellowest	71.7	0.37	nd
250 mg mid-range	72.7	0.28	nd
250 mg palest	71.8	0.29	nd
250 mg surface	72.1	0.28	nd

Note: Contents of degradation products 1 and 2 are given as % w/w and are determined by HPLC analysis.

Tablets had been exposed to 115 klx for 24 h (Dr. Höhnle SOL 2).

Certain tablets were selected as representative of the exposed tablets; others were taken as the palest or yellowest samples obtained during the test.

nd = not detected.

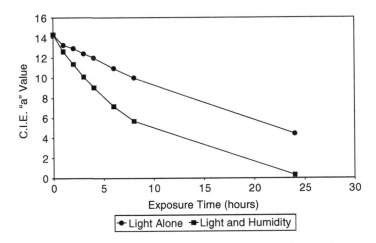

Figure 13.2 The effect of light and humidity on the color fastness of a film coat containing the red dye Carmine.

The final case study concerns another marketed product that also contained a nonphotolabile drug, presented as a range of active tablet strengths of equal tablet weight. The different strengths were distinguished by the color of the film coat. Photostability testing was performed on the coated tablets by exposure to simulated sunlight (Dr. Höhnle SOL 2) for various times at either ambient or elevated humidity (75% r.h.). The effect on one film coat containing the red dye Carmine is shown in Figure 13.2. It is clear that color loss is much increased in the presence of high humidity, a factor not generally controlled in photostability tests. In fact, there is a danger that total color loss may result (redness "a" value reduced to zero), which could lead to confusion between tablets of different strengths. In order to avoid this possibility, moisture-resistant packaging was employed along with the use of a higher level of dye in the film coat to allow some loss of dye without perceived loss of color. As noted for the topical formulations above, color stability of film coats can also sometimes be improved by the use of sunscreening additives in the film coat formulation (Hajratwala, 1974a, b, 1977, 1985).

Whilst discussing tablets it is worth noting that excipients used in these formulations are not always inert and can themselves be subject to photochemically induced changes. This is particularly important when the excipient is intended to control the rate of release of the active component from the formulation. Careful choice of excipient is required when significant light exposure is anticipated (Maggi et al., 2003).

13.4 CONCLUSIONS

The case studies presented in this chapter illustrate a consistent strategy for evaluating light stability of a compound during formulation development. This includes studies to evaluate mechanistically the effects of formulation and packaging on the stability of the final product. The challenge to the product under likely in-use conditions is examined as an important aspect of understanding the photostability of the drug substance and its formulations.

ACKNOWLEDGMENTS

The authors wish to acknowledge S. A. Hancock, A. J. Goodall, M. E. Morris, and J. Oduro–Yeboah for their help in valuable discussion and preparing this presentation. They also wish to acknowledge T. Franklin, M. Lau, B. Hall, S. Ritchie, R. Denton, I. Gower, and the late M. Churchill for much of the practical work contained in these studies.

REFERENCES

Allwood, M.C. and Plane, J.H. (1986) The wavelength dependent degradation of vitamin A exposed to ultraviolet radiation, *Int. J. Pharm.*, 31, 1–7.
Aman, W. and Thoma, K. (2002) The influence of formulation and manufacturing process on the photostability of tablets, *Int. J. Pharm. Tech.*, 243, 33–41.
Anderson, N.H., Johnston, D., McLelland, M.A., and Munden, P. (1991a) Photostability testing of drugs, *Manuf. Chemist*, 62, 25.
Anderson, N.H., Johnston, D., McLelland, M.A., and Munden, P. (1991b) Photostability testing of drug substances and drug products in UK pharmaceutical laboratories, *J. Pharm. Biomed. Anal.*, 9, 443–449.
Asker, A.F. and Harris, C.W. (1988) Influence of certain additives on the photostability of physostigmine sulphate solutions, *Drug Dev. Ind. Pharm.*, 14, 733–746.

Asker, A.F. and Habib, M.J. (1989) Influence of certain additives on the photostatibility of colchicine solutions, *Drug Dev. Ind. Pharm.*, 15, 845–849.

Augustine, M.A., Bernstein, D.F., Narurkar, A.N., and Sheen, P.-C. (1986) Studies on the light stability of flordipine tablets in amber blister packaging material, *Drug Dev. Ind. Pharm.*, 12, 1241–1247.

Bechard, S.R., Kwong, E., and Ouraishi, O. (1992) Film coating: effect of titanium dioxide concentration and film thickness on the photostability of nifidipine, *Int. J. Pharm.*, 87, 133–139.

Bhadresa, B. and Sugden, J.K. (1981) Light transmittance through amber glass medicine bottles, *Pharm. Acta Helv.*, 56, 122.

Colardyn, F., De Muynck, C., Remon, J.P., and Vandenbossche, G.M.R. (1993) Light stability of molsidomine in infusion fluids, *J. Pharm. Pharmacol.*, 45, 486–488.

Drew, H.D. (1998) Photostability of drug substances and drug products: a validated reference method for implementing the ICH photostability guidelines, in Albini, A. and Fasani, E. (Eds.) *Drugs: Photochemistry and Photostability,* Cambridge: Royal Society of Chemistry, p. 227–246.

EMEA guideline CPMP/SWP/398/01 CPMP Note for guidance on photosafety testing, adopted June 2002.

Hajratwala, B.R. (1974 a) Influence of sunscreening agents on colour stability of tablets coated with certified dyes F D & C Red No 3, *J. Pharm. Sci.*, 63, 129–132.

Hajratwala, B.R. (1974b) Influence of sunscreening agents on colour stability of tablets coated with certified dyes F D & C Blue No 1, *J. Pharm. Sci.*, 63, 1927–1930.

Hajratwala, B.R. (1977) Influence of sunscreening agents on colour stability of tablets coated with certified dyes F D & C Yellow No 6, *J. Pharm. Sci.*, 66, 107–109.

Hajratwala, B.R. (1985) Stability of colours, *S.T.P. Pharma*, 1, 539–544.

Helboe, P. (1998) The elaboration and application of the ICH guideline on photostability: a European view, in Albini, A. and Fasani, E. (Eds.) *Drugs: Photochemistry and Photostability*, Cambridge: Royal Society of Chemistry, pp. 243–246.

Jansson, I., Lundgren, P., and Nyquist, H. (1980) Studies on the physical properties of tablets and tablet excipients. II. Testing of light stability of tablets, *Acta Pharm. Suec.*, 17, 148–156.

Lin, H.S., Chan, S.Y., Ng, Y.Y., and Ho, P.C. (2000) 2-Hydroxypropyl-beta-cyclodextrin increases aqueous solubility and photostability of all-trans retinoic acid, *J. Clin. Pharm. Ther.*, 25, 265–269.

Maggi, L., Ochoa, M.E., Fasani, E., Albini, A., Seagle, L., and Conte, U. (2003) Photostability of extended-release matrix formulations, *Eur. J. Pharm. Biopharm.*, 55, 99–105.

Moore, D.E. (1987) Principles and practice of drug photodegradation studies, *J. Pharm. Biomed. Anal.*, 5, 441–453.

Nema, S., Waskuhn, R.J., and Beussink, D.R. (1995) Photostability testing: an overview, *Pharm. Technol.*, 19, 170–185.

Nyquist, H. (1984) Light stability testing of tablets in the Xenotest and the Fadeometer, *Acta Pharm. Suec.*, 21, 245–252.

Quattrocchi, O.A. and Gador, S.A. (2001) Photostability of drugs and formulations, *Revista SAFYBIA*, 40, 3–11.

Spielmann, H., Balls, M., Dupuis, J., Pape, W.J., Pechovitch, G., de Silva, O., Holzhütter, H.-G., Clothier, R., Desolle, P., Gerberick, F., Liebsch, M., Lovell, W.W., Maurer, T., Pfannenbecker, U., Potthast, J.M., Csato, M., Sladowski, D., Steiling, W., and Brantom, P. (1998) The international EU/COLIPA *in vitro* phototoxicity validation study, *Toxicol. In Vitro*, 12, 305–327.

Takada, H., Mori, H., Yamazaki, F., Goto, S., and Niwa, C. (1998) Photostability testing of various ophthalmic solutions and its application for medication instructions, *Byoin Yakugaku*, 24, 601–610.

Teraoka, R., Konishi, Y., and Matsuda, Y. (2001) Photochemical and oxidative degradation of the solid-state tretinoin tocoferil, *Chem. Pharm. Bull.*, 49, 368–372.

Thatcher, S.R., Mansfield, R.K., Miller, R.B., Davis, C.W., and Baertschi, S.W. (2001a) Pharmaceutical photostability: a technical guide and practical interpretation of the ICH guideline and its application to pharmaceutical stability. I, *Pharm. Technol. North Am.*, 25(3), 98–110.

Thatcher, S.R., Mansfield, R.K., Miller, R.B., Davis, C.W., and Baertschi, S.W. (2001b) Pharmaceutical photostability: a technical guide and practical interpretation of the ICH guideline and its application to pharmaceutical stability. II, *Pharm. Technol. North Am.*, 25(4), 50–62.

Thoma, K. and Klimek, R. (1991) Photostabilization of drugs in dosage forms without protection from packaging materials, *Int. J. Pharm.*, 67, 169–175.

Thoma, K. and Kerker, R. (1992) Photostability of drugs. V. The photostabilization of glucocoticoids, *Pharm. Ind.*, 54, 551–554.

Thoma, K. and Kübler, N. (1997) Influence of excipients on the photodegradation of drug substances, *Pharmazie*, 52, 122–129.

Torres Suarez, T.A. and Camacho, M.A. (1994) Photolability evaluation of the new cytostatic drug mitonafide, *Arznm. Forsch.*, 44, 81–83.

Tønnesen, H.H. (1991) Photochemical degradation of components in drug formulations. I. An approach to the standardization of degradation studies, *Pharmazie*, 46, 263–265.

Tønnesen, H.H. and Moore, D.E. (1993) Photochemical degradation of components in drug formulations. II. Selection of radiation sources in light stability testing, *Pharm. Technol.*, 5, 27–34.

Vandenbossche, G.M.R., De Muynck, C., Colardyn, F., and Remon, J.P. (1993) Light stability of molsidomine in infusion fluids, *J. Pharm. Pharmacol.*, 45, 486–488.

CHAPTER 14

Photostability of Parenteral Products

Solveig Kristensen

CONTENTS

14.1 INTRODUCTION

Photochemical reactivity of drug formulations is an important aspect to consider during development, production, storage, and use of pharmaceutical preparations. However, photochemical stability of drug substances is rarely as well documented as thermal stability of the compounds. For instance, in order to obtain a high sterility assurance level of the product, a parenteral preparation is sterilized in its final container if possible. Steam sterilization at minimum temperature of 121°C for

15 min is recommended for sterile liquid preparations, according to the *European Pharmacopoeia* (2002). Accordingly, the effect of temperature on stability of the drug substance and the formulation must be evaluated to ensure optimal drug stability, to estimate product shelf-life, and to determine storage conditions. Photochemical reactions can also influence the shelf-life of a drug and therefore should be an essential consideration in the evaluation of storage conditions.

Guidelines for testing the photochemical stability of new drug substances and products were established in 1998 (ICH, 1997). However, most drugs used today were already present on the market before 1998 and are not covered by the guidelines. Moreover, the ICH guidelines do not include in-use conditions, which can be an important aspect for drugs administered by the parenteral routes. Ready-to-use parenteral products are liquid preparations, with the exception of implants, and the drugs are diluted, dissolved, or added to infusion liquids by the producer or prior to administration. Liquid preparations are usually far more photolabile than solid formulations of the corresponding drug substances.

Parenterals often consist of dilute aqueous solutions; thus, a large fraction of the incident optical irradiation (UV irradiation and visible light) will penetrate through the drug formulation. The optical irradiation will reach sensitive molecules in the solution, and photons can be absorbed by the drug substance, excipients, or impurities in the formulation. Absorption may be followed by decomposition of the drug substance or the formulation by primary or secondary photochemical reactions (see Chapter 2).

In busy hospital wards, the nursing staff often removes the outer cartons from parenteral products, and the preparations are stored without any protection against optical irradiation. In addition, *ex tempore* preparations are produced in the hospital pharmacy without outer protection and not usually protected when distributed to the wards (e.g., by use of aluminum foil) unless specifically instructed. According to the *European Pharmacopoeia* (2002), containers for parenteral preparations are to be made when possible from materials (usually glass or plastic materials) that are sufficiently transparent to permit visual inspection of the contents. As a consequence, the containers will offer no protection, or in some cases only limited protection, against photochemical decomposition of the drug substance or the formulation.

At least in Europe, hospital wards are equipped with plenty of windows, and drugs administered as infusions are exposed to optical irradiation at the wards during storage and administration. The exposure can last for several hours or even days. The infusion bags are usually hanging unprotected in the room, often in front of a window or close to indoor lighting. Sunlight penetrating window glass as well as optical irradiation from light bulbs and fluorescent tubes can cause severe photochemical degradation.

In Europe, greater emphasis is put on health care at home. The hospital pharmacies are under more pressure to provide these extended services, which include provision of home parenteral nutrition, home dialysis, home chemotherapy, i.v. antibiotic therapy at home, and patient-controlled analgesia (Hutchinson and Graham, 1998; Hutchinson, 1998). UV and light exposure is thus hard to control. In certain cases patients take their infusion devices outdoors, an opportunity also provided by some hospitals, especially for children. However, sensitive preparations

Table 14.1   Formulation Properties and Terms Concerning Administration of Parenteral Products Essential for the Photochemical Stability

Formulation Properties and Administration Terms	Photochemical Consequences
Liquid formulation	Increased reactivity due to high mobility of photochemically excited molecules and reactive species
Dilute solution	Increased transmission and absorption of optical irradiation by individual molecules
Transparent container	Penetration of optical irradiation
Product not stored in an outer packaging	Formulation unprotected against optical irradiation
Administration carried out in front of a window, close to indoor lighting, or outdoors	Exposure to optical irradiation during administration
Long-term administration	Exposure to a high cumulative dose of optical irradiation

exposed to UV and visible irradiation for a long time will absorb a high cumulative dose of optical irradiation and be especially vulnerable to photochemical degradation.

As a consequence, knowledge of photochemical stability and possible photochemical stability problems is of great importance in regard to parenteral preparations, especially infusions (see Table 14.1). Few systematic studies have been performed. Several factors can influence photochemical stability of dilute solutions and disperse systems, which constitute the main part of parenteral formulations. This chapter is focused on the different parameters that, theoretically, may influence photochemical stability of parenterals. Of course, the best way to stabilize the preparations is by preventing exposure to optical irradiation, an option not often accomplished in practice. Thus, it is valuable to know the factors that can influence photochemical stability of the products, as well as to prepare protocols for the management of parenterals at each hospital ward based on this knowledge.

The formulation of parenteral products involves careful consideration of the proposed route of administration and the volume of the injection. Injections are administered to the body by many routes: into various layers of the skin, the subcutaneous and muscle tissue, into arteries or veins, into or around the spinal cord, or directly into various organs (e.g., the heart or the eye). The volume to be injected can range from microliters, typically diagnostic agents administered intradermally or insulin administered subcutaneously, to several liters administered intravenously as infusions. The route of administration and the volume to be injected affect the composition of the formulation.

The most common way to formulate injections is to dissolve the drug in an aqueous vehicle. Excipients are added to control solubility, osmotic pressure, pH, stability, specific gravity, and preservation. Injections may also be formulated as oily solutions, disperse systems like suspensions and emulsions (aqueous and oily), and liposomal dispersions. Drug formulation is an essential factor in photochemical stability of the drug substance. Excipients or impurities in the formulation can also participate in photochemical reactions, leading to decomposition of the drug substance or the formulation.

A photochemical reaction is always initiated by absorption of optical irradiation. The ozone layer in the atmosphere transmits sunlight in the ranges of 290 to 320 nm (UVB), 320 to 400 nm (UVA), and 400 to 800 nm (visible light), which reaches the Earth's surface. These photons have sufficient energy to interact with electrons in organic molecules, i.e., drugs, excipients, or impurities in the formulation. Most drugs are able to absorb UV radiation; colored compounds can also absorb visible light. Absorption of photons of certain wavelengths leads to an excitation of electrons into molecular orbitals of higher energy levels. The result is an excited, or energy-rich, organic molecule.

The physicochemical properties of the excited state and ground state molecules can be totally different, e.g., steric configuration, electron distribution, and pKa values. Thus, photochemical reaction mechanisms cannot necessarily be predicted from knowledge about thermal reactions of the ground state molecule. The excited state must be considered as a distinct compound. The excited molecule may dissipate energy as heat or luminescence, by chemical reactions, or by energy transfer to neighboring molecules. Photochemical reaction mechanisms are complex and involve the formation of free radicals. Several photodegradation pathways are common for drugs in solution, e.g., oxidation, dechlorination, hydrolysis, *cis–trans* isomerization, decarboxylation, and rearrangements. It is essential to note that photodecomposition of drugs is not likely to be discernible by visual observation of color change.

Some organic molecules are able to absorb optical irradiation and pass on the increased energy to other molecules, which then degrade; these processes are called photosensitized or secondary photochemical reactions. The absorbing molecules, which are denoted photosensitizers, are not decomposed in the photochemical reactions. Excipients, solvents, or small amounts of impurities may act as photosensitizers that initiate the photochemical reaction in the formulation. The photosensitizers may be present in concentrations not detectable by conventional analytical methods, such as UV-visible absorption spectroscopy or HPLC with UV detection.

Molecular oxygen is often involved in photochemical reactions. There are two mechanisms of photosensitized oxidation. In the type I process, the excited organic molecule reacts with substances in the formulation to give free radicals or radical ions. Reaction of these radicals with oxygen gives oxygenated products. Reactive forms of oxygen may also be produced, i.e., superoxide anions ($O_2 \cdot -$), hydroxyl radicals ($OH \cdot$), and hydrogen peroxide (H_2O_2).

In the type II process, the excited molecule will pass on the increased energy to molecular oxygen to form singlet molecular oxygen ($^1O_2$), an electronically excited state that is highly reactive. Singlet oxygen can react with drugs or excipients in the formulation to form oxygenated products. The reaction is so-called self-sensitized when the photosensitizer is attacked by singlet oxygen and decomposed. Thus, exclusion of molecular oxygen from the parenteral preparation may protect the product from photochemical decomposition. On the other hand, molecular oxygen is able to quench photochemical degradation reactions by absorbing the excitation energy and thereby bringing the excited drug molecule back to the ground state. In such cases, the presence of oxygen may be essential for the photochemical stability of the product, assuming that the drug molecule does not react with the singlet oxygen formed in this process. (See Chapter 2 and Chapter 3 in this textbook for an introduction to fundamental principles of photochemistry and photochemical reaction mechanisms.)

A photochemical degradation product can be therapeutically inactive or less active than the parent drug. Complete loss of pharmacological activity is possible after exposure to optical irradiation. However, a photodegradation product can also have a different or higher pharmacological activity than the parent compound or it can be toxic. Photochemical reactions may further lead to destabilization or degradation of complex formulations, e.g., emulsions or suspensions. Coalescence of the disperse phase in emulsions is an example of a possible effect that can be harmful, depending on the actual route of administration. Parenteral preparations are often administered by injection directly into the blood stream. Thus, photochemical stability of parenterals also becomes an issue of patient safety, as well as an aspect to consider during development of the formulation and determination of drug shelf-life and storage conditions.

14.2 INFLUENCE OF THE VEHICLE

The composition of the formulation (the vehicle) can have a vital influence on the photochemical reactivity of the drug substance. The pharmacologically active compound can be photochemically stable in one environment and highly reactive in another. Composition of the formulation will influence the absorption properties of the drug and the mechanisms by which energy is dissipated from the excited molecule. Excipients that are demonstrated to stabilize one drug substance can have the opposite effect on another compound. Excipients in the formulation may also be photoreactive and initiate degradation of the active substance by photosensitized reactions.

Thus, evaluation of photoreactivity should be included in the preformulation phase during drug development, in order to optimize photochemical stability of the drug. In principle, all excipients in a parenteral formulation can influence the photochemical stability of the drug substance. Therefore, the pharmacist can possibly affect photoreactivity by changing the formulation properties (see Table 14.2).

14.2.1 Drug Concentration

The rate of a photochemical reaction depends on the overlap integral between the incident irradiation and the molar absorptivity of the reacting species (see Chapter 3). Thus, the concentration of the dissolved drug will affect the photochemical degradation rate of the substance. Photochemical reactions can be described by several models of reaction kinetics, depending on the reaction mechanisms involved and the concentration of the reacting substances.

First-order reactions are predicted for very dilute solutions, as may be the case for intravenous preparations. The incident optical irradiation is transmitted through the formulation and the photons can reach and be absorbed by any of the molecules in the formulation. The photochemical decomposition of dilute solutions with a

Table 14.2   Formulation Parameters That Can Influence Photochemical Stability of Parenteral Preparations

Drug substance	Drug vehicle
Physicochemical properties	Polarity
Drug concentration	Amphiprotic properties
	Viscosity
Excipients	pH
Solvent and cosolvents	Ionic strength
Surfactants	Oxygen content
Buffer salts and concentrations	Metal ions
Antioxidants	Organic impurities (from raw materials
Preservatives	or container)
Chelators	
Cyclodextrins	**Container transparency**
Liposomes/particles	Material
Other additives	Thickness
	Color
	Opacity

maximum total absorbance < 0.4 will often follow first-order degradation kinetics (see Chapter 3). At small overlap integrals due to limited absorption in the UV and visible regions, first-order kinetics can be observed even at high concentrations of the drug. By first-order kinetics the degradation rate ($-d[C]/dt$) depends on the concentration of the reacting drug substance (C) (Connors, 1990):

$$-d[C]/dt = k_1 \cdot [C] \qquad (14.1)$$

or in its integrated form:

$$\ln C_t = \ln C_0 - k_1 \cdot t \qquad (14.2)$$

where:

    t = time
    C_0 = initial drug concentration
    C_t = drug concentration at the time t
    k_1 = first-order rate constant

When the drug has a substantial absorption in the UV and visible regions (maximum absorbance ≥ 2), solutions of the compound will absorb most of the incident irradiation (see Chapter 3). The drug acts as a filter and prevents transmission of photons through the formulation. The rate limiting factor is then the intensity of the incident irradiation, and the photochemical degradation rate will follow zero-order degradation kinetics. Photochemical reactions can take place at the surface of the product; the decomposition proceeds at a constant rate independent of the concentration of the reactant (Connors, 1990):

$$-d[C]/dt = k_0 \qquad (14.3)$$

or in integrated form:

$$C_t = C_0 - k_0 \cdot t \tag{14.4}$$

where k_0 = zero-order rate constant. Concentrates for injections and concentrates for infusions (*European Pharmacopoeia*, 2002) are examples of parenteral formulations likely to follow zero-order reaction kinetics during photochemical decomposition. Photochemically stable concentrates may show an increased degree of photodecomposition after dilution, when the extent of decomposition is expressed as a percentage of the initial concentration of the diluted solution. An apparently decreased photochemical degradation rate as function of an increased drug concentration was demonstrated for ciprofloxacine in citrate buffer, but this was only because the degradation was expressed in terms of the percentage of initial drug remaining at various times (Torniainen et al.,1996). If the decomposition was expressed as the amount, i.e., milligrams or millimoles of drug degraded, it would have been clear that the amount was essentially the same across all concentrations except the lowest. This is a zero-order kinetic situation in which the rate-determining step is the rate of photon absorption (see Chapter 3).

Degradation processes sensitized by an excipient or impurity present in the formulation or by the drug itself (self-sensitization) will follow second-order reaction kinetics if the concentrations (or the overlap integrals) are low. Reactions between an excited drug molecule and molecular oxygen or an excipient also follow second-order kinetics. The photochemical degradation rate is determined by the concentrations of the two reacting species (C and X) (Connors, 1990):

$$-d[C]/dt = -d[X]/dt = k_2 \cdot [C] \cdot [X] \tag{14.5}$$

where k_2 = second-order rate constant. If the initial concentrations of the two reacting species are identical ($C_0 = X_0$), integration leads to:

$$1/C_t = 1/C_0 + k_2 \cdot t \tag{14.6}$$

Photochemical decomposition of the drug by self-sensitization or a dimerization reaction involving two molecules of the drug substance will follow Equation 14.6. However, in most cases the initial concentrations of two different reacting species will not be identical ($C_0 \neq X_0$), giving a more complex expression after integration and rearrangement into a linear form (Connors, 1990):

$$\ln\left(C_t/X_t\right) = \left(C_0 - X_0\right) \cdot k_2 \cdot t + \ln\left(C_0/X_0\right) \tag{14.7}$$

Photochemical degradation of minoxidil in solution is likely sensitized by the impurity chlorominoxidil from the drug synthesis. An increase in the drug concentration is followed by an increased concentration of the photosensitizer, leading to a rise in the photochemical degradation rate (Chinnian and Asker, 1996).

14.2.2 Solvents and Cosolvents

In general, parenteral dosage forms are relatively simple from a formulation point of view, and mostly consist of aqueous solutions of the drug. *Water for injection* (*European Pharmacopoeia*, 2002) is the most common solvent, but may be combined with a cosolvent to improve solubility or stability of the products. Ethanol or propylene glycol, used separately or in combination, occur in more than 50% of parenteral cosolvent systems. Fixed oils of vegetable origin are used to dissolve drugs of low aqueous solubility and provide sustained drug delivery. Glycerol, polyethylene glycols (PEGs), benzyl benzoate, and *N,N*-dimethylacetamide are other cosolvents commonly used for parenteral preparations (Nema et al., 2002).

The absorption spectrum of a compound determines the wavelength range to which the drug may be sensitive (see Chapter 2). The vehicle will influence the electronic absorption spectra of molecules by solvation, an effect denoted *solvato-chromism*. The term *solvation* is quantitatively defined as the energy of interaction between a dissolved molecule (solute) and the solvent (Suppan and Nagwa, 1997). The interaction can be nonspecific in character, which does not imply any fixed stoichiometry or geometry of the molecules. Nonspecific interactions depend on the polarity of the solute molecule and the solvent. Interactions may also be specific as in hydrogen bonding. Hydrogen bonding can play a most important role in the photophysical properties of many molecules, by its effect on electronic absorption spectra and by leading to extended lifetimes of excited states.

Amphiprotic (proton donor and acceptor) properties of the drug (solute) and the solvent are essential for the formation of hydrogen bonds. Changing solvent polarity by addition or substitution with cosolvents can lead to a solvatochromic shift of the absorption spectrum of the drug. Such shifts result from the difference of solvation energies between the two electronic states involved in the absorption transition, i.e., the ground state and the excited state of the molecule (Suppan and Nagwa, 1997). The absorption shift can be blue (i.e., to lower wavelengths of higher energy) or red (i.e., to higher wavelengths of lower energy), depending on the nature of the pho-tophysical transition, and the physicochemical properties of the solvent and the drug in the ground state and the excited state (e.g., polarity and amphiprotic properties). Solvatochromic shifts of absorption spectra observed after the introduction of more polar solvents can also be due to a rupture of intramolecular hydrogen-bonded structures during formation of competing drug–solvent interactions.

Many drug molecules are acids and bases and can exist in protonated and deprotonated forms; these species may have different absorption spectra. The total absorption spectrum will be a combination of these individual spectra, which will depend on the pK value of the drug *vs.* the pH of the solution (Suppan and Nagwa, 1997). An acid-base equilibrium reaction is a combination of two steps, ionization and dissociation (Pecsok et al., 1976):

$$DH + SH \underset{ionization}{\rightleftharpoons} D^-SH_2^+ \underset{dissociation}{\rightleftharpoons} D^- + SH_2^+ \qquad \begin{array}{l} DH = acidic\ drug \\ SH = solvent \end{array} \qquad (Scheme\ 1)$$

$$DH + SH \underset{\text{ionization}}{\rightleftharpoons} DH_2^+ S^- \underset{\text{dissociation}}{\rightleftharpoons} DH_2^+ + S^- \qquad \begin{array}{l} DH = \text{basic drug} \\ SH = \text{solvent} \end{array} \qquad \text{(Scheme 2)}$$

The extent of the ionization step depends on the relative strength of the conjugate acid–conjugate base pairs. The amphiprotic properties of the solvent have an essential effect on the equilibrium constant of this reaction step. The extent of the dissociation step is influenced by the polarity of the solvent, increasing with the dielectric constant of the solvent. In water, all products of acid–base reactions of moderate to low concentrations are essentially completely dissociated into solvated ions (Pecsok et al., 1976). The dissociation step is suppressed by addition or substitution with cosolvents of lower polarity, e.g., alcohols in aqueous formulations. The ion–pair aggregates may have absorption spectra different from the dissociated species. Thus, the amphiprotic properties and polarity (expressed as the dielectric constant) of the solvent are essential for the acid–base equilibrium of the drug and thus the absorption spectrum of the compound. This subject is further discussed in Section 14.2.3.

As a result of protic and polar properties, solvents may influence the reaction mechanisms of the excited state of the drug molecule (D*) and thus the photochemical stability of the compound. The fluorescence of many aromatic molecules is quenched by protic solvents such as water, alcohols, amines, and acids. The effect results from protonation of the drug (D) in the excited state (D*):

$$D \xrightarrow{\text{hv}} D* \qquad\qquad \text{Excitation}$$

$$\text{(Scheme 3)}$$

$$D* + H^+ \rightarrow (DH)^+ \rightarrow D + H^+ \quad \text{Protonation and quenching}$$

A general mechanism also exists for the quenching action of polar solvents on compounds that are more polar in the excited state (D*) than in the ground state (D). Thus, the effective transition energy (D* → D) diminishes with increasing solvent polarity. However, there are cases of increased luminescence caused by polar solvents, resulting from physical alterations of the excited state (Suppan and Nagwa, 1997). The electronically excited state of a drug molecule (D*) can act as an electron donor or acceptor, resulting in the formation of highly reactive free radical ions. The nature of the reaction will depend on the redox properties of D* and the participating compound, which may be the solvent or excipients in the formulation. The electron transfer process is favored in strongly polar solvents such as water because the electron is readily stabilized by the polar solvent (see Chapter 2).

Some solvents are able to filter parts of the optical spectrum — an effect that may provide some protection of the dissolved or suspended drug. However, absorption of photons may lead to formation of free radicals from solvent molecules, resulting in chain reactions that are well documented for unsaturated fats and oils (Florence and Attwood, 1998). Drugs dissolved or suspended in such vehicles may participate in the free radical chain reactions, leading to photosensitized decomposition of the drug induced by the solvent.

Many of the reactive species generated by absorption of optical irradiation, i.e., excited states, singlet oxygen, and free radicals, are very short lived with half-lives in the range of nanoseconds (several singlet excited states); microseconds (singlet oxygen); and milliseconds (many triplet excited states and free radicals) (Bayley et al., 1987). The lifetimes of these reactive intermediates and the viscosity of the medium determine the mean distances that the species will diffuse before they are inactivated by decay to the ground state or by chemical reactions.

In water, short-lived singlet excited states of organic molecules are essentially immobile, while singlet oxygen can diffuse a limited distance (about 33 nm) prior to deactivation (Bayley et al., 1987). These intermediates can participate in chemical reactions at the site of absorption with molecules located in the near vicinity. Triplet excited states and free radicals of small organic molecules can diffuse several micrometers in water and have a larger potential to deactivate by chemical reactions with other substances in the formulation (Bayley et al., 1987).

Adding viscous cosolvents, e.g., glycerol, to the parenteral formulation will reduce the mean diffusion path of the reactive intermediates and other reactive species. The diffusion-controlled rate constant (k_D) of a given chemical reaction is inversely proportional to the viscosity of the formulation (Suppan and Nagwa, 1997):

$$k_D = 8RT/3000 \, \eta \qquad (14.8)$$

where:

  R = universal gas constant
  T = temperature (Kelvin)
  η = viscosity (Poise)

For example, the second-order diffusional rate constant in water at 20°C is $6.5 \cdot 10^9$ dm^{3}mol^{-1}s^{-1}, while the constant is reduced to $4.3 \cdot 10^6$ dm^{3}mol^{-1}s^{-1} in glycerol at the same temperature. The reason is the much higher viscosity of glycerol (15 P at 20°C) compared to water (0.010 P at 20°C) (Suppan and Nagwa, 1997).

A wide variety of molecules act as collisional quenchers. Molecular oxygen is one such quencher, possessing diffusion-controlled reaction rates (see Section 14.2.4). In the presence of a quenching agent, the lifetime (τ) of a reactive intermediate is inversely proportional to the collisional quenching constant (K_q) and the concentration of the quencher, [Q], when the quenching reaction is diffusion controlled (Lakowicz, 1983):

$$\tau = \tau_0 \big/ \left(1 + K_q[Q]\right) \qquad (14.9)$$

where τ_0 = lifetime in the absence of quenching.

Thus, an increase in viscosity, leading to a reduced quenching rate constant (Equation 14.8), will result in an increased lifetime of the reactive intermediate (Equation 14.9). Depending on the nature of the diffusion-controlled reaction and the reactivity of the products formed, a rise in viscosity by addition of cosolvents like glycerol to the parenteral preparation can lead to improved or reduced photochemical stability of the formulation. Increased viscosity will stabilize the product if the diffusion-controlled reaction leads to formation of reactive intermediates and a subsequent degradation of the drug substance or excipients. Photochemical destabilization is observed when the diffusion-controlled reaction leads to a decay to the ground state and no further attack on the drug substance or excipients in the preparation. The effect of molecular oxygen on photochemical stability is discussed in Chapter 2 and in Section 14.2.4.

14.2.3 Buffers and pH

The photochemical reactivity of an ionizable drug will depend on the degree of protonation, which influences the molecular electronic distribution and charge of the compound, as well as formation of intramolecular bonds and interactions with the vehicle. These factors are essential for the chemical reactivity as well as the described absorption properties of the drug (Section 14.2.2). Acidification of the medium will lead to protonation of nitrogen atoms in amines and blocking of the easily excited n-electrons in several organic molecules. The antimalarial drug primaquine is photochemically stabilized by these mechanisms, which also leads to a red shift of the absorption spectrum (Kristensen et al., results to be published). Photochemical reactivity as a function of pH has been published for several drugs, e.g., ciprofloxacin, minoxidil, chloroquine, and sulfamethoxazole (Torniainen et al., 1996; Chinnian and Asker, 1996; Nord et al., 1997a; Moore and Zhou, 1994).

The pH of parenteral preparations should be close to physiological pH when possible. Injections intended for intrathecal, peridural, or intracisternal administration must be adjusted to physiological pH due to the risk of aseptic meningitis caused by nonphysiological solutions (Ford, 1988). Large-volume parenterals intended for intravenous administration ought to have a pH close to physiological conditions, but deviation is accepted if the preparation is not buffered. Unbuffered saline or glucose solutions are frequently used as dissolution media for intravenous infusions (see Table 14.3). The pH of the preparation will thus depend on the pH of the solution and the concentration and pKa value of the dissolved drug.

Buffers are used to adjust the pH of formulations in order to optimize stability and solubility, and to produce biocompatible products. Phosphate, citrate, and acetate are the most common buffer salts used in parenteral products (Nema et al., 2002). Buffer salts can catalyze the photochemical decomposition of drugs. The photochemical degradation rate of daunorubicin is increased in the presence of phosphate, citrate, or acetate salts (Islam and Asker, 1995). A correlation exists between degradation rate and concentration of the three buffer systems investigated. Experiments indicate that divalent phosphate ions (HPO_4^{2-}) have the highest catalytic effect on the photochemical decomposition rate. Photochemical stability of daunorubicin can be optimized by selecting the less destabilizing buffer (i.e., acetate) and keeping the buffer capacity as low as possible.

Buffer salts can influence photochemical reactivity by several mechanisms and deplete or increase photochemical stability of the product. Salts may influence photochemical stability by initiating solvation of the molecular ground state or excited state to various extents. Solvation can lead to absorption shifts into the red

TABLE 14.3   Physicochemical Properties of Selected Solutions Commonly Used for Infusions

Formulation Parameter	Glucose (50 mg/ml)	Sodium Chloride (9 mg/ml)	Ringer-Acetate	Macrodex® (60 mg/ml with sodium chloride)
Solvent	Water for injection	Water for injection	Water for injection	Water for injection
Sugar	Glucose			
Salts		NaCl	NaCl; Na-acetate; KCl; CaCl$_2$; MgCl$_2$	NaCl
pH adjustor			HCl	
Polymer				Dextran 70
Osmolality (mosmol/kg)	290	290	270	300
pH	4	5	6	4–7
Buffer capacity	NO	NO	YES	NO

(low-energy) or blue (high-energy) area in the UV and visible regions. Buffer salts are able to participate in acid-base reactions with molecules also in the excited state — reactions that are not easy to predict.

The pKa value can vary greatly with the electronic state of the molecule. Examples are phenols, which are bases in the ground state (pKa ~ 9) but become strong acids in the lowest singlet excited state (pKa ~ 2) (Suppan and Nagwa, 1997), and the amino acid tyrosine, which has pKa > 10 in the ground state and pKa ≤ 4 in the excited state (Lakowicz, 1983). The proton donor and proton acceptor properties of the buffers will have influence on the acid–base reactions between the salts and the drug in the ground and excited states. The lifetime of the drug excited state is a critical factor that will influence probability for a proton exchange (see Section 14.2.2).

Some salts are able to form complexes with drug molecules in the formulation, and thus influence photochemical stability. Citrates are common buffers that can have a dual role as chelating agents. Recent studies show an interaction between primaquine and citrate in aqueous solutions. The photochemical stabilization of primaquine occurring in the presence of citrate is probably caused by the observed interaction between the drug and the buffer. The mechanisms involved are under investigation (Kristensen et al., to be published).

14.2.4 Oxygen Content

Molecular oxygen is present in all solvents at atmospheric conditions. The molecule is an efficient collisional quencher, which diffuses rapidly in most solvents. Molecular oxygen is a triplet in the ground state, and thus spin-matched with triplet-excited photosensitizers (e.g., excited drug molecules). Singlet molecular oxygen ($^1O_2$) is easily formed by electronic energy transfer from excited triplet photosensitizers (e.g., drugs) to ground state molecular oxygen with a subsequent regeneration of the photosensitizer.

Many compounds are capable of acting as sensitizers for singlet oxygen formation because of the small energy gap between the triplet ground state and the singlet excited state of the molecule (see Chapter 2). Singlet molecular oxygen is deactivated with emission of luminescence in the infrared region, or by physical or chemical quenching reactions with the drug or excipients in the formulation. The photochemical stability of a preparation is increased by collisional quenching with molecular oxygen when singlet oxygen is deactivated without attack on any components in the formulation. This is demonstrated for chloroquine, which is photochemically stabilized in the presence of molecular oxygen (Nord et al., 1997a).

Molecular oxygen is a very good scavenger of free radicals. The molecule adds rapidly to free radicals or free radical ions formed in photochemical reactions, subsequently leading to free radical chain reactions. The result can be decomposition of the drug substance or destruction of the formulation, unless scavengers of free radicals are added to protect the preparation (see Section 14.2.5). The photosensitized oxidation reactions very often lead to generation of other reactive oxygen species in addition to singlet oxygen (i.e., hydroxyl radicals, OH·, superoxide radicals, O_2·− and peroxyl radicals, RO_2·) and the toxic and reactive substance hydrogen peroxide (H_2O_2) (see Chapter 2).

These reactive species will likely attack and decompose the drug or excipients in the formulation. This was demonstrated for primaquine (Kristensen et al., 1998), which can be photochemically stabilized by an inert atmosphere. Photo-oxidation reactions of type I (free radical) or type II (singlet oxygen) mechanisms can take place simultaneously in a competitive fashion. Oxygen concentration and the properties of the vehicle are factors influencing the distribution between the two processes. Free radical reactions are favored by polar vehicles such as water.

The use of an inert atmosphere in the container can be an approach to limit photosensitized oxidative decomposition of the preparation. Small-volume parenterals, such as ampoules and multidose vials, have a large headspace gas to formulation volume ratio. In nonpurged packages, this is offering an ample reservoir of molecular oxygen. Purging the preparation with an inert gas (e.g., nitrogen or argon) during production and filling into ampoules that are immediately sealed will considerably suppress the oxygen content and limit the reactions involving oxygen. However, large-volume parenterals intended for intravenous administration (intravenous infusions) should not be depleted of their oxygen content prior to injection. Photosensitive intravenous infusions decomposed by oxidative reaction mechanisms should be produced *ex tempore* and administered immediately. When held in a transparent container, the contents may be protected by an outer colored envelope or by wrapping with aluminum foil (see Section 14.3).

14.2.5 Antioxidants

Antioxidants are added to the formulation to prevent oxidation of active substances or excipients in the finished product. They should only be included in a formulation if their use cannot be avoided, e.g., by purging the product with an inert gas (see Section 14.2.4) (CPMP, 1998). Compatibility of antioxidants with the drug and the packaging system must be thoroughly evaluated, and the chosen concentration must be justified.

Antioxidants are divided into categories based on their main mechanism of action (Nema et al., 2002; CPMP, 1998). The *free radical scavengers* act as terminators of free radical chain reactions. The chain process is inhibited by the formation of relatively nonreactive free antioxidant radicals. Butylated hydroxytoluene (BHT) and butylated hydroxy anisole (BHA) are two such radical scavengers that are widely used as antioxidants in semi- or nonaqueous vehicles because of their low aqueous solubility. Addition of free radical scavengers can provide protection against photosensitized oxidation reactions of the type I (free radical) category and against reactive oxygen species formed during the photosensitized reactions.

Reducing agents such as sulfite, bisulfite, and metabisulfite are preferentially oxidized due to a lower redox potential than the drug or other excipients in the formulation. Salts of sulfite, bisulfite, and metabisulfite are oxygen scavengers with a high aqueous solubility; they are the most common antioxidants in aqueous parenteral products. Another important compound widely used in aqueous parenteral products is ascorbic acid/ascorbate, which may also serve as a chelating agent and buffer in the same formulation. Reducing agents are not necessarily regenerated *in vitro* and may be consumed during the shelf-life of the product. Consumption of ascorbic acid/ascorbate can lead to a change of formulation properties due to the additional functions of the excipient. Moreover, L-ascorbic acid is reported to be a photolabile compound in aqueous solutions due to decomposition through free radical reaction mechanisms (Ho et al., 1994), and should be stored protected from light (*European Pharmacopoeia*, 2002; Parfitt, 1999). Thus, ascorbic acid is a potential photosensitizer in parenteral formulations.

The *antioxidant synergists* enhance the effect of antioxidants. Chelating agents like ethylenediaminetetraacetic acid (EDTA) are commonly added to parenterals. They may reduce oxidative damage by forming complexes with oxidative metal ion catalysts. Chelating agents are further discussed in Section 14.2.6.

Sulfite (SO_3^{2-}), bisulfite (HSO_3^-), and metabisulfite ($S_2O_5^{2-}$) are antioxidants that have been used in aqueous parenteral preparations for decades. They will be present in equilibrium and potentially initiate radical chain reactions. Due to several reports of incompatibility and toxicity, they are considered especially undesirable in drug formulations (Munson et al., 1977; Enever et al., 1977). Nevertheless, these compounds constitute the majority of antioxidants used in parenteral products (Nema et al., 2002). Traditionally, the sulfites are considered to be oxygen scavengers, thus their wide application as antioxidants. Sulfites may lead to reduced photochemical decomposition of drugs, illustrated by the photochemical stabilization of daunorubicin in the presence of sodium sulfite (Islam and Asker, 1995). However, reactive hydroxyl radicals (OH·) are demonstrated to be formed in the reaction between sulfite and superoxide ($O_2^{·-}$) (Stewart and Tucker, 1985). In theory, one superoxide radical, which may be produced in the photochemical reactions, can give rise to two highly reactive hydroxyl radicals in the presence of sulfite, thus accelerating decomposition of the drug. This may be partly the reason for the increased photochemical decomposition of epinephrine infusions observed in the presence of sodium metabisulfite (Brustugun et al., 1999). However, photochemical decomposition of epinephrine accelerated by metabisulfite has a complex reaction pattern. Present studies indicate that a photosensitizer is formed *in situ* by thermal (dark) reactions in the epinephrine infusion. Metabisulfite is directly involved in the production of the photosensitizer, which seems to initiate degradation by photochemical formation of singlet oxygen. The reactions are dependent on the oxygen content and the concentration of metabisulfite (Brustugun et al., 2004a). Addition of sulfites to pharmaceutical products can thus lead to photochemical destabilization by the formation of reactive free radicals and photosensitizers in the formulation. Sulfites should clearly not be selected as stabilizers in pharmaceutical products unless the influence on photochemical stability is thoroughly evaluated.

14.2.6 Metal Ions and Chelating Agents

Aqueous parenteral preparations can contain trace amounts of heavy metal ions in concentrations sufficient to catalyze oxidative reactions. Aqueous parenterals are produced with the use of *Water for injection*, which complies with the limit test for heavy metals (*European Pharmacopoeia*, 2002). This is, however, no guarantee for exclusion of metal ions. Heavy metal contamination brought into the formulation by excipients is also a problem, especially for sugars, phosphate, and citrate (Nema et al., 2002). Heavy metals may also be extracted from the container by the preparation (*European Pharmacopoeia*, 2002; see Section 14.3). Moreover, trace elements like zinc, copper, manganese, and chromium constitute important components in several parenteral nutrition formulas (Trissel, 2001).

Metal ions can accelerate or inhibit photochemical reactions by influencing the formation of free radicals or by forming complexes with organic molecules in the formulation. The reaction mechanisms and the overall effect on photochemical stability will depend on the metal ion species, the drug, and the composition of the formulation. The influence of various metal ions on photochemical reactivity can be illustrated by photochemical decomposition of nitrofurazone (Shahjahan and Enever, 1996b). The photochemical degradation rate of the drug substance is increased in the presence of ferric ions (Fe^{3+}), while cupric ions (Cu^{2+}) have the opposite effect, leading to photochemical stabilization.

The cupric ion is a quencher of excited states and can regenerate the ground state of an excited molecule by an exchange of electrons (Lakowicz, 1983) — possibly resulting in reduced photochemical degradation in the presence of cupric ions, as observed for nitrofurazone (Shahjahan and Enever, 1996b). Ferric ions and ferrous ions (Fe^{2+}) can provoke a whole series of radical reactions in the presence of hydrogen peroxide (H_2O_2), a process called the Fenton reaction (Halliwell and Gutteridge, 1985). Hydrogen peroxide is often formed during photochemical reactions and may thus be present in the formulation (see Chapter 2). The Fenton reaction leads to formation of the highly reactive hydroxyl radical (OH·), which decomposes organic molecules.

Cuprous ions (Cu^+) also react with hydrogen peroxide to make hydroxyl radicals (Halliwell and Gutteridge, 1985). The presence of iron salts and cuprous ions in the formulation can lead to an accelerated photochemical degradation of the drug by these reaction mechanisms. Metal ions can also participate in redox reactions with the drug or excipients in the preparation, depending on the redox properties of the species involved. Such reactions may further influence photochemical stability of the product, e.g., by the formation of photosensitizers.

Chelating agents serve to complex heavy metals, and may improve or reduce photochemical stability of the product by influencing the previously described reactions. Only a limited number of chelating agents are used in parenteral preparations; salts of EDTA (ethylenediaminetetraacetic acid) are the most frequently used substances (Nema et al., 2002). The photochemical stability of an aqueous preparation of riboflavin is observed to increase as a function of EDTA concentration (Asker and Habib, 1990). However, complexation with EDTA can lead to photochemical destabilization of the product because EDTA-chelated iron salts react faster in certain radical reactions than the unchelated ions (Halliwell and Gutteridge, 1985). Citric acid, tartaric acid, ascorbic acid, and some amino acids can also serve as chelating agents in parenteral preparations. These compounds have additional functions that may influence photochemical stability of the product, e.g., by serving as buffers or antioxidants.

14.2.7 Tonicity Adjusters, Preservatives, Bulking Agents, Protectants

Parenteral formulations often contain excipients considered to be chemically stable and inert; however, all excipients in a formulation may influence the photochemical stability of the product. Dextrose and sodium chloride are used to adjust tonicity in the majority of parenteral formulations. Sodium chloride can affect photochemical processes by influencing solvation of the photoreactive molecules (see Section 14.2.3). The ionic strength is reported to affect the photochemical decomposition rate of minoxidil until a saturation level is reached (Chinnian and Asker, 1996). The photostability of L-ascorbic acid (vitamin C) in aqueous solution is enhanced in the presence of dextrose, probably caused by the scavenging effect of the excipient on hydroxyl radicals mediated by the photolysis of ascorbic acid; sucrose, sorbitol, and mannitol have the same effect (Ho et al., 1994). Monosaccharides (dextrose, glucose, maltose, and lactose), disaccharides (sucrose and trehalose), and polyhydric alcohols (inositol, mannitol, and sorbitol) are examples of commonly used lyo-additives in parenterals. These excipients may also affect photochemical stability of the products after reconstitution.

The amino acid histidine is used as a bulking agent for lyophilization and can additionally serve as a buffer (histidine/histidine hydrochloride) and stabilizer in the formulation (Nema et al., 2002). Histidine is a rather efficient quencher of singlet oxygen and a scavenger of hydroxyl radicals (Halliwell and Gutteridge, 1985). Thus, the presence of histidine in the parenteral formulation can be very important for the photochemical stability of the product.

Methylene blue is an ingredient listed in the *Inactive Ingredient Guide* published by the FDA (1996). The compound is accepted as an excipient in parenteral preparations. However, methylene blue is a well-documented generator of singlet oxygen and superoxide radicals in the presence of light (Kuramoto and Kitao, 1982). The dark blue compound is an efficient absorber of visible light, and acts as an effective photosensitizer. Methylene blue dissolved in aqueous solution generates photoinduced reactive oxygen species also when the preparation is stored in brown glass vials. Primaquine is decomposed in the presence of methylene blue when stored in colored glass containers during irradiation (Kristensen et al., 1998). Methylene blue is thus not a recommended excipient in any kind of drug formulation unless optical radiation is fully excluded from the preparation during production, storage, and use.

Benzyl alcohol and the parabens (methyl-, ethyl-, propyl-, and butyl parahydroxybenzoates) are the most common antimicrobial preservatives present in multidose aqueous parenteral formulations (Nema et al., 2002). Methyl paraben is reported to decrease the aerobic photodegradation of phosphate-buffered riboflavin phosphate (vitamin B_2) solutions by more than 70%, and the photostability of riboflavin is increased as a function of methyl paraben concentration. The photostabilizing effect can be attributed to the phenolic group of methyl paraben acting as a radical scavenger, thus playing a significant role through lowering the apparent quantum efficiency of riboflavin (Asker and Habib, 1990).

14.2.8 Disperse Formulations

A broad category of polymeric and surface active compounds can be added to parenteral preparations. Their function is to impart viscosity or act as suspending agents in the formulation (e.g., carboxy methyl cellulose (sodium), acacia, and hydrolyzed gelatine); to act as solubilizing, wetting, or emulsifying agents (e.g., polysorbates (Tweens); sodium dodecyl sulphate; PEG castor oils; lecithin; and egg yolk phospholipids); and, in certain cases, to form gels. Polysorbates can be contaminated by residual amounts of peroxides (Nema et al., 2002). Peroxides are reactive species and may participate in photochemically initiated free radical chain reactions that can occur in the formulation (see Chapter 2).

Photochemical stability of suspensions and emulsions is a rather complicated area. The optical properties of a disperse system (transmission of photons through the formulation and spread of optical irradiation) will depend on the size of the particles or droplets in the disperse phase, the fractional relationship between the disperse and homogenous phases, flocculation in the system, and physicochemical properties of the disperse and homogenous phases. The photochemical stability of a drug formulated as an emulsion will partly depend on the photochemical reactivity of the drug in the lipophilic and hydrophilic phases. The distribution of the drug between the two immiscible phases is an essential aspect to consider as part of an evaluation. Influence of the solvent properties on photochemical reactivity is covered in Section 14.2.2.

Photochemical stability of the solid compound is an important aspect when the drug is formulated as a suspension. Photochemical stability of a drug in the solid state can depend on the polymorphic/pseudo polymorphic form of the compound, which is demonstrated for chloroquine, mefloquine, and furosemide (Nord et al., 1997b; Tønnesen et al., 1997; De Villiers et al., 1992). The crystal structure, molecular conformations, and surface of the particles can thus influence photoreactivity of a suspended drug. When formulated as a suspension, the drug should be in the form of the stable polymorph. Transformation can occur between different crystal forms in the presence of a liquid, often accompanied by caking of the crystals (Martin, 1993). Photochemical stability of solids is further discussed in Chapter 16.

Surface active compounds can form micelles in the formulation. Drugs may diffuse into various layers of the micelles, leading to a change of the microenvironment surrounding the drug. The environmental changes include reduced polarity, increased molecular oxygen level, and altered chemical properties of the surrounding medium; they are likely to influence photochemical stability of the drug. Photochemical stability of nitrofurazone in an aqueous solution is reduced in the presence of nonionic surfactants or PEG 1000, which are added to improve solubility and obtain therapeutic concentrations of the drug (Shahjahan and Enever, 1996a, b). The stability decreases as a function of increased lipophilicity or increased viscosity of the vehicle. In this case it is possible to optimize photochemical stability of the product by using the most appropriate solubilizer (i.e., resulting in a preparation of high polarity and low viscosity).

The chemical properties of the micelles will have an influence on the result obtained. Irradiation of riboflavin in the presence of tryptophan leads to formation of cytotoxic photoproducts. The presence of neutral micelles favors photoproduct generation, which is hindered by ionic micelles (Bueno et al., 1999). Solubilization of an ionic drug in ionic micelles will alter the apparent pKa of the drug by up to two units, thereby shifting the pH dependence of the drug's photostability. This was demonstrated for chlorpromazine (a basic drug) and frusemide (an acidic drug) with respect to their photosensitization of free radical and singlet oxygen reactions in micellar solutions of cetrimide (a cationic surfactant) and sodium dodecyl sulfate (an anionic surfactant) (Moore and Burt, 1981).

Surfactants are able to extract organic additives from plastic materials. Cremophor EL is a mixture of nonionic surfactants used as a vehicle in various intravenous injections. It is incompatible with polyvinyl chloride containers due to extraction of phthalates from the polymer (Parfitt, 1999; Nuijen et al., 1999, 2001). Because even trace amounts of organic molecules may act as photosensitizers in the formulation, extraction of organic additives from the container should be avoided (see Section 14.3).

Molecules of low molecular weight (e.g., drugs) are likely to adsorb to polymers present in the formulation. Adsorption occurs by the formation of weak (localized) interactions, hydrogen bonds, or ionic bonds between molecules and polymers. Microcrystalline cellulose, which is an important pharmaceutical excipient, has been demonstrated to influence chemical and photochemical reaction mechanisms of adsorbed compounds (Wilkinson et al., 1991). Interactions between drugs and polymeric compounds and the subsequent influence on photoreactivity are further discussed in Chapter 15.

Preparations for total parenteral nutrition (TPN) are complex formulations intended for administration by the intravenous route. TPN preparations are formulated as aqueous solutions or hydrophilic (oil-in-water) emulsions; they may contain amino acids, carbohydrates, fatty acids, emulsifiers, electrolytes, trace metals, vitamins, and minerals (Hutchinson, 1998; Trissel, 2001). In certain cases, drugs are added to the preparations prior to administration. The environment is thus rather heterogeneous, and the photochemical stability of different components can vary from formulation to formulation and be hard to predict. It is necessary to perform experiments based on studies of the actual composition to obtain correct information concerning photostability of the formulation or components present.

Several vitamins are known to be photolabile, and the photochemical stability of these compounds is influenced by TPN composition. The photochemical stability depends on composition of the amino acid solutions as well as the presence of lipids in the preparations (i.e., the formation of emulsions). Photochemical decomposition of the lipophilic vitamin A is reduced in admixtures containing lipids, possibly due to diffusion of the vitamin into the lipophilic phase. On the other hand, the hydrophilic vitamin riboflavin is protected by emulsification, probably because the opaque emulsion will reduce the optical transmission of the preparation to some extent (Smith et al., 1988). However, emulsification protects neither the water-soluble vitamin C nor the lipohilic vitamins A and K1 from photochemical degradation, which illustrates the complexity of photochemical reactions in heterogeneous media (Smith et. al., 1988; Billionrey et al., 1993).

The droplet size distribution of the emulsions may change as a consequence of photochemical reactions in TPN formulations. Physical stability of the emulsion is an important issue for patient safety because coalescence of the disperse phase and a subsequent increase in globule size could result in thrombosis *in vivo* (Ford, 1988). Thus, stability testing of TPN emulsions should also include size distribution analyses after exposure to irradiation, as described by Williams et al. (1990). Ideally, the emulsion should be formulated so that the disperse droplets have a size distribution corresponding to the chylomicra (500 to 1000 nm), which are the natural transport systems for fat through the blood stream (Ford, 1988). The size of the disperse droplets should not be affected by the storage temperature or exposure to optical irradiation. However, it is important to note that addition of any substance (e.g., a drug) to a photochemically stable TPN preparation may alter the photoreactivity and thus the photochemical stability of the formulation.

14.2.9 Drug Carriers

Parenteral products can be formulated using drug carriers. The purpose is to improve solubility of poorly soluble drugs and thus enhance bioavailability and affect pharmacokinetic behavior, improve physical and chemical properties (e.g., increase stability), and obtain drug targeting of the active substance. Incorporation of drugs into cyclodextrins, liposomes, or other drug carrier systems can lead to a change in photochemical stability as an additional effect; the drug is in equilibrium between the carrier and the dispersion medium. Interactions of a weak or strong character are established between the drug and the carrier, e.g., hydrophobic interactions, hydrogen bonding, or electrostatic interactions (Bortolus and Monti, 1996). The chemical microenvironment surrounding the drug is changed after incorporation. All these parameters can influence photochemical behavior.

Cyclodextrins will easily form inclusion complexes with several lipophilic drugs, leading to improved pharmaceutical and chemical properties such as solubility, chemical reactivity, pKa value, diffusion properties, and spectral properties (Bortolus and Monti, 1996). As a result of the complexation, photochemical stability of the included drug may be improved or reduced (Ammar and El-Nahhas, 1995; Chen et al., 1996; Mielcarek, 1996). Inclusion of riboflavin into γ-cyclodextrins increases photochemical stability sixfold (Loukas et al., 1995). Factors essential for the resulting effect are the size and the chemical properties of the cyclodextrin, physicochemical properties of the drug, stoichiometry and stability of the inclusion complex, and three-dimensional structure of the inclusion complex.

Development of parenteral formulations of all-*trans*-retinoic acid is possible by using 2-hydroxypropyl-β-cyclodextrin, and inclusion into the carrier also leads to an improved photostability of the drug (Lin et al., 2000). Aqueous cyclodextrin solutions are shown to extract plastic additives when stored in containers of polypropylene (Peiris et al., 1999; see Section 14.3). The result may be photosensitized decomposition of the drug. Photochemical stability of drug–cyclodextrin inclusion complexes are further discussed in Chapter 16.

Photochemical stability of photolabile drugs can be improved by inclusion into liposomes. Liposomes are vesicles consisting of one or several membrane-like alternating bilayers of phospholipids forming aqueous compartments. Water-soluble and lipid-soluble drugs may be accommodated in the aqueous compartment and the lipid layer phases, respectively. Certain macromolecules can insert their hydrophobic regions into the lipid bilayers with the hydrophilic portions extending into the water phase (Florence and Attwood, 1998). Entrapment of the photolabile vitamin riboflavin into the aqueous phases of multilamellar vesicles increases the photochemical half-life four times under given experimental conditions (Loukas et al., 1995).

Multicomponent liposomes reveal optimal UV protection in combination with the γ-cyclodextrin complex of the vitamin present in the aqueous phases and lipid-soluble protectors (UV absorbers and antioxidants) present in the lipid phases (Loukas et al., 1995). The experimental photochemical half-life of riboflavin can be increased up to 266-fold by these complex liposome formulations. The liposomal composition is also important for the stabilizing effect obtained as incorporation of riboflavin into neutral and negatively charged liposomes increases photochemical stability, while association with positively charged liposomes leads to a decreased photochemical stability (Habib and Asker, 1991). Incorporation of retinol into multilamellar liposomes is demonstrated to extend the shelf-life of retinol under various conditions of pH, temperature, and light exposure (Lee et al., 2002).

14.3 INFLUENCE OF THE CONTAINER

A container for pharmaceutical use is the article that contains and is in direct contact with the product, including the closure. The container is considered part of the formulation (*European Pharmacopoeia*, 2002), provides protection and delivery of the product, and should not interact physically or chemically with the contents. Glass and plastic are the most commonly used materials and closures are made by elastomers such as rubber or silicone (*European Pharmacopoeia*, 2002).

Parenteral preparations are filled into various types of containers, depending on the nature of the product. Single-dose injections are filled into glass ampoules sealed by fusion or *ex tempore* into plastic syringes. Multidose injections are delivered in glass vials sealed with rubber closures with mechanical properties suitable for multiple piercing. Concentrates and powders for injections or infusions are also

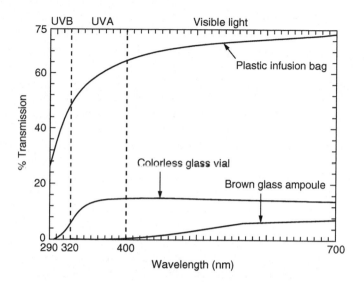

Figure 14.1 Optical properties of pharmaceutical containers.

dispensed in glass vials. Bags or bottles of plastic materials are used as containers for aqueous solutions for parenteral infusion and TPN emulsions. It is recommended that containers for preparations for parenteral use permit a visual inspection of the contents (*European Pharmacopoeia*, 2002). Clear glass is highly transparent to UV and visible radiation (Figure 14.1) and does not protect the contents from exposure even in the UVB region (see Chapter 2). The *European Pharmacopoeia* allows the use of colored glass for substances known to be photosensitive and describes limits of transmission of optical irradiation (290 to 450 nm) in colored light-protecting glass containers intended for parenteral preparations (*European Pharmacopoeia*, 2002). However, the stabilizing effect of amber glass is not always satisfactory for highly photolabile drugs, as demonstrated for molsidomine (Thoma and Kübler, 1996).

The chemical stability of glass containers for pharmaceutical use is expressed by the hydrolytic resistance, i.e., the resistance to release of water-soluble mineral substances, evaluated by titrating the released alkalinity. According to the *European Pharmacopoeia* (2002), aqueous preparations for parenteral use are to be dispensed into glass containers of high hydrolytic resistance, while nonaqueous preparations and powders for parenteral use can be filled into glass containers of moderate hydrolytic resistance. It is obvious that release of alkaline substances may influence photochemical stability by an increase in pH (see Section 14.2.3).

Plastic bags or bottles are commonly used as containers for aqueous parenteral solutions. Plastic containers are most frequently composed of polyethylene, polypropylene or poly(vinyl chloride) (*European Pharmacopoeia*, 2002). More than 60% of the plastic containers for parenteral preparations are made of the polyolefins (Fischer, 2002). Containers made of poly(ethylene-vinyl acetate) (EVA bags) are suitable as containers for TPN preparations because they do not have added plasticizers, which easily diffuse into fat components (*European Pharmacopoeia*, 2002; Hutchinson, 1998). The containers must be sufficiently transparent to allow the appearance of the contents to be examined at any time. Plastic materials transmit most of the optical irradiation in the UVB, UVA, and visible region (Figure 14.1) and provide even less protection against photochemical degradation than clear glass containers. However, plastic containers may be enclosed in a protective (colored) envelope to provide adequate stability of preparations sensitive to optical irradiation.

Alternatively, the container may contain an opacifying agent, e.g., titanium dioxide. The use of multilayered bags was demonstrated to inhibit photochemical decomposition of vitamin E in TPN fat emulsions (Allwood and Martin, 2000). PVC films discolor on irradiative exposure due to photochemical degradation of the polymer (Hollande and Laurent, 1997). Plastic containers for parenteral use may contain several additives, e.g., antioxidants, stabilizers, plasticizers, lubricants, impact modifiers, and coloring matter when justified and authorized. In an appendix, the *European Pharmacopoeia* presents a list of plastic additives that may be used (*European Pharmacopoeia*, 2002). The additives should not be extracted by the contents in such quantities as to alter efficacy or stability of the product or to present any risk of toxicity (*European Pharmacopoeia*, 2002). However, organic additives extracted in concentrations below the detection limits of the analytical methods authorized by the *European Pharmacopoeia* may be sufficient to initiate photosensitized reactions in the formulation.

Exposure of low-density polyethylene containers to sunlight leads to photodecomposition of the polymer and a subsequent extraction of organic compounds, including phthalates, into water (Calvosa et al., 1994). Migration of organic additives and degradation products from plastic containers into parenteral preparations is further demonstrated by Sarbach et al. (1996) — a phenomenon likely to increase when the preparation is heat sterilized in the container. Surfactants or cyclodextrins present in the preparation can extract organic additives from plastic containers (Nuijen et al., 1999, 2001; Peiris et al., 1999). The *European Pharmacopoeia* defines the maximum concentration of extractable metals from polymeric plastic materials and elastomeric closures (*European Pharmacopoeia*, 2002), but even trace amounts of metal ions can have a significant influence on photochemical stability of the product.

Plastic containers, e.g., infusion bags and syringes, are regularly presterilized by the use of ethylene oxide. A limit value for the content of ethylene oxide is given by the *European Pharmacopoeia* (2002), but small amounts of this highly reactive substance extracted into the contents may be sufficient to influence stability, e.g., by the formation of photosensitizers or reactions with the drug substance. The presence of ethylene oxide was demonstrated to accelerate degradation of cisplatin in isotonic sodium chloride (Zieske et al., 1991).

The International Conference on Harmonization (ICH) has established guidelines for photostability testing of new drug substances and products (ICH, 1997). Photochemical stability of the formulation must be evaluated in the final form of the product. However, the guidelines do not cover products that were on the market before 1998.

14.4 *EX TEMPORE* PREPARATIONS OF PARENTERAL PRODUCTS

Parenteral preparations are regularly prepared aseptically a short time or immediately prior to administration. Compounds susceptible to hydrolysis or oxidative decomposition in solution are preferentially stored as dry powders, concentrates under an inert atmosphere, or in combination with stabilizers. Concentrates for injections or infusions (*European Pharmacopoeia*, 2002) are diluted prior to administration, usually with sterilized *Water for injection* (*European Pharmacopoeia*, 2002) or sterile, isotonic solutions of sodium chloride, glucose, dextran, or buffer (see Table 14.3). Powders for injections or infusions (*European Pharmacopoeia*, 2002) are dissolved or suspended in the same media. Vitamins are aseptically added *ex tempore* to TPN preparations due to poor stability and the risk of precipitation (Hutchinson, 1998), as are trace metals that may influence the stability of the TPN formulation. A limited number of drugs may also be dissolved in the TPN infusion prior to administration (Hutchinson, 1998).

According to the ICH guidelines for testing photostability, drug substances available as powders or concentrates for injections or infusions need only be tested prior to their final dilution or reconstitution (ICH, 1997). As previously described, photochemically stable powders for injections or infusions can become photolabile after dissolution (Section 14.2.1). This is also the case for concentrates for injections or infusions, which in dilution may become photolabile due to transmission of optical irradiation through the sample and a change of the relative photochemical degradation rate (Section 14.2.1).

Concentrates or powders for injections or infusions may be diluted, dissolved, or suspended *ex tempore* in several different media prior to administration. The most common solutions for this purpose are listed in Table 14.3. These are sterile, isotonic, aqueous solutions, but differ with respect to pH, ionic strength, buffer capacity, and chemical composition. As described in Section 14.2, these physicochemical properties may be highly important for the photochemical stability of a dissolved compound; choice of medium can be critical and photochemical behavior can change dramatically in different media.

A study of catecholamines dissolved and irradiated according to the ICH guidelines (UVA and visible irradiation) in various infusion solutions can serve as an illustration (Brustugun et al., 2004b). Adrenaline, dopamine, and isoprenaline are photolabile in all the isotonic aqueous media investigated, i.e., glucose 50 mg/ml; sodium chloride 9 mg/ml; Macrodex® 60 mg/ml with sodium chloride; and Ringer-acetate (Table 14.3). The drugs are decomposed to various extents depending on the properties of the dissolution media and chemical properties of the drugs. Adrenaline is formulated as a concentrate for injections and infusions in combination with the antioxidant sodium metabisulfite. As described in Section 14.2.5, the presence of metabisulfite decreases photochemical stability of adrenaline. The photolability of adrenaline in the presence of metabisulfite depends on the physicochemical properties of the medium.

The initial pH of the infusion solutions has a significant effect, even though the actual pH range (4 to 7; see Table 14.3) will not influence the protonation of adrenaline to a great extent (pKa values = 8.69 and 9.90). The pH of the medium seems to be important for the photochemical reaction mechanisms because the acidification of the medium, which occurs during photodegradation, leads to an autoprotection against further photodecomposition (Brustugun et al., 2004b). Ringer-acetate (pH = 6) destabilizes adrenaline photochemically, probably because the presence of buffer inhibits acidic autoprotection.

Viscosity is a property undocumented by the producer, but is important for the photochemical stability of adrenaline due to decreased degradation in viscous media. However, the most interesting phenomenon is the influence of glucose, which has a significant destabilizing effect on the photochemical stability of adrenaline. The effect is especially pronounced in the absence of metabisulfite. The reaction mechanisms are under study in our laboratory. It is clear from this work that the medium selected for parenteral administration of adrenaline will influence the effective dose administered and the formation of degradation products unless the product is protected from irradiation during production, storage, and administration (Brustugun et al., 2004b).

14.5 STORAGE AND ADMINISTRATION OF PARENTERAL PRODUCTS

As described in the previous sections, plastic and glass containers used for the production of parenterals offer little or no protection against optical irradiation (Figure 14.1). It is therefore important to protect photolabile products from sunlight and artificial light sources by using an outer, nontransparent container or a colored outer bag, or wrapping with aluminum foil.

It is highly recommended that parenteral products be stored in the original outer container. Information concerning storage must be given to hospital ward personnel and also should be included on the product label. Parenteral preparations produced *ex tempore* (e.g., infusion bags) are usually not protected by an outer container. These products should not be stored on an open bench or in a case without a lid, but rather be kept in a dark locker or covered by a nontransparent material (e.g., aluminum foil). Of course, storage in a refrigerator will protect the product against irradiation.

Administration of injections is not a great issue with respect to photostability because the administration time is rather short (seconds to minutes). Infusions, however, are administered over a long period of time. Administration of a single infusion bag can be extended to several days. Exposure to optical irradiation is thus a problem because of sunlight shining through the windows as well as indoor lightning. In some cases, the patient may also want to carry the infusion set outdoors during the administration. The preparation can then easily be protected by use of an outer protective envelope or an aluminum foil wrap. Although visual inspection of the content will still be easy to perform, in busy wards this is rarely a prioritized task. Wrapping the administration set can be recommended in certain cases, for instance, when the administered drug is extremely photosensitive (e.g., nifedipine; Thoma and Klimek, 1985a, b) or when toxic photodegradation products are known to be formed during exposure (e.g., chloramphenicol; Mubarak et al., 1982; de Vries et al., 1984). For parenteral administration of dacarbazine, the use of opaque infusion tubing for protection of the photolabile drug is recommended (El Aatmani et al., 2002).

Today, at least in Europe, an increasing number of patients are provided with health care services at home (Hutchinson and Graham, 1998). Such services include supply of home parenteral nutrition (HPN), home dialysis, i.v., antibiotic therapy at home, and patient-controlled analgesia (PCA) (Hutchinson and Graham, 1998; Hutchinson 1998). Standard bags of TPN are prepared in certain hospital pharmacies, sealed into a dark-colored outer plastic bag, and stored in a refrigerator for several weeks (Hutchinson, 1998). Information on correct storage and administration of TPN preparations at home should be provided by the pharmacist and health care personnel involved, with a focus also on the protective effect of the outer colored bag.

Important concerns are the photochemical stability of vitamins, as described previously (Section 14.2.8), and drugs in PCA devices. The shelf-life of morphine solutions dispensed in plastic syringes used in PCA devices is reduced from 6 weeks to 1 week during light exposure. The structural analogue pethidine is photochemically stable in the same syringes (Strong et al., 1994), which illustrates the need for exact information to the patient on storage conditions. Medicinal home treatment is an area expected to increase in the future due to the focus on patient convenience as well as on economic issues. Adequate education of the health care personnel involved is a prerequisite for correct handling of pharmaceutical parenteral products during their storage and use.

REFERENCES

Allwood, M.C. and Martin, H.J. (2000) The photodegradation of vitamins A and E in parenteral nutrition mixtures during infusion, *Clin. Nutr.*, 19, 339–342.

Ammar, H.O. and El-Nahhas, S.A. (1995) Improvement of some pharmaceutical properties of drugs by cyclodextrin complexation. 2. Colchicine, *Pharmazie*, 50, 269–272.

Asker, A.F. and Habib, M.J. (1990) Effect of certain stabilizers on photobleaching of riboflavin solutions, *Drug. Dev. Ind. Pharm.*, 16, 149–156.

Bayley, H., Gasparro, F., and Edelson, R. (1987) Photoactivable drugs, *Trends Pharmacol. Sci.*, 8, 138–143.

Billionrey, F., Guillaumont, M., Frederich, A., and Aulagner, G. (1993) Stability of fat-soluble vitamin A (retinol palmitate), vitamin E (tocopherol acetate), and vitamin K1 (phylloquinone) in total parenteral nutrition at home, *JPEN J. Parenter. Enteral. Nutr.*, 17, 56–60.

Bortolus, P. and Monti, S. (1996) Photochemistry in cyclodextrin cavities, in Neckers, D.C., Volman, D.H., and von Bünau, G. (Eds.) *Advances in Photochemistry*, Vol. 21, New York: John Wiley & Sons, pp. 4–5.

Brustugun, J., Tønnesen, H.H., Klem, W., and Kjønniksen, I. (1999) Photodestabilization of epinephrine by sodium metabisulfite, *PDA J. Pharm. Sci. Technol.*, 54, 136–143.

Brustugun, J., Kristensen, S., and Tønnesen, H.H. (2004a) Photostabilty of epinephrine — the influence of bisulfite and degradation products, *Pharmazie,* accepted.

Brustugun, J., Kristensen, S., and Tønnesen, H.H. (2004b) Photostability of symphatomime-thic agents in commonly used infusion media in the absence and presence of bisulfite, *Pharmazie*, submitted.

Bueno, C.A., Silva, E., and Edwards, A.M. (1999) Incorporation and photodegradation of flavin and indole derivatives in anionic, cationic and neutral micellar dispersions, *J. Photochem. Photobiol. B-Biol.*, 52, 123–130.

Calvosa, L., Chiodini, G., Coretii, W., Donaggio, P., Orlandi, M., Paratici, V., and Rindone, B. (1994) Taste and odor development in water in polyethylene containers exposed to direct sunlight, *Water Res.*, 28, 1595–1600.

Chen, C.-Y., Chen, F.-A., Wu, A.-B., Hsu, H.-C., Kang, J.-J., and Cheng, H.-W. (1996) Effects of hydroxypropyl-β-cyclodextrin on the solubility, photostability and *in vitro* permeability of alkannin/shikonin enantiomers, *Int. J. Pharm.*, 141, 171–178.

Chinnian, D. and Asker, A.F. (1996) Photostability profiles of minoxidil solutions, *PDA J. Pharm. Sci. Technol.*, 50, 94–98.

Connors, K.A. (1990) *Chemical Kinetics*, New York: VCH Publishers, pp. 17–22.

CPMP (1998) Note for guidance on inclusion of antioxidants and antimicrobial preservatives in medicinal products, The European Agency for the Evaluation of Medicinal Products, CPMP/CVMP/QWP/115/95, January.

De Villiers, M.M., van der Watt, J.G., and Lötter, A.P. (1992) Kinetic studies of the solid-state photolytic degradation of two polymorphic forms of furosemide, *Int. J. Pharm.*, 88, 275–283.

de Vries, H., Beijersbergen van Henegouwen, G.M.J., and Huf, F.A. (1984) Photochemical decomposition of chloramphenicol in a 0.25% eyedrop and in a therapeutic intraocular concentration, *Int. J. Pharm.*, 20, 265–271.

El Aatmani, M., Poujol, S., Astre, C., Malosse, F., and Pinguet, F. (2002) Stability of dacarbazine in amber glass vials and polyvinyl chloride bags, *Am. J. Health Syst. Pharm.*, 59, 1351–1356.

Enever, R.P., Li Wan Po, A., and Shotton, E. (1977) Factors influencing decomposition rate of amytriptyline hydrochloride in aqueous solution, *J. Pharm. Sci.*, 66, 1087–1089.

European Pharmacopoeia (2002), Strasbourg: Directorate for the Quality of Medicines of the Council of Europe, pp. 233–283, 405–407, 548–550, 674–675, 2132–2133.

FDA (Food and Drug Administration) (1996) *Inactive Ingredient Guide*, Division of Drug Information Resources, CDER, January.

Fischer, B. (2002) Form-fill-seal manufacturing of polyolefin bags for parenteralia, *Pharm. Ind.*, 64, 892–898.

Florence, A.T. and Attwood, D. (1998) *Physicochemical Principles of Pharmacy*, Great Britain: MacMillan Press Ltd., pp. 103, 237–238.

Ford, J.L. (1988) Parenteral products, in Aulton, M.E. (Ed.) *Pharmaceutics. The Science of Dosage Form Design*, Edinburgh: Churchill Livingstone, pp. 359–380.

Habib, M.J. and Asker, A.F. (1991) Photostabilization of riboflavin by incorporation into liposomes, *J. Parenter. Sci. Technol.*, 45, 124–127.

Halliwell, B. and Gutteridge, J.M.C. (1985) *Free Radicals in Biology and Medicine*, Oxford: Clarendon Press, pp. 18, 26, 52, 124.

Ho, A.H.L., Puri, A., and Sugden, J.K. (1994) Effect of sweetening agents on the light stability of aqueous solutions of L-ascorbic acid, *Int. J. Pharm.*, 107, 199–203.

Hollande, S. and Laurent, J.L. (1997) Study of discoloring change in PVC, plasticizer and plasticized PVC films, *Polym. Degrad. Stab.*, 55, 141–145.

Hutchinson, S.L. (1998) Hospital at home: the alternative care setting, in Winfield, A.J. and Richards, R.M.E. (Eds.) *Pharmaceutical Practice*, London: Churchill Livingstone, pp. 274–288.

Hutchinson, S.L. and Graham, D. (1998) Specialized services from a hospital pharmacy, in Winfield, A.J. and Richards, R.M.E. (Eds.) *Pharmaceutical Practice*, London: Churchill Livingstone, pp. 254–273.

ICH Q1B (1997) Photostability testing of new drug substances and products, *Fed. Reg.*, 62, 27115–27122.

Islam, M.S. and Asker, A.F. (1995) Photoprotection of daunorubicin hydrochloride with sodium sulfite, *PDA J. Pharm. Sci. Technol.*, 49, 122–126.

Kristensen, S., Nord, K., Orsteen, A.-L., and Tønnesen, H.H. (1998) Photoreactivity of biologically active compounds. XIV. Influence of oxygen on light induced reactions of primaquine, *Pharmazie*, 53, 98–103.

Kuramoto, N. and Kitao, T. (1982) Contribution of superoxide ion to the photophading of dyes, *JSDC*, 98, 159–162.

Lakowicz, J.R. (1983) *Principles of Fluorescence Spectroscopy*, New York: Plenum Press, pp. 179, 243, 359–360.

Lee, S.C., Yuk, H.G., Lee, D.H., Lee, K.E., Hwang, Y.I., and Ludescher, R.D. (2002) Stabilization of retinol through incorporation into liposomes, *J. Biochem. Mol. Biol.*, 35, 358–363.

Lin, H.S., Chean, C.S., Ng, Y.Y., Chan, S.Y., and Ho, P.C. (2000) 2-Hydroxypropyl-beta-cyclodextrin increases aqueous solubility and photostability of all-trans-retinoic acid, *J. Clin. Pharm. Ther.*, 25, 265–269.

Loukas, Y.L., Jayasekera, P., and Gregoriadis, G. (1995) Novel liposome-based multicomponent systems for the protection of photolabile agents, *Int. J. Pharm.*, 117, 85–94.

Martin, A. (1993) *Physical Pharmacy*, Philadelphia: Lea & Febiger, pp. 33–36.

Mielcarek, J. (1996) Photochemical stability of the inclusion complexes of nicardipine with α-, γ-cyclodextrin, methyl-β-cyclodextrin and hydroxypropyl-β-cyclodextrin in the solid state and in solution, *Pharmazie*, 51, 477–479.

Moore, D.E. and Burt, C.D. (1981), Photosensitization of drugs in surfactant solutions, *Photochem. Photobiol.*, 34, 431–439.

Moore, D.E. and Zhou, W. (1994) Photodegradation of sulfamethoxazole — a chemical system capable of monitoring seasonal changes in UVB intensity, *Photochem. Photobiol.*, 59, 497–502.

Mubarak, S.I.M., Standford, J.B., and Sudgen, J.K. (1982) Some aspects of the photochemical degradation of chloramphenicol, *Pharm. Acta Helv.*, 57, 226–230.

Munson, J.W., Hussain, A., and Bilous, R. (1977) Precautionary note for use of bisulfite in pharmaceutical formulations, *J. Pharm. Sci.*, 66, 1775–1776.

Nema, S., Brendel, R.J., and Washkuhn, R.J. (2002) Excipients — their role in parenteral dosage forms, in *Encyclopedia of Pharmaceutical Technology*, New York: Marcel Dekker, Inc., pp. 1164–1187.

Nord, K., Orsteen, A.-L., Karlsen, J., and Tønnesen, H.H. (1997a) Photoreactivity of biologically active compounds. X. Photoreactivity of chloroquine in aqueous solution, *Pharmazie*, 52, 598–603.

Nord, K., Andersen, H., and Tønnesen, H.H. (1997b) Photoreactivity of biologically active compounds. XII. Photostability of polymorphic modifications of chloroquine diphosphate, *Drug Stab.*, 1, 243–248.

Nuijen, B., Bouma, M., Henrar, R.E.C., Manada, C., Bult, A., and Beijnen, J.H. (1999) Compatibility and stability of aplidine, a novel marine-derived depsipeptide antitumor agent, in infusion devices, and its hemolytic and precipitation potential upon i.v. administration, *Anticancer Drugs*, 10, 879–887.

Nuijen, B., Bouma, M., Manada, C., Jimeno, J.M., Lazaro, L.L., Bult, A., and Beijnen, J.H. (2001) Compatibility and stability of the investigational polypeptide marine anticancer agent kahalalide F in infusion devices, *Invest. New Drugs*, 19, 273–281.

Parfitt, K. (Ed.) (1999) *Martindale. The Complete Drug Reference*, London: Pharmaceutical Press, pp. 1327, 1365.

Pecsok, R.L., Shields, L.D., Cairns, T., and McWilliam, I.G. (1976) *Modern Methods of Chemical Analysis*, New York: John Wiley & Sons, pp. 415–416.

Peiris, D.M., Mohanty, D.K., and Sharma, A. (1999) Undesirable reaction of aqueous cyclodextrin solutions with polypropylene, *Microchem. J.*, 62, 266–272.

Sarbach, C., Yagoubi, N., Sauzieres, J., Renaux, C., Ferrier, D., and Postaire, E. (1996) Migration of impurities from a multilayer plastic container into a parenteral infusion fluid, *Int. J. Pharm.*, 140, 169–174.

Shahjahan, M. and Enever, R.P. (1996a) Photolability of nitrofurazone in aqueous solution. I. Quantum yield studies, *Int. J. Pharm.*, 143, 75–82.

Shahjahan, M. and Enever, R.P. (1996b) Photolability of nitrofurazone in aqueous solution. II. Kinetic studies, *Int. J. Pharm.*, 143, 83–92.

Smith, J.L., Canham, J.E., and Wells, P.A. (1988) Effect of phototherapy light, sodium bisulfite, and pH on vitamin stability in total parenteral nutrition admixtures, *JPEN J. Parenter. Enteral. Nutr.*, 12, 394–402.

Stewart, P.J. and Tucker, I.G. (1985) Prediction of drug stability. III. Oxidation and photolytic degradation, *Aust. J. Pharm.*, 15, 111–117.

Strong, M.L., Schaaf, L.J., Pankaskie, M.C., and Robinson, D.H. (1994) Self-lives and factors affecting the stability of morphine sulphate and meperidine (pethidine) hydrochloride in plastic syringes for use in patient-controlled analgesic devices, *J. Clin. Pharm. Ther.*, 19, 361–369.

Suppan, P. and Nagwa, G. (1997) *Solvatochromism*, London: The Royal Society of Chemistry, pp. 1–95.

Thoma, K. and Klimek, R. (1985a) Untersuchungen zur Photoinstabilität von Nifedipin. I. Mitteilung: Zersetsungskinetik und Reactionsmechanismus, *Pharm. Ind.*, 47, 207–215.

Thoma, K. and Klimek, R. (1985b) Untersuchungen zur Photoinstabilität von Nifedipin. II. Mitteilung: Einfluß von Milieubedingungen, *Pharm. Ind.*, 47, 319–327.

Thoma, K. and Kübler, N. (1996) Einfluß der Wellenlänge auf die Photozersetzung von Arzneistoffen, *Pharmazie*, 51, 660–664.

Tønnesen, H.H., Skrede, G., and Martinsen, B.K. (1997) Photoreactivity of biologically active compounds. XIII. Photostability of mefloquine hydrochloride in the solide state, *Drug Stab.*, 1, 249–253.

Torniainen, K., Tammilehto, S., and Ulvi, V. (1996) The eggect of pH, buffer type and drug concentration on the photodegradation of ciprofloxacin, *Int. J. Pharm.*, 132, 53–61.

Trissel, L.A. (2001) *Handbook on Injectable Drugs*, Bethesda: American Society of Health-System Pharmacists, pp. 1373–1386.

Wilkinson, F., Leicester, P.A., Ferreira, L.F.V., and Freires, V.M.M.R. (1991) Photochemistry on surfaces: triplet–triplet energy transfer on microcrystalline cellulose studied by diffuse reflectance transient absorption and emission spectroscopy, *Photochem. Photobiol.*, 54, 599–608.

Williams, M.F., Hak, L.J., and Dukes, G. (1990) *In vitro* evaluation of the stability of ranitidine hydrochloride in total parenteral nutrient mixtures, *Am. J. Hosp. Pharm.*, 47, 1574–1579.

Zieske, P.A., Koberda, M., Hines, J.L., Knight, C.C., Sriram, R., Raghavan, N.V., and Rabinow, B.E. (1991) Characterization of cisplatin degradation as affected by pH and light, *Am. J. Hosp. Pharm.*, 48, 1500–1506.

CHAPTER **15**

Photoactivated Drugs and Drug Formulations

Jan Karlsen and Hanne Hjorth Tønnesen

CONTENTS

15.1 INTRODUCTION

The efficacy of a pharmaceutical product depends to a large extent on its dosage form. Although new and powerful drugs continue to be developed, increasing attention is being given to the methods by which the active substances are administered. A number of advancements have been made recently in the development of new techniques for drug delivery. Novel drug delivery systems are directed toward a controlled release rate, a sustained duration of therapeutic action, and/or a targeting of the delivery to a specific tissue. The advantage of advanced drug delivery systems over traditional systems is the ability to deliver a drug more selectively to a specific site, thus resulting in easier, more accurate and less frequent dosing; less variations in systemic drug concentrations; absorption more consistent with the site and mechanisms of action; and reduction in toxic metabolites.

Advanced drug delivery strategies are typically based on one of the following principles: biological, chemical, physical, or mechanical responsive systems. According to these principles, the drug release rate can be activated by an external stimuli (e.g., temperature changes, magnetism, ultrasound, electrical effects, irradiation) or by controlled variables within the body that reflect the physiological needs (e.g., change in pH, ionic strength, glucose, or urea concentration) (Kost and Langer, 2001). Circadian rhythms are established for almost all body functions (e.g., blood pressure, viscosity, and flow; heart rate; stroke volume; body temperature; gastric pH; and plasma concentration of various substances such as hormones) (Vyas et al., 1997). Thus, the assumption that physiological processes and biological functions display constancy over time is no longer valid. It is therefore becoming increasingly important to design systems that can synchronize drug delivery with biological rhythms.

Photoactivation is an attractive option for triggering drug delivery. The method provides a broad range of adjustable parameters (e.g., wavelength, intensity, duration, spatial and temporal control) that can be optimized to suit a given application. Modern laser systems provide an energy confinement that makes them suitable for a number of biomedical applications, including surgical, diagnostic, and therapeutic purposes. Laser excitation methods can further be used for activating processes like membrane phase transitions, enzymatic reactions, cross-linking, or cleavage reactions.

Photoactivated drug delivery includes different approaches. The method can be used to control the release rate of the active principle from a dosage form (i.e., a carrier system), to activate a drug molecule already present at the site of action in an inactive form (e.g., a photosensitizer or a prodrug), or to combine the two (i.e., photocontrolled drug release and drug activation). Although photoactivation of sensitizers is well established — for instance, in treatment of cancer (photodynamic therapy, PDT) or psoriasis (PUVA-therapy) (Chapter 9) — the application of photoactivated carrier systems offers an underdeveloped opportunity. This chapter focuses on recent literature describing phototriggering mechanisms that may ultimately find utility in drug delivery applications.

15.2 PHOTOACTIVATED DOSAGE FORMS

15.2.1 Photosensitive Hydrogels

A hydrogel consists of a cross-linked hydrophilic polymer swollen with water. The three-dimensional network formed is able to imbibe large amounts of water or biological fluids and thus has many similarities with a biological tissue. The polymer can be synthetic or natural, possibly modified. Synthetic materials can be prepared to respond to various physiological stimuli such as pH, ionic strength, and temperature. Hydrogels have attracted an increasing interest in pharmacy and medicine due to their biocompatibility (ascribed to the high water content), biofunctionality (due to their soft consistency), and bioadhesive properties. Drug molecules can be incorporated in the three-dimensional network and hydrogel-based delivery devices may be used for oral, ocular, topical, rectal, and parenteral application (Peppas et al., 2000; Bos et al., 2001).

The structure and mechanical properties of the hydrogels are important for the loading of drug molecules and the sensitivity to stimuli (e.g., radiation). Changing the degree of cross-linking and copolymerization have been utilized to achieve the desired properties of the gels (Peppas et al., 2000). A photosensitive gel system can be obtained by a combination of monomers, a photosensitive component, and a cross-linker. A photosensitive co-monomer can be incorporated to increase release from the matrix, while a photosensitive cross linker can be applied to reduce the drug release (Tomer and Florence, 1993). Water-soluble polymers that can generate cations by UV irradiation have been developed to control the nonbiospecific interaction with endothelial cell membranes (Nakayama and Matsuda, 2003).

A number of gel systems for pharmaceutical applications are based on polysaccharides (e.g., cellulose derivatives, dextrans, alginate, hyaluronic acid, and chitosan). Methylene blue, thionine, Rose Bengal, and riboflavine are examples of photosensitizers used in combination with polysaccharide hydrogels. Photosensitized degradation of hyaluronic acid has been thoroughly studied, and the degradation mechanism seems to depend on the experimental conditions (Andly and Chakrabarti, 1983; Frati et al., 1997). It has been demonstrated that cationic dye-sensitized degradation of sodium hyaluronate involves photoinduced electron transfer, while hydroxyl radicals and singlet oxygen do not participate in the reaction. On the other hand, anionic dye-sensitized degradation of hyaluronic acid seems to proceed through a different mechanism (Kojima et al., 2001). It has further been demonstrated that carboxyl groups play an important role in the degradation of polysaccharide polymers like hyaluronic acid, alginate, or polygalacturonic acid (Scott and Tigwell, 1973; Kojima et al., 2001; Baldursdottir et al., 2003). Polysaccharides without carboxyl groups, such as methyl cellulose, do not show photodegradation under the same conditions (Kojima et al., 2001). The reaction may be dependent on pH; it has been reported that radical-induced degradation of neutral (though not necessarily of acidic) polysaccharides should be most efficient at high and low pH (Gilbert et al., 1984).

Light-activated delivery systems would be especially useful in accessible regions of the body, e.g., eye or skin, to allow triggering of drug release by natural light. Visible light-induced degradation of cross-linked hyaluronic acid gels has been examined as a potential drug delivery system to the eye (Yui et al., 1993). However, the most significant weaknesses of the radiation-sensitive hydrogels are a slow response time and leaching of the reactants during the swelling–deswelling cycles. Development of fast-acting hydrogels and alternative sensitizers is thus a challenge. In theory, the hydrogels can be sensitive to UV, visible, and infrared radiation. The latter can be used in the absence of chromophores due to the high IR absorbency of water, which generates heat. This principle is applied in the transformation of a poly(N-isopropylacrylamide) gel by irradiation with a CO_2 laser (Qiu and Park, 2001). A change in viscosity of specially thermosensitive gels can also be obtained by combining the trisodium salt of copper chlorophyllin (i.e., a sensitizer that generates heat) and visible light.

In addition to acting as potential drug delivery devices, such hydrogels are postulated to have an application in the development of photoresponsive artificial muscles and memory devices (Qiu and Park, 2001). Selective tumor targeting and accumulation of cytotoxic drugs in the tumor have been demonstrated for certain drug–polymer conjugates in combination with light, which makes these systems particularly promising in cancer treatment (Shiah et al., 2000). Photoresponsive, ion-selective membranes can be made from polyacrylamide hydrogels containing a compound that undergoes dissociation into a cation and an hydroxyl anion on UV irradiation (e.g., triphenylmethane lecoderivatives) (Kodzwa et al., 1999). Production of photocations during irradiation increases the transport of anionic compounds through the gel.

A different approach to the phototriggering principle is obtained when radiation is used as a tool in formation of the drug delivery device, thereby reducing drug release rather than activating the drug release process by degrading the carrier system. The drug formulation remains fluid in the dark and can be injected into the tissue. The administration site is then exposed to the adequate radiation. This leads to a rapid gelation and the *in situ* formation of a drug reservoir that slowly will release the active principle when the light is turned off.

Photopolymerization is already used extensively in dentistry to form sealants on teeth. However, recent attention has focused on materials that may be implanted in a minimally invasive manner for use in orthoscopic and plastic surgery. *In situ* photopolymerization seems promising in wound healing and for reconstruction of soft tissues (Silverman et al., 1999; Collier et al., 2001; Smeds and Grinstaff, 2001; Ward and Peppas, 2001), and a transdermal photopolymerization technique has recently been developed (Elisseeff et al., 1999a, b, 2000). An injected macromer solution was demonstrated to photopolymerize by UV radiation transmitted through the skin. The system was responsive to low UV doses and short exposure times (2 to 3 min) and has proved efficient for encapsulation of cells and drug molecules. The increased control in spatial resolution with photopolymerization may allow design of scaffold architecture on the micrometer or nanoscale level — making the principle particularly feasible for tissue engineering (Elisseeff, 2000).

15.2.2 Photosensitive Microcapsules

Microencapsulation is a technique that involves encapsulation of small particles of a drug, or a solution of a drug, in a polymer film or coat. The resulting units have a size typically in the range 200 to 6000 µm. The drug is usually released by swelling of the polymeric barrier followed by diffusion. Analogue to hydrogels (see preceding discussion), exposure of microcapsules to radiation can lead to an opening of the capsule wall, thereby facilitating drug delivery, or to an induction of polymerization, thereby changing the mechanical properties of the capsule wall to slow down the release process.

Photochemically controlled delivery systems have been prepared by incorporating a substance (azobisisobutyronitrile, AIBN) in microcapsules prepared from polyamide or other polymer-based membranes, including double wall systems. AIBN will photochemically emanate nitrogen gas. The pressure build-up in the capsule due to the release of nitrogen during exposure will cause immediate rupture of the capsule wall and release the contents (Mathiowitz et al., 1981; Mathiowitz and Cohen, 1989a, b, c, d, e). The method could be applied for microcapsules of different sizes and membrane flexibility. A photorelease of the capsule content was obtained at high- and low-light intensity, and even sunlight was demonstrated to induce release of a model compound.

Microencapsulation is one of the most important techniques for immobilizing biologically active materials in biomedical applications. The technique is applied to circumvent the problem of host rejection by immune response. The ideal delivery system should have an "appropriate" mechanical strength, it should be biocompatible and should allow diffusion of nutrients and growth factors while retaining migration of antibodies and cells across the membrane. It has been demonstrated that the mechanical strength of alginate-based microcapsules was significantly improved by coating the capsules with a photosensitive polymer (Chang et al., 1999; Lu et al., 2000). Cross-linking of the coating was induced by light exposure. Because the wavelengths necessary to induced polymerization were outside the UV range where most bioactive entities absorb, the capsule wall could be photopolymerized without harming its content.

15.2.3 Photosensitive Liposomes

Liposomes make a convenient vehicle for drug delivery of many conventional drugs (e.g., anticancer, antimicrobial, antifungal) (Gregoriadis, 1995; Hillery, 1998a). Liposomes are formed spontaneously when phospholipids are dispersed in aqueous media. The resulting structures consist of an inner aqueous space enclosed by a phospholipid bilayer. The liposomes can be unilamellar or multilamellar with many concentric bilayers and aqueous compartments. Lipid-soluble drugs will be incorporated within the lipid bilayer while water-soluble molecules will be entrapped within the aqueous interior of the liposome. Surfactants with dialkyl chains can pack in a similar manner to the phospholipids; vesicle formation by synthetic cationic surfactants such as cetyl-trimetyl-ammonium chloride in the presence of PEG-yltated surfactants and octanol has been reported (Lasic, 1998).

The toxicity of ionic surfactants does restrict their use as drug-delivery vehicles, however. Nonionic surfactants are less toxic and therefore the neutral polyoxyethylene ethers or the combination of cholesterol with certain glyceryl-containing surfactants demonstrated to form vesicles is more interesting from a pharmaceutical point of view. The resultant nonionic vesicles have been termed niosomes (Hillery, 1998b).

The stability of liposomes *in vivo* depends on their size, charge, and chemical composition. Liposomes may contain sterols, glycolipids, proteins, polymers, organic acids, and bases, depending on the type of vesicle required. Sterically stabilized long-circulating liposomes described as "stealth" liposomes have been

Table 15.1   Examples of Direct and Mediated
            Methods Used for Photoactivation
            of Liposomes

Direct Mechanisms	Mediated Mechanisms
Photoisomerization	Photo-oxidation
Photopolymerization	Photoinduced hydrolysis
Photofragmentation	Photothermal reactions

developed (Hillery, 1998b).These units seem to accumulate passively at disease sites with a porous vasculature, thereby facilitating drug targeting. Targeting can also be obtained by conjugating liposomes with antibodies or by injecting liposomes directly into target sites.

Liposomes can further be used for topical drug delivery. However, even when drug-loaded liposomes can be targeted successfully to a given cell type, the drug bioavailability may still be low. Liposomes are not necessarily taken up into the cells and passive release of the encapsulated agent may be too slow to achieve an optimal therapeutic effect. Efforts have therefore been directed toward the development of active triggering mechanisms that can promote maximum efficacy (Shum et al., 2001). Photoactivation can lead to fusion with cells or tissues and/or exchange of membrane-bound constituents. This will alter the clearance rate and may also facilitate transport of lipophilic agents from the vesicle into the cellular membrane.

Photoactivation can further allow release of entrapped materials from aqueous space. The liposomes may carry materials of low toxicity that are converted to more potent materials upon radiation (e.g., prodrugs). The methods applied for photoactivation of liposomes are based on reactions that destabilize the liposomal bilayer. The mechanisms can be direct or mediated by oxygen, heat, specific ions, or a change in pH (Table 15.1). Many efforts are underway to develop two-photon schemes or near-infrared-sensitive liposomes to optimize the effectiveness of the system. The toxicity of the photosensitive units is a problem that must be looked into to improve the utility of these phototriggered systems.

15.2.3.1   Direct Activation Mechanisms

15.2.3.1.1   Photoisomerization

A change in molecular structure in response to external stimuli (e.g., radiation) may interfere with the organization of the liposomal bilayer. Synthetic phospholipids bearing azobenzene pendant groups (e.g., Bis-AzoPC) exist as mixtures of *cis* and *trans* isomers. Under normal daylight conditions, approximately 90% of the molecules are in the *trans* (*E*) form. This form represents linear molecules of low polarity (Morgan et al., 1995a). Exposure to UVA radiation transforms the compounds into the *cis* (*Z*) isomer, which is nonlinear with a higher dipole moment. This change interferes with the bilayer packing and the result is that the liposomes become leaky (Figure 15.1). Entrapped molecules can thereby diffuse out (Bisby et al., 2000a).

This effect requires, however, that the membranes exist in the gel phase. In the more fluid liquid crystalline phase, the packing defects arising from the *trans–cis*

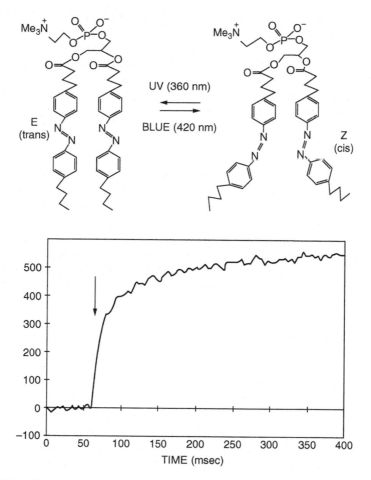

Figure 15.1 *Cis–trans* photoisomerization of Bis-AzoPC and the release of calcein from DPPC/Bis-AzoPC liposomes after flash photolysis at 355 nm detected as an increase in fluorescence (arbitary units). The arrow indicates the point of the laser flash. (Redrawn from Morgan, C.G. et al., 1995a and Bisby, R.H. et al., 2000. With permission.)

isomerization are rapidly annealed. One of the advantages of this system is a very fast release of the drug down to a millisecond time scale (Morgan et al., 1995b). The release should preferentially be triggered by one laser pulse or a flash lamp. Continuous exposure will lead to a gradual lateral equilibrium and seems to be ineffective at low sensitizer concentration (Bisby et al., 1999).

Exposure to white light causes reversion of the process and halts leakage and vesicle fusion unless the liposomes contain cholesterol (Morgan et al., 1995a; Bisby et al., 2000b). A release of an aqueous marker could be demonstrated by incorporation of retinoic acid in the liposome membrane. Only 30 to 120 sec of irradiation was necessary to stimulate 100% leakage from the model systems (Pidgeon and Hunt, 1983). On–off switching of drug release from liposomes can potentially be obtained by incorporating a spiropyran group at the hydrophopic terminus of a single chain lipid (Ohya et al., 1998). Spiropyrans isomerize reversibly between two different states upon visible light and UV radiation. The polar twitter ionic merocyanine (MC) form is present upon UV exposure, while the nonionic spiropyran (SP) form exists under exposure to visible light. This process is reversible. The formation of an ionic twitter destroys the amphiphilic characteristics of the molecule, leading to a perturbation of the liposome membrane.

15.2.3.1.2 Photopolymerization

Photoinduced destabilization of liposomes can be induced by polymerization of the bilayers. Lipids such as phosphatidylcholines that contain a photoreactive group in the hydrophobic tail can react to form a cross-linked polymer network within a bilayer, thus resulting in the formation of lipid domains (Lamparski et al., 1992; Armitage, 1993). A number of reactive moieties sensitive to various parts of the spectrum have been utilized to induce photopolymerization. Dienoyl, sorbyl, and styryl groups are sensitive to short-wavelength UV radiation, while ionic cyanine dyes have been used in combination with visible light (Clapp et al., 1997; O'Brien et al., 1998; Miller et al., 2000; Mueller et al., 2000). The possibility for using visible light stimuli rather than UV radiation increases the utility of the liposome system. The depth of light penetration increases and the risk of damaging the drug incorporated in the liposome or components in the surrounding tissue is reduced.

Special attention has been given to the release of drugs from the sterically stabilized, long-circulating PEG-liposomes (Lamparski et al., 1992; Bondurant and O'Brian, 1999; Mueller et al., 2000; Bondurant et al., 2001). The photosensitive lipid bis-SorbPC (1,2-bis[10-(2′,4′-hexadienoyloxy)decanonyl]-*sn*-glycero-3-phosphocholine) has been incorporated in PEG liposomes of various composition. The observed increase in liposome permeability was about 200-fold upon exposure to UV radiation in the presence of oxygen on a time scale of minutes to hours (Bondurant et al., 2001).

The photoinduced destabilization of the lipid membrane was apparently a consequence of cross linking of the bis-SorbPC by a successive photoaddition process leading to defects in the bilayer structure (Figure 15.2). This created separate domains of polymerized bis-SorbPC lipid and nonpolymerized PEG-lipid, respectively. The effect was independent on the charge and mole fraction of the PEG-lipid in the formulation. By including distearoyl indocarbocyanine (DiIC(18)3) as a photosensitizer in selected bis-SorbPC containing PEG-liposome formulations, the leakage could be induced by visible light (Mueller et al., 2000).

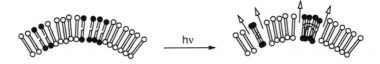

Figure 15.2 Photoinduced cross linking leading to separate domains of polymerized and nonpolymerized lipids. Defects in the bilayer due to polymer shrinkage result in content release.

15.2.3.1.3 Photofragmentation

1,2-Dioleoylphosphatidylethanol (DOPE)-derivatives that form bilayers can undergo photofragmentation reactions and thereby induce liposome leakage. This was demonstrated for the compound NVOC–DOPE, which rapidly decomposes by exposure to radiation above 300 nm (Zhang and Smith, 1999). Photolysis resulting in the generation of ions (e.g., carboxylate ions or zwitterions) in the hydrophobic region of the bilayer will also introduce disintegration of the liposome wall (Sunamoto et al., 1982; Kusumi et al., 1989)

15.2.3.2 Mediated Activation Mechanisms

15.2.3.2.1 Photo-oxidation

Photosensitive liposomes may contain a photosensitizer that can induce peroxidation of the membrane lipids by type I and/or type II photodynamic sensitization (Srinath and Jain, 1994). In the type I process, the electronically excited sensitizer reacts directly with the lipids. The result is the formation of radicals that can react further with molecular oxygen or other components present at the reaction site. In the type II process, the electronically excited sensitizer reacts directly with molecular oxygen via energy transfer to produce singlet oxygen.

Formation of reactive oxygen species (ROS) by either of these processes may result in oxidation of the phospholipids, leading to a change in membrane permeability (Gerasimov et al., 1997). The photosensitizer should preferentially absorb in the visible or near-infrared region of the spectrum to obtain optimal light penetration depths and minimal interference with biological chromophores or incorporated drugs. This is the case for bacteriochlorophyll *a*, which has been used to phototrigger the release of calcium ions and the zinc and tin phthalocyanine derivatives applied in the study of semisynthetic plasmenylcholine liposomes (Thompson et al., 1996; Wymer, 1998). In the latter case, the photo-oxidation resulted in a multistep release process (Figure 15.3) (Thompson et al., 1996).

An alternative approach is to use the drug substance as a sensitizer for phospholipid oxidation to induce release, assuming that the drug molecule is not altered in the process. Light-induced leakage was demonstrated in the presence of the highly phototoxic phenothiazines (Copeland et al., 1976). It should be kept in mind, however, that reactive oxygen (ROS)-mediated release may involve competing reactions between the sensitizer and the phospholipds, leading to the degradation of both.

15.2.3.2.2 Photoinduced Hydrolysis

Intramolecular photoinduced electron transfer (PET) may lead to the generation of acid (i.e., protons) from an anthracene-based sulphonium salt. An amphiphilic anthracene-based photoacid generator (An-PAG) has been synthesized and was reported to induce acid-catalyzed hydrolytic degradation of certain lipids upon UV exposure (Shum et al., 2001). At present, the application of this technique is limited by the short wavelength required to trigger the response.

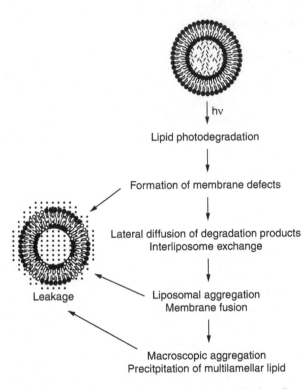

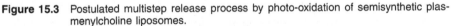

Figure 15.3   Postulated multistep release process by photo-oxidation of semisynthetic plasmenylcholine liposomes.

15.2.3.2.3   Photothermal Reactions

Deactivation of an excited state molecule may follow a nonradiative pathway. If the absorbed energy is released as heat, a local temperature increase may induce chemical changes to nearby molecules. This principle is applied in photothermally induced cell killing and could also have potential in the development of future phototriggered liposomal drug delivery systems (Busetti et al., 1999; Shum et al., 2001). Light-targeted delivery of drugs to the eye has successfully been reported (Zeimer and Goldberg, 2001). The drug was encapsulated in a heat-sensitive liposome preparation and injected intravenously. Drug release at the site of choice was obtained by noninvasively warming up the targeted tissue with a light pulse directed through the pupil of the eye.

15.3   PHOTOACTIVATED DRUG SUBSTANCE

15.3.1   Caged Compounds

The term "caged compounds" is used to describe synthetic molecules whose biological activity is controlled by light, usually by photolytic conversion from an

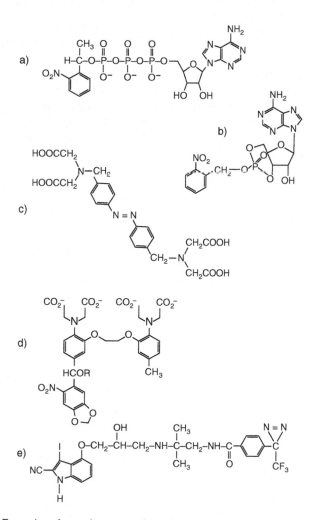

Figure 15.4 Examples of caged compounds and other photoactivatable drugs. (a) and (b): on photolysis the *o*-nitrobenzyl protecting group detaches and releases ATP and cAMP, respectively; (c) photosensitive chelator. The *cis* form of the molecule binds Zn^{2+} but the *trans* form does not; (d) frequently used photosensitive calcium chelators. Nitr-2 (R = CH_3), Nitr-5 (R = H); (e) ICYP-diazirine, a photoaffinity reagent that becomes covalently attached to β-adrenergic receptors upon photolysis.

inactive to an active form. In nearly all the caged molecules, the active species is released by photocleavage of a single covalent bond. A variety of caged compounds exists, including caged nucleotides, chelators, neurotransmittors, and drugs (Marriott, 1998) (Figure 15.4). A caged drug molecule can also be described as a light-activated prodrug. Reactions of a time scale ~10^{-3} sec and longer are usually of little interest from a photochemical point of view, but are definitely important in the study of biological processes.

Photolabile protecting groups offer means to deliver bioactive materials rapidly to well-defined target sites; they thus enable biologists to follow a physiological process in real time. This capability is particularly valuable when microscopic spatial gradients are desired or when rapid mechanical mixing is impractical, e.g., inside intact cells, tissue, or proteins. Caged compounds therefore have found a wide application in biochemistry and cell biology (Marriott, 1998).

Frequent applications of these activatable bioagents are found in the photorelease of enzymes, peptides and proteins, nucleotides, calcium and calcium chelators, neurotransmitters, and other second messengers (Gurney and Lester, 1987; Kaplan and Somlyo, 1989; Adams and Tsien, 1993; Hagen et al., 1998; Pelliccioli and Wirz, 2002). Photoactivation of gene expression and a number of other applications are also under investigation (Pelliccioli and Wirz, 2002). The photoremovable protecting group or caging group must satisfy certain criteria to allow for a biological application. Basically, it should render the biomolecule inert until activation occurs and then release the biomolecule at high yield and sufficient speed. The process should take place at wavelengths unlikely to cause damage to the biological environment. Any photoproducts other than the desired biomolecule should be inert to the biological system.

Knowledge of the actual release rate and, preferably also, the reaction mechanism under physiological conditions are prerequisites to interpreting data obtained in biological studies. Release rate and efficiency may depend strongly on the environment, e.g., pH, ionic strength, and medium lipophilicity. The released active principle and the protective group should possess adequate solubility at the application site and pass selected membranes in order to reach the target or to be eliminated. The release mechanism normally falls within two categories. The photorelease can be a direct elimination from the excited singlet or triplet state of the photoactivatable molecule, or the driving force can be the light-induced generation of reactive ground state intermediates that have a low energy barrier for substrate release.

The chemical structure of some protective groups is given in Figure 15.5. Derivatives of 2-nitrobenzyl are by far the most frequently used photolabile protective groups in caged compounds. More than 40 nitrobenzyl protected biochemicals are now on the market (Pelliccioli and Wirz, 2002). The nitrobenzyl derivatives, however, have their disadvantages, such as slow release rate upon excitation and toxic and strongly absorbing degradation products. Efforts are therefore made in order to obtain efficient and fast systems based on alternative structures like RHN^-, RO^-, or RS^-, which are regarded as poor leaving groups. Most of the existing protective groups absorb in the UV region of the spectrum (Figure 15.5). The excitation light thereby has a low tissue penetration depth and is likely to cause damage to biological molecules. Two-photon photolysis (i.e., simultaneous absorption of two IR photons of energy equivalent to the UV excitation) to overcome these problems is now under investigation.

15.3.2 Photodynamic Therapy

In the conventional applications of photodynamic therapy (PDT), light-activated chemicals (photosensitizers) are used to destroy fast-growing cells and tissues. The commonly used sensitizers are based upon porphyrin-like molecules, e.g., porphyrins, chlorins, bacteriocholins, and phthalocyanines. These substances collect in

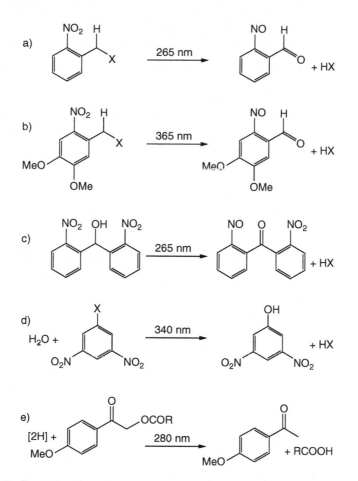

Figure 15.5 Examples of photolabile protecting groups used for caged compounds. (a) 2-nitrobenzyl (NB); (b) 4,5-dimethoxy-2-nitrobenzyl (DMNB); (c) 2,2'-dinitrobenzhydryl (DNB); (d) 3,5-dinitrophenyl; (e) 4-methoxyphenacyl.

rapidly proliferating cells when administered parenterally. Upon exposure to light, the sensitizer initiates a cascade of molecular reactions that can destroy those cells and the tissues they compose. The destructive reactions are partly based on the sensitized formation of singlet oxygen and partly on the degradation of the sensitizer, leading to free radical formation. Photohermal destruction of neoplastic cells can also result, leading to stimulation of the immunological defense system against residual and metastatic tumor cells (Chen et al., 1997). The concept of photodymamic therapy is further discussed in Chapter 9.

Numerous clinical trials of PDT have been carried out (phases I to IV). Light-activated drugs used in PDT have potential in cancer treatment located at a number of sites (e.g., esophagus, lungs, skin, head and neck, brain, mouth, and bladder) (Sibata et al., 2000). Other targets for the therapy include troublesome bacterial infections, Barrett's esophagus, arthritis, atherosclerotic plaques in coronary arteries, and abnormal blood vessels in the retinas of people with age-related macular degeneration (the leading cause of adult blindness) (Trauner and Hasan, 1996; Moore, 1998; Overholt et al., 1999; Overhaus et al., 2000; Hinnen et al., 2002; Lane, 2003).

15.3.2.1 Role of Delivery Vehicles

The future progress of photodynamic therapy will require input from pharmacists for improvement of the drug formulation. Ideally, the photosensitizer should be retained selectively at the target site, whether this is a tumor or a bacterially infected area. In practice, normal tissues, especially in skin and eye, also retain the photosensitizer present in the systemic circulation. The patient may become sensitive to natural sunlight for weeks or months after treatment. Drug targeting by means of an optimized drug delivery system is therefore essential to improve sensitizer selectivity (Konan et al., 2002). Thus far, targeting has been achieved by using liposomes, oil emulsions, microspheres, micellar nanoparticles, proteins, and monoclonal antibodies in parenteral administration of sensitizers (Hasan, 1992; Klyashchitsky et al., 1994; Hoebeke, 1995; Jori, 1992, 1996; Love et al., 1996; Reddi, 1997; Valduga et al., 1998; Sibata et al., 2000; Houlton, 2003). In most cases the system under study (e.g., a specific sensitizer liposome–antibody conjugate studied in a selected tumor model) had a favorable targeting effect, but so far it has not been reported that one specific delivery system can be valid for general application (Reddi, 1997).

It is also apparent that little effort is made to optimize the vehicle for topically administered sensitizers (e.g., dermal, oral cavity). For such applications, one is frequently presented to simple aqueous solutions of ethanol or DMSO. Topical preparations thereby offer a challenge to the formulation expert. Application to the oral cavity and larynx would benefit from bioadhesive formulations to increase the contact time between sensitizer and tissue. A well-designed vehicle could allow topical administration of sensitizers to tumors located close to the skin surface and thus offer an alternative to the present systemic administration.

The selection of vehicle not only facilitates targeting to specific tissues but will also influence localization of the sensitizer within the cells (Jori, 1992, 1996). Liposome-, lipoprotein-, or emulsion-delivered sensitizers tend to be released inside tumor cells and may cause damage at the level of lysosomes and endoplasmatic reticulum. Photosensitizers dissolved in aqueous solutions tend to cause damage to cytoplasmic and mitochondrial membranes, whereas albumin-carried substances are mainly deposited in the extracellular matrix. Features of photosensitizer medication that can be affected by the vehicle are given in Table 15.2.

In order to minimize side effects further, the drug should be effective at a low dose. This requires that the drug molecule have a high molar absorbtivity at a wavelength that penetrates the tissue and a high quantum yield for singlet oxygen formation. Because hemoglobin absorbs light up to 577 nm, it is necessary to irradiate the tissue at wavelengths above 600 nm to ensure significant penetration. For wavelengths >850 to 900 nm, the photons may not have sufficient energy to induce the desired reactions. The generation of singlet oxygen via the type II mechanism is then energetically unfavorable because the triplet-state energies of the sensitizers are too low. The "therapeutic window" for PDT is therefore confined to radiation between 600 and 800 nm (Parrish, 1981; Sibata et al. 2000).

Table 15.2   Features of Photosensitizer Medication That Can Be Affected by the Vehicle

Requirements for Optimal Photosensitizer Effect	Problems Related to Drug Administration
Retains selectively in the tumor	Widely distributed to the skin and eye
Fast clearance	Patient remains photosensitive for weeks
Absorbs light at 600–800nm	Low molar absorbtivity >600 nm; change in photophysical characteristics within vehicle
Effective at low concentration	Low quantum yield for singlet oxygen formation; change in photophysical characteristics within vehicle
Monomolecular form	Aggregate formation in aqueous media
Molecule as zwitterion	Low solubility in hydrophilic and lipophilic media
Photolabile *in vivo*	Degradation during storage and use

The use of two-photon infrared excitation could offer an alternative (equivalent to visible single photon excitation). The multiphoton absorption process, however, tends to show a nonlinear intensity dependence that can introduce further complications to the regime (MacRobert et al., 1989). Some photosensitizers (e.g., hematoporphyrin derivatives) have a low molar absorbtivity in the red region and therefore special emphasis is made on development of red-light absorbing sensitizers. Such molecules are characterized by a polycyclic chemical structure in which the energy gap between the ground state and the first excited singlet state is reduced by the extensively delocalized electron cloud, thereby allowing electronic excitation by low energy light (Jori, 1990). A pharmaceutical formulation contains excipients (e.g., surfactants, polymers) and will definitely introduce a change in the microenvironment (e.g., polarity, viscosity, refractive index), which can influence the spectral characteristics and photoreactivity of the sensitizer (Udal'tsov, 1997). A topical preparation should preferentially be transparent and enhance the optical transmissiveness of the tissue by hydration (Meserol, 1996). These aspects must be taken into account in the formulation process.

Aggregation of sensitizer molecules has a deleterious effect on the sensitizing properties. Formation of aggregates shortens the lifetimes of the singlet and triplet excited states and may induce a hypochromic shift in the absorption spectrum (Jori, 1990). Aggregation of sensitizer molecules usually occurs in aqueous media and is therefore a problem in parenteral solutions. Aggregation can be prevented by the introduction of charged substituents leading to electrostatic repulsion between the molecules; the presence of ligands that sterically inhibit aggregation; or by incorporating the sensitizer in a vehicle that solubilizes the molecules (Jori, 1990).

It has been demonstrated that liposomes, micelles, certain oil emulsions, and macromolecules (e.g., serum albumin) will keep the sensitizer molecules in a monomeric form (Jori, 1992; Hoebeke, 1995). The fact that most sensitizers are heavily charged and in many cases present as zwitterions, combined with their extensive polycyclic nature, makes some of these compounds particularly difficult to incorporate in a simple emulsion or liposome preparation. The zwitterions, in particular, can be almost insoluble in hydrophilic and lipophilic media. Alternative formulation approaches may therefore be required for this group of sensitizers.

For any drug product, the active principle within the formulation (e.g., the sensitizer) and the formulation (e.g., liposomes) should be stable during storage and use. In the case of commonly used photosensitizers, however, it can be advantageous that the compounds are labile *in vivo* (MacRobert, 1989; Sibata et al., 2000). Photodegradation *in vivo* can lead to formation of free radicals that can play an active role in the photoinduced cell toxicity. Bleaching the sensitizer near the surface can further allow light to penetrate deeper into the tissue, and degradation of the substance can facilitate clearance from the body. It is therefore important that the drug delivery system stabilizes the drug *in vitro,* thereby also preventing the drug inducing degradation of the vehicle (e.g., photo-oxidation of phospholipids in liposomes) without interfering with its *in vivo* stability.

15.4 OTHER APPROACHES TO PHOTOINDUCED DRUG DELIVERY

The application of photodynamic therapy (PDT) in treatment of bacterial infections will definitely continue to attract the attention of scientists. It is considered unlikely that bacteria develop resistance against singlet oxygen; therefore, PDT may become a useful weapon to battle infections. For treatment of infections in the oral cavity, a "smart toothpaste" containing a photosensitizer that is selectively absorbed by bacteria is under development. This can be used in combination with a "laser toothbrush" designed recently (Moore, 1998). Sensitizers can further be incorporated into bandages for disinfection of wounds or in cellophane films used to wrap food (Moore, 1998).

Another approach to overcome development of resistance is to reduce the amount of antibiotics present in the environment. Self-destructive antibiotic drugs have been synthesized. Their structure contains a functionality that unmasks upon exposure to artificial light or sunshine. The unmasking generates a reactive functionality that destroys the molecular structure and abolishes the antibiotic property. Thus, as soon as the compound is eliminated from the body and gets into the environment, it will be decomposed by natural daylight eliminating the antibacterial activity (Henry, 2000; Lee et al., 2000).

REFERENCES

Adams, S.R. and Tsien, R.Y. (1993) Controlling cell chemistry with caged compounds, *Annu. Rev. Physiol.,* 55, 755–784.

Andley, U.P. and Chakrabarti, B. (1983) Role of singlet oxygen in the degradation of hyaluronic acid, *Biochem. Biophys. Res. Commun.,* 115, 894–901.

Armitage, B., Klekota, P.A., Oblinger, E., and O'Brien, D.F. (1993) Enhancement of energy transfer on bilayer surfaces via polymerization-induced domain formation, *J. Am. Chem. Soc.,* 115, 7920–7921.

Baldursdottir, S.G., Kjøniksen, A.-L., Karlsen, J., Nystrøm, B., Roots, J., and Tønnesen, H.H. (2003) Riboflavin-photosensitized changes in aqueous solutions of alginate. Rheological studies, *Biomacromolecules,* 4, 429–436.

Bisby, R.H., Mead, C., Mitchell, A.C., and Morgan, C.G. (1999) Fast laser-induced solute release from liposomes sensitized with photochromic lipid: effects of temperature, lipid host, and sensitizer concentration, *Biochem. Biophys. Res. Commun.,* 262, 406–410.

Bisby, R.H., Mead, C., and Morgan, C.G. (2000a) Active uptake of drugs into photosensitive liposomes and rapid release on UV photolysis, *Photochem. Photobiol.,* 72, 57–61.

Bisby, R.H., Mead, C., and Morgan, C.G. (2000b) Wavelength-programmed solute release from photosensitive liposomes, *Biochem. Biophys. Res. Commun.*, 276, 169–173.

Bondurant, B. and O'Brian, D.F. (1999) Photoinduced destabilization of sterically stabilized liposomes, *Polymer Preprints Am.*, 40, 353–354.

Bondurant, B., Mueller, A., and O'Brien, D.F. (2001) Photoinitiated destabilization of sterically stabilized liposomes, *Biochim. Biophys. Acta*, 1511, 113–122.

Bos, G.W., Verrijk, R., Franssen, O., Bezemer, J.M., Hennink, W.E., and Crommelin, D.J.A. (2001) Hydrogels for the controlled release of pharmaceutical proteins, *Pharm. Technol. North Am.*, 25, 110–120.

Busetti, A., Soncin, M., Reddi, E., Rodgers, M.A.J., Kenny, M.E., and Jori, G. (1999) Photothermal sensitization of amelanotic melanoma cells by Ni(II)-octabutoxy-naphthalocyanine, *J. Photochem. Photobiol. B Biol.*, 53,103–109.

Chang, S.J., Lee, C.H., and Wang, Y.J. (1999) Microcapsules prepared from alginate and a photosensitive poly (L-lysine), *J. Biomater. Sci. Polymer Edn.*, 10, 531–542.

Chen, W.R., Adams, R.L., Carubelli, R., and Nordquist, R.E. (1997) Laser-photosensitizer assisted immunotherapy: a novel modality for cancer treatment, *Cancer Lett.*, 115, 25–30.

Clapp, P.J., Armitage, B.A., and O'Brien, D.F. (1997) Two-dimensional polymerization of lipid bilayers: visible-light-sensitized photoinitiation, *Macromolecules*, 30, 32–41.

Collier, J.H., Hu, B.-H., Ruberti, J.W., Zhang, J., Shum, P., Thompson, D.H., and Messersmith, P.B. (2001) Thermally and photochemically triggered self-assembly of peptide hydrogels, *J. Am. Chem. Soc.*, 123, 9463–9464.

Copeland, E.S., Alving, C.R., and Grenan, M.M. (1976) Light-induced leakage of spin label marker from liposomes in the presence of phototoxic phenothiazines, *Photochem. Photobiol.*, 24, 41–48.

Elisseeff, J., Anseth, K., Sims, D., McIntosh, W., Randolph, M., and Langer, R. (1999a) Transdermal photopolymerization for minimally invasive implantation, *Proc. Natl. Acad. Sci. U.S.A.*, 96, 3104–3107.

Elisseeff, J., Anseth, K., Sims, D., McIntosh,W., Randolph, M., Yaremchuk, M., and Langer, R. (1999b) Transdermal photopolymerization of poly (ethylene oxide)-based injectable hydrogels for tissue-engineered cartilage, *Plastic Reconstr. Surg.*, 104, 1014–1022.

Elisseeff, J., McIntosh, W., Anseth, K., Riley, S., Ragan, P., and Langer, R. (2000) Photoencapsulation of chondrocytes in poly(ethylene oxide)-based semi-interpenetrating networks, *J. Biomed. Mater. Res.*, 51, 164–171.

Frati, E., Khatib, A.-M., Front, P., Panasyuk, A., Aprile, F., and Mitrovic, D.R. (1997) Degradation of hyaluronic acid by photosensitized riboflavin *in vitro*. Modulation of the effect by transition metals, radical quenchers, and metal chelators, *Free Rad. Biol. Med.*, 22, 1139–1144.

Gerasimov, O.V., Schwan, A., and Thompson, D.H. (1997) Acid-catalyzed plasmenylcholine hydrolysis and its effect on bilayer permeability: a quantitative study, *Biochim. Biophys. Acta*, 1324, 200–214.

Gilbert, B.C., King, D.M., and Thomas, C.B. (1984) The oxidation of some polysaccharides by the hydroxyl radical: an E.S.R. investigation, *Carbohydr. Res.*, 125, 217–235.

Gregoriadis, G. (1995) Engineering liposomes for drug delivery: progress and problems, *TIBTECH*, 13, 527–537.

Gurney, A.M. and Lester, H.A. (1987) Light-flash physiology with synthetic photosensitive compounds, *Physiol. Rev.*, 67, 583–617.

Hagen, V., Dzeja, C., Bendig, J., Baeger, I., and Kaupp, V.B. (1998) Novel caged compounds of hydrolysis-resistant 8-Br-cAMP and 8-Br-cGMP: photolabile NPE esters, *J. Photochem. Photobiol. B Biol.*, 42, 71–78.

Hasan, T. (1992) Photosensitizer delivery mediated by macromolecular carrier systems, in Henderson, B.W. and Dougherty, T.J. (Eds.) *Photodynamic Therapy. Basic Principles and Clinical Applications*, New York: Marcel Dekker.

Henry, C.M. (2000) Antibiotic resistance, *Chem. Eng. News*, 78, 41–58.

Hillery, A.M. (1998a) Liposomal drug delivery. II. Clinical applications, *Pharm. J.*, 261, 712–715.

Hillery, A.M. (1998b) Liposomal drug delivery. I. Introduction, *Pharm. J.*, 261, 626–629.

Hinnen, P., de Rooij, F.W.M., Hop, W.C.J., Edixhoven, A., van Dekken, H., Wilson, J.H.P., and Siersema, P.D. (2002) Timing of 5-aminolaevulinic acid-induced photodynamic therapy for the treatment of patients with Barrett's oesophagus, *J. Photchem. Photobiol. B Biol.*, 68, 8–14.

Hoebeke, M. (1995) The importance of liposomes as models and tools in the understanding of photosensitization mechanisms, *J. Photochem. Photobiol. B Biol.*, 28, 189–196.

Houlton, S. (2003) Blocking the way forward, *Manuf. Chem.*, 74, 27–30.

Jori, G. (1990) Photosensitized processes *in vivo*: proposed phototherapeutic applications, *Photochem. Photobiol.*, 52, 439–443.

Jori, G. (1992) Low-density lipoproteins-liposome delivery systems for tumor photosensitizers *in vivo*, in Henderson, B.W. and Dougherty, T.J. (Eds.), *Photodynamic Therapy. Basic Principles and Clinical Applications*, New York: Marcel Dekker.

Jori, G. (1996) Tumor photosensitizers: approaches to enhance the selectivity and efficiency of photodynamic therapy, *J. Photochem. Photobiol. B Biol.*, 36, 87–93.

Kaplan, J.H. and Somlyo, A.P. (1989) Flash photolysis of caged compounds: new tools for cellular physiology, TINS, 12, 54,–59.

Klyashchitsky, B.A., Nechaeva, I.S., and Ponomaryov, G.V. (1994) Approaches to targeted photodynamic tumor therapy, *J. Contr. Rel.*, 29, 1–16.

Kodzwa, M.G., Staben, M.E., and Rethwisch, D.G. (1999) Photoresponsive control of ion-exchange in leucohydroxide containing hydrogel membranes, *J. Membr. Sci.*, 158, 85–92.

Kojima, M., Takahashi, K., and Nakamura, K. (2001) Cationic dye-sensitized degradation of sodium hyaluronate through photoinduced electron transfer in the upper excited state, *Photochem. Photobiol.*, 74, 369–377.

Konan, Y.N., Gurney, R., and Allémann, E. (2002) State of the art in the delivery of photosensitizers for photodynamic therapy, *J. Photochem. Photobiol. B Biol.*, 66, 89–106.

Kost, J. and Langer, R. (2001) Responsive polymeric delivery systems, *Adv. Drug Del. Rev.*, 46, 125–148.

Kusumi, A., Nakahama, S., and Yamaguchi, K. (1989) Liposomes that can be disintegrated by photo-irradiation, *Chem. Lett.*, 433–436.

Lamparski, H., Liman, U., Barry, J.A., Frankel, D.A., Ramaswami, F.V., Brown, M.F., and O'Brien, D.F. (1992) Photoinduced destabilization of liposomes, *Biochemistry*, 31, 685–694.

Lane, N. (2003) Light on medicine, *Sci. Am.*, 288, 26–33.

Lasic, D.D. (1998) Novel applications of liposomes, *TIBTECH*, 16, 307–321.

Lee, W., Li, Z.-H., Vakulenko, S., and Mobashery, S. (2000) A light-inactivated antibiotic, *J. Med. Chem.*, 43, 128–132.

Love, W.G., Duk, S., Biolo, R., Jori, G., and Taylor, P.W. (1996) Liposome-mediated delivery of photosensitizers: location of zinc (II)-phthalocyanine within implanted tumors after intravenous administration, *Photochem. Photobiol.*, 63, 656–661.

Lu, M.Z., Lan, H.L., Wang, F.F., and Wang, Y.J. (2000) A novel cell encapsulation method using photosensitive poly(allylamine α-cyanocinnamylideneacteate), *J. Microencapsulation*, 17, 245–251.

MacRobert, A.J., Bown, S.G., and Phillips, D. (1989) What are the ideal photoproperties for a sensitizer? *Ciba Foundation Symposium 146*, New York: John Wiley & Sons.

Marriott, G. (Ed.) (1998) *Caged Compounds, Methods in Enzymology,* 291, New York: Academic Press.

Mathiowitz, E., Raziel, A., Cohen, M.D., and Fischer, E. (1981) Photochemical rupture of microcapsules: a model system, *J. Appl. Polym. Sci.,* 26, 809–822.

Mathiowitz, E. and Cohen, M.D. (1989a) Polyamide microcapsules for controlled release. I. Characterization of the membranes, *J. Membr. Sci.,* 40, 1–26.

Mathiowitz, E. and Cohen, M.D. (1989b) Polyamide microcapsules for controlled release. II. Release characteristics of the microcapsules, *J. Membr. Sci.,* 40, 27–41.

Mathiowitz, E. and Cohen, M.D. (1989c) Polyamide microcapsules for controlled release. III. Spontaneous release of azobenzene, *J. Membr. Sci.,* 40, 43–54.

Mathiowitz, E. and Cohen, M.D. (1989d) Polyamide microcapsules for controlled release. IV. Effects of swelling, *J. Membr. Sci.,* 40, 55–65.

Mathiowitz, E. and Cohen, M.D. (1989e) Polyamide microcapsules for controlled release. V. Photochemical release, *J. Membr. Sci.,* 40, 67–86.

Meserol, P.M. (1996) Article of Manufacture for the Photodynamic Therapy of Dermal Lesion, U.S. Patent 5505726.

Miller, C.R., Clapp, P.J., and O'Brien, D.F. (2000) Visible light-induced destabilization of endocytosed liposomes, *FEBS Lett.,* 467, 52–56.

Moore, P. (1998) Lethal weapon, *New Sci.,* 158, 40–43.

Morgan, C.G., Yianni, Y.P., Sandhu, S.S., and Mitchell, A.C. (1995a) Liposome fusion and lipid exchange on ultraviolet irradiation of liposomes containing a photochromic phospholipid, *Photochem. Photobiol.,* 62, 24–29.

Morgan, C.G., Bisby, R.H., Johnson, S.A., and Mitchell, A.C. (1995b) Fast solute release from photosensitive liposomes: an alternative to "caged" reagents for use in biological systems, *FEBS Lett.,* 375, 113–116.

Mueller, A., Bondurant, B., and O'Brien, D.F. (2000) Visible-light-stimulated destabilization of PEG-liposomes, *Macromolecules,* 33, 4799–4804.

Nakayama, Y. and Matsuda, T. (2003) Photo-control of the interaction between endothelial cells and photo-cation generatable water-soluble polymers, *J. Contr. Rel.,* 89, 213–224.

O'Brien, D.F., Armitage, B., Benedicto, A., Bennett, D.E., Lamparski, H.G., Lee, Y.-S., Srisiri, W., and Sisson, T.M. (1998) Polymerization of performed self-organized assemblies, *Acc. Chem. Res.,* 31, 861–868.

Ohya, Y., Okuyama, Y., Fukunaga, A., and Ouchi, T. (1998) Photo-sensitive lipid membrane perturbation by a single chain lipid having terminal spiropyran group, *Supramol. Sci.,* 5, 21–29.

Overhaus, M., Heckenkamp, J., Kossodo, S., Leszczynski, D., and LaMuraglia, G.M. (2000) Photodynamic therapy generates a matrix barrier to invasive vascular cell migration, *Circ. Res.,* 86, 334–340.

Overholt, B.F., Panjehpour, M., and Haydek, J.M. (1999) Photodynamic therapy for Barrett's esophagus: follow-up in 100 patients, *Gastrointest. Endosc.,* 49, 1–7.

Parrish, J.A. (1981) New concepts in therapeutic photomedicine: photochemistry, optical targeting and the therapeutic window, *J. Invest. Dermatol.,* 77, 45–50.

Pellicci023, A.P. and Wirz, J. (2002) Photoremovable protecting groups: reaction mechanisms and applications, *Photochem. Photobiol. Sci.,* 1, 441–458.

Peppas, N.A., Bures, P., Leobandung, W., and Ichikawa, H. (2000) Hydrogels in pharmaceutical formulations, *Eur. J. Pharm. Biopharm.,* 50, 27–46.

Pidgeon, C. and Hunt, C.A. (1983) Light sensitive liposmes, *Photochem. Photobiol.,* 37, 491–494.

Qiu, Y. and Park, K. (2001) Environment-sensitive hydrogels for drug delivery, *Adv. Drug. Del. Rev.,* 53, 321–339.

Reddi, E. (1997) Role of delivery vehicles for photosensitizers in the photodynamic therapy of tumors, *J. Photochem. Photobiol. B Biol.,* 37, 189–195.

Scott, J.E. and Tigwell, M.J. (1973) Periodate-induced viscosity decreases in aqueous solutions of acetal- and ether-linked polymers, *Carbohyd. Res.,* 28, 53–59.

Shiah, J.-G., Sun, Y., Peterson, C.M., Straight, R.C., and Kopecek, J. (2000) Antitumor activity of N-(2-hydroxypropyl)methacrylamide copolymer-mesochlorin e_6 and adriamycin conjugates in combination treatments, *Clin. Cancer Res.,* 6, 1008–1015.

Shum, P., Kim, J.-M. and Thompson, D.H. (2001) Phototriggering of liposomal drug delivery systems, *Adv. Drug Del. Rev.,* 53, 273–284.

Sibata, C.H., Colussi, V.C., Oleinick, N.L., and Kinsella, T.J. (2000) Photodynamic therapy: a new concept in medical treatment, *Braz. J. Med. Biol. Res.,* 33, 869–880.

Silverman, R.P., Elisseeff, J., Passaretti, D., Huang, W., Randolph, M.A., and Yaremchuk, M.J. (1999) Transdermal photopolymerized adhesive for seroma prevention, *Plastic Reconstr. Surg.,* 103, 531–535.

Smeds, K.A. and Grinstaff, M.W. (2001) Photocrosslinkable polysaccharides for *in situ* hydrogel formation, *J. Biomed. Mat. Res.,* 54, 115–121.

Srinath, P. and Jain, N.K. (1994) Signal sensitive liposomes, *Indian Drugs,* 31, 284–290.

Sunamoto, J., Iwamoto, K., Mohri, Y., and Kominato, T. (1982) Liposomal membranes. 13. Transport of an amino acid across liposomal bilayers as mediated by a photoresponsive carrier, *J. Am. Chem. Soc.,* 104, 5502–5504.

Thompson, D.H., Gerasimov, O.V., Wheeler, J.J., Rui, Y., and Anderson, V.C. (1996) Triggerable plasmalogen liposomes: improvement of system efficiency, *Biochim. Biophys. Acta,* 1279, 25–34.

Tomer, R. and Florence, A.T. (1993) Photo-responsive hydrogels for potential responsive release applications, *Int. J. Pharm.,* 99, R5–R8.

Trauner, K.B. and Hasan, T. (1996) Photodynamic treatment of rheumatoid and inflammatory arthritis, *Photochem. Photobiol.,* 64, 740–750.

Udal'tsov, A.V. (1997) Characterisitcs of donor-acceptor complexes formed in porphyrin-polymer systems and their photoactivation in electron transfer photoreaction, *J. Photochem. Photobiol. B Biol.,* 37, 31–39.

Valduga, G., Reddi, E., Garbisa, S., and Jori, G. (1998) Photosensitization of cells with different metastatic potentials by liposome-delivered Zn(II)-phthalocyanine, *Int. J. Cancer,* 75, 412–417.

Vyas, S.P., Sood, A., Venugopalan, P., and Mysore, N. (1997) Circadian rhythm and drug delivery design, *Pharmazie,* 52, 815–820.

Ward, J.H. and Peppas, N.A. (2001) Preparation of controlled release systems by free-radical UV polymerizations in the presence of a drug, *J. Contr. Rel.,* 71, 183–192.

Wymer, N.J., Gerasimov, O.V., and Thompson, D.H. (1998) Cascade liposomal triggering: light-induced Ca^{2+} release from diplasmenylcholine liposomes triggers PLA_2-catalyzed hydrolysis and contents leakage from DPPC liposomes, *Bioconjugate Chem.,* 9, 305–308.

Yui, N., Okano, T., and Sakurai, Y. (1993) Photo-responsive degradation of heterogeneous hydrogels comprising crosslinked hyaluronic acid and lipid microspheres for termporal drug delivery, *J. Contr. Rel.,* 26, 141–145.

Zeimer, R. and Goldberg, M.F. (2001) Novel ophthalmic therapeutic modalities based on noninvasive light-targeted drug delivery to the posterior pole of the eye, *Adv. Drug Del. Res.,* 52, 49–61.

Zhang, Z.-Y. and Smith, B.D. (1999) Synthesis and characterization of NVOC-DOPE, a caged photoactivatable derivative of dioleoylphosphatidylethanolamine, *Bioconjugate Chem.,* 10, 1150–1152.

CHAPTER 16

Formulation Approaches for Improving Solubility and Its Impact on Drug Photostability

Hanne Hjorth Tønnesen

CONTENTS

16.1 INTRODUCTION

Solubility is one of the critical parameters determining bioavailability of a drug. It is estimated that more than 40% of potential drug candidates fail in the drug development process due to poor bioavailability (Lipper, 1999). With the exception of drug transport facilitated by endocytosis, the drug should be present in soluble form to be absorbed. According to Fick's first law, the absorption process mediated by passive transport is then a direct function of the concentration gradient over the biological membrane and the permeability of the membrane (Equation 16.1) (Löbenberg and Amidon, 2000).

$$J_w = P_w \times C_w \qquad (16.1)$$

where J_w is the transport across the membrane, P_w is the permeability of the membrane, and C_w is the concentration of dissolved drug at the absorption site.

For many drugs that cross a biological membrane rather easily, the onset of drug levels will be determined by the time required for the dosage form to release its contents and for the drug substance to dissolve. For solid dosage forms, the dissolution rate may be so slow that the active compound is not released within the time that it is in transit past the absorptive sites. The drug dissolution rate is primarily determined by the aqueous solubility; it has been demonstrated that compounds with aqueous solubilities lower than 100 μg/ml often present dissolution limitations to absorption (Hörter and Dressman, 1997). For parenteral preparations, the dose can become a significant formulation challenge due to limitations in aqueous solubility and administration volume. Thus, the formulation process often requires strategies to improve the aqueous solubility of the drug substance in order to achieve delivery of a correct dose or to improve bioavailability.

The choice of a particular solubilization method will depend on the drug's stability, biocompatibility of the vehicle, and the solubility efficiency of the system. The two approaches for enhancing drug solubility are (1) manipulation of the solid compound (e.g., by selection of a different crystalline form) or (2) modification of the drug formulation (e.g., by altering the viscosity or polarity of the medium). Any of the techniques used for drug solubilization are likely to have an impact on the photostability of the drug molecule. This should be kept in mind, particularly in cases where reformulation is applied on a previously well characterized drug substance.

16.1.1   Why the Formulation Process Can Influence Drug Photostability

In order to undergo a photochemical reaction, a molecule must contain a chromophore that can absorb the incident radiation. A chromophore consists of π and n electrons. Absorption of radiation leads to a transition of one of these electrons from the ground state (S_0) to the excited singlet state and subsequently to a triplet state (^1S and ^3T, respectively). A number of reactions may occur from these excited states, leading to, for instance, degradation of the drug molecule; interactions with excipients; or a change in physicochemical properties of the formulation (e.g., color, viscosity). By introducing the pure drug substance into a complex formulation, the character of the lowest singlet and triplet states can be changed, thereby changing reactivity and types of reactions. The formulation approaches applied to improve drug solubility may influence the environment of the molecule and change the drug photostability by one or more of the following:

- Alteration in drug absorption spectrum
- Change in excited state lifetime of the drug
- Change in aggregate formation of drug molecules within the formulation

Each of these factors will have an impact on essential processes like energy or charge transfer, diffusion–controlled reactions, and interactions between molecules (Table 16.1).

Table 16.1 Physicochemical Approaches for Increasing Solubility and Resulting Changes in Product Characteristics Influencing Drug Photostability

Parameter to be Altered	Physicochemical Approach	Resulting Change Likely to Influence Photostability
Particle surface area	Solid dispersion	Restricted molecular mobility
		Exposure of individual molecules
		Transparent matrix
		Photoreactive matrix
	Particle size reduction	Polymorphism
Solubility	Crystal structure	Polymorphism
	Hydrophilicity	Nonhomogenous environment
		Arranged medium
	Solubilization	Nonhomogenous environment
		Arranged medium

16.2 SOLID STATE

The two essential components governing the solubility of an organic solute in water are the molecular structure and the crystal structure. The activity coefficient of the solute in solution describes the effect of molecular structure on the aqueous solubility. Assuming that the molecular entities are not to be altered during the drug development process, the activity coefficient can be changed by modifying the polarity and/or viscosity of the drug formulation. On the other hand, the crystal structure determines melting point and enthalpy of solution. The melting point reflects the energy needed to free a solute molecule from its crystal and is strictly a function of the pure crystal and independent of the solvent. The higher the melting point is, the lower the solubility.

The various crystal modifications of a substance will possess different melting points and thereby have different solubilities (Brittain, 2002). Only one solid phase is thermodynamically stable for a given set of environmental conditions. The most stable form has the lowest free energy and therefore the lowest solubility. Metastable forms can theoretically be used to improve solubility, but the kinetics of the transformation back to the stable crystal modification must then be taken into account. In suspension, solvent-mediated transition is generally too rapid to give a product with an acceptable shelf-life. In solid dosage forms, it may be possible, however, to utilize a less stable crystal modification. Due to the decreased molecular mobility in the solid state the transition rate can be low.

The molecular mobility can be further restricted by using appropriate excipients. It is obvious that the unknowing use of a metastable crystal form can lead to solubility and stability problems. A change in crystal modification can occur during processing or as a result of a photochemical process (Briggner et al., 1994; Ahmed et al., 1996; Nord et al., 1997). The data available on photochemical stability of drugs in the solid state that include information on the effect of formulation principle and excipients are at present rather limited.

16.2.1 Photochemical Reactions in the Solid State

The photodegradation pathway of a drug in the solid state does not necessarily follow the degradation in solution. The difference arises mainly from a lack of molecular mobility, which restricts diffusion-controlled reactions and interactions between molecules, thereby influencing the reactant conformation. The photochemical properties of molecules in the solid state or embedded in a solid matrix depend greatly on the organization (if any) and nature of the solid lattice. In a number of cases, it has been possible to correlate the reactivity with packing of the reactant molecule in the crystal (Hadjoudis et al., 1986).

Solids can be classified according to their organizational structure and, in the context of photochemical reactions, be classified as glasses, polymers, and crystals. Furthermore, a solid-state photochemical reaction only takes place in a thin layer of the solid surface, while in a dilute solution all the molecules are equally exposed to the radiation. In a tablet, the radiation has been estimated to reach a penetration depth of 0.03 cm, and the faded layer apparently does not increase upon further exposure (Carstensen, 1974). A study of nifedipine in the solid state demonstrated that the degradation rate was inversely proportional to the thickness of the powder bed (Marciniec and Rychcik, 1994).

A considerable increase in photodecomposition of a number of drug substances is reported under continuous mixing of solid samples during irradiation (Takács and Reisch, 1986; Takács et al., 1990). A solid-state photoreaction is influenced by the shape, size, and texture of the particles or drug device because these parameters will influence the absorption, reflection, and scattering properties of the sample. The ratio of absorbed to scattered radiation, θ, relates to the fraction of absorbed to reflected radiation (R) as given in Equation 16.2 (Carstensen, 1974):

$$\theta = (1 - R^2)/2R \tag{16.2}$$

A study on furosemide demonstrated that separate particles (size 22 μm) degraded more rapidly than agglomerates of smaller particles (individual size 2.5 μm) due to a larger surface area exposed in the first case (de Villiers et al., 1993). The presence of excipients or coatings will further influence the solid state degradation as discussed next. Solid state degradation is a heterogenous process; as a result, photodegradation in the solid state differs from that in the liquid state in a mechanistic sense and also in its kinetics.

First-order kinetics normally applies to the photodecomposition of a drug substance in dilute solution (see Chapter 3). In cases in which the rate-limiting factor is the intensity of the incident radiation, the reaction will follow pseudo-zero-order kinetics. Unless the reaction is followed to the extent of >50% conversion, it is difficult to decide the reaction order clearly. It may be particularly difficult to follow a reaction this far in the solid state. A solid-state reaction would be expected to follow pseudo-zero-order kinetics due to the optical properties of most solid samples. However, the powder bed or formulation in which the reaction takes place is usually thicker than a monomolecular layer, leading to varying conditions (i.e., intensity of radiation) throughout the sample.

Because the radiation in most cases will not penetrate the entire sample, the concentration of the reactant is unlikely to approach zero at infinite time. A plot of remaining concentration *vs.* time will therefore level off at a value greater than zero. This should be taken into account when selecting the kinetic model for studies of solid-state degradation (Sande, 1996). The solid-state degradation will in some cases appear to consist of a series of consecutive processes with different mechanisms and rates (Carstensen, 1974). Such a stepwise change in reaction rate is most likely caused by an alteration in sample surface and fading of subsequent layers. The concept of reaction order may not be useful for photodecomposition in the solid state (De Villiers et al., 1992).

As discussed in Chapter 3, the use of rate constants can be helpful for comparative purposes under the same irradiation conditions; however, the quantitative expression of photochemical rate should be given in the terms of the quantum yield of the reaction. To determine the quantum yield in the solid state is unfortunately not a straightforward process. As a result, most studies in the literature refer to an apparent reaction order calculated from the degradation of a minor fraction of the sample. Such data should preferentially be supported by some detail of the reaction mechanism.

Various techniques have been applied in studies of solid samples. A direct method based on infrared spectroscopy has been used in the study of pharmacologically active compounds such as nitroimidazole derivatives (chemotherapeutics) and nifedipine (Marciniec and Rychcik, 1994; Marciniec et al., 1997). This approach can be applied in kinetic and quantitative examinations of the degradation process. An alternative method based on diffuse reflectance spectrophotometry takes into account the surface reflectance of the sample (Zhan et al., 1995). In recent years, Raman spectroscopy has become a powerful tool for the study of various processes in the solid state. The technique allows mapping of the concentration of one specific component within a sample, e.g., an active ingredient in a tablet formulation. This method may therefore become a useful tool in future studies of solid-state photodegradation (Opel and Venturini, 2002).

The fact that molecular movements in the solid state are restricted and that the sample permeability to oxygen and moisture can be low may lead to an alteration in product formation compared to the drug in solution. This is illustrated by formation of new rotamers in the solid-state degradation of tolrestat; favoring of the redox product in the solid-state degradation of nifedipine compared to the photo-oxidation product observed in solution; or by diversity in degradation products of santonins in solution and in the solid state (Reisch et al., 1986; Lee and Lee, 1990; Marciniec and Rychcik, 1994). These observations emphasize that results obtained on the drug substance in solution are not necessarily valid for the solid state.

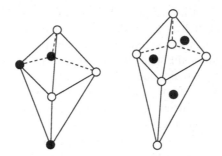

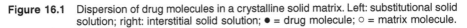

Figure 16.1   Dispersion of drug molecules in a crystalline solid matrix. Left: substitutional solid solution; right: interstitial solid solution; ● = drug molecule; ○ = matrix molecule.

16.2.2 Dispersion of Drugs in Carriers

The dispersion of a poorly water-soluble drug in a strongly hydrophilic carrier increases the dissolution rate by increasing the drug surface area and optimizing the wetting, thus improving bioavailability of the substance. The resulting system can consist of very fine drug crystals in a carrier (i.e., a eutectic combination) or be a dispersion of drug molecules in a solid medium (i.e., solid solution). The benefit of the latter in the context of improved solubility is that the drug no longer has a crystal structure, and therefore no crystalline bonding to be broken before dissolution. The solid carrier can be in a crystalline or an amorphous state (e.g., a glass). In a crystalline solid solution, the drug molecules can be incorporated by fitting into the interstices between the carrier molecules or by substitution of carrier molecules in the crystal lattice (Figure 16.1) (Leuner and Dressman, 2000).

Within an amorphous carrier, the drug molecules are dispersed irregularly (Figure 16.2). The carriers most commonly used for preparing solid dispersions are polyethylene glycol (PEG), polyvinyl pyrrolidone (PVP), cellulose derivatives, acrylates, urea, citric acid, and sugars. Whether the resulting product is a crystalline or an amorphous solid solution or a eutectic mixture is dependent on the individual combination of drug and carrier and on the preparation method. For example, chlorpropamide mixed with urea in the proportion 89:11 is reported to form a eutectic composition (Ford and Rubinstein, 1977). Higher concentrations of chlorpropamide result in a solid solution of urea in chlorpropamide, whereas solid solutions do not form at less than 89%. Produced by the hot melt method (Leuner and Dressman, 2000), mixtures of >50% chlorpropamide in urea, including the eutectic, existed as solid glass (Ford and Rubinstein, 1977).

Thus far, little information is available on the photochemical behavior of drug molecules in solid dispersions; however, the solid dispersion type of formulation is likely to influence the photoreactivty of the molecules in some way. The molecules are, in a sense, as "unprotected" as in a solution, but their movements are heavily restricted. Compared to a traditional powder mixture or granulate, the carrier system is often highly transparent to radiation. In theory, lactose, mannitol, sugar, starches, and PVP are excipients with the potential to participate in photochemical processes

Figure 16.2   Dispersion of drug molecules in an amorphous carrier.

(Chapter 2). They have abstractable hydrogens and are thereby susceptible to free radical attack. However, the reactions will be restricted by the rigidity of the matrix.

Common excipients also frequently contain impurities that are likely to participate in photochemical reactions. For example, PVP may contain peroxides; lactose may contain aldehydes and reducing sugars; PEG may contain aldehydes, peroxides, and organic acids; while glyoxal may be detected in cellulose derivatives (Crowly and Martini, 2001). Citric acid, which is capable of forming a transparent glass, can bind trace amounts of Fe^{3+}. The resulting iron–citrate complex has an absorption spectrum that extends into the visible region (i.e., up to 450 nm) and can thereby act as a "filter" to protect the drug or as a sensitizer to induce photodegradation. Citric acid can also facilitate the reduction of Fe^{3+} to Fe^{2+}, which is an important step in the Fenton reaction. Whether a solid dispersion will have a photostabilizing or -destabilizing effect on the drug depends on the dominating factors: the increase in exposure area (single molecules *vs.* particles), system transparency and sensitizing properties, or low reactivity due to restricted mobility.

16.2.2.1   Solid Glasses

In the glass form, the immobilized molecules have retained the structure of the liquid state but there is no special ordering of solvent molecules around a solute. Diffusion of molecules in rigid glasses is negligible within the lifetime of the excited state. The main reactions are unimolecular dissociations and isomerization. Although the processes are rather similar to liquid reactions, the fragments cannot separate through diffusion. Therefore, the reactants are often restored due to recombination of the fragments (Suppan, 1994).

16.2.2.2   Crystals

The crystal form is an ordered molecular structure. The crystals can consist of the pure drug substance or the drug substance can be incorporated in a carrier, which is in the crystalline form, as discussed earlier. In such systems, similar molecules are forced to stay relatively close and in fixed orientations. Upon excitation, the energy of one excited molecule can be delocalized and transferred from one molecule to another throughout the lattice, leading to damage of molecules relatively distant from the site of absorption (Suppan, 1994). Bimolecular reactions can also take

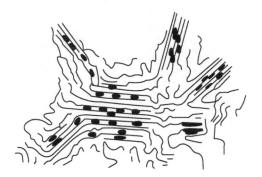

Figure 16.3   Dispersion of drug molecules in a semicrystalline polymer.

place when molecules are kept in close contact in a crystal lattice, and some geometrical spacing is ideal for photodimerization (Reisch et al., 1994a). Photoreactivity for a number of pure drug substances in the crystal form has been examined (Reisch and Müller, 1984; Reisch et al., 1989a, b, 1992a, b, 1993, 1994b; Ekiz–Gücer and Reisch, 1991a, b; Ekiz–Gücer et al., 1991a, b; Takács et al., 1991) but few data are available on solid dispersions in crystalline matrixes.

16.2.2.3 Polymeric Systems

A number of solid drug formulations, including solid dispersions, contain polymers. In such systems, the drug molecules are usually dispersed at random among the polymer chains, which may be semicrystalline in nature (Figure 16.3). Apparently the crystalline regions of a polymer lack oxygen, while the diffusion of small molecules in the amorphous part of the system is much higher than would be predicted from the internal viscosity of the medium (Rabek, 1996). Transfer of excitation energy can occur in polymeric structures in a similar manner to what is observed in crystals. In addition to this, bimolecular interactions between polymer and drug can take place. In the absence of diffusion or migration as in rigid polymer matrices, interaction is only possible within a certain "sphere of action." For short-range interactions, this distance is about 1 to 1.5 nm (Rabek, 1996).

Radiation can also be absorbed by the carrier. The polymer may contain chromophores as part of its molecular structure or as low molecular weight impurity chromophores (e.g., traces of solvents, additives, coextracts from natural sources). Such chromophores can act as sensitizers leading to the degradation of the drug. Furthermore, the polymers may contain reactive groups (e.g., carbonyl in methyl acrylates) that can play an active role in photoinduced reactions between the drug and the carrier. For example, incorporation of the drug suprofen in hydrogels based on two macromolecules (PAHy and PHEA) clearly changed the photoreactivity of the drug substance. This was demonstrated as a significant reduction in the drug photosensitizing activity (Giammona et al., 1998).

A study on nitrobenzyl compounds demonstrated a similarity between the reaction products and mechanism in organic solvent and in a poly(methyl methacrylate) matrix, although the quantum yield for the photoreaction was substantially lower in the solid state (Reichmanis et al., 1985). The low quantum yield was ascribed to the decreased conformational mobility in the matrix, whereas interactions with the polymer seemed to be of little importance. Another study has showed that Eudragit (RS100 and RL100) has a strong inhibitory effect on the photodegradation of diflunisal present as solid dispersions in these polymers (Pignatello et al., 2001). It was suggested that an electrostatic interaction between the drug and the polymer influenced the electron-trapping reaction essential for the photodegradation. It was further postulated that the host organic functional groups had a scavenging effect toward the free radicals generated during the drug degradation process.

16.2.3 Polymorphism

Solid state property differences derived from the existence of alternate crystal forms can lead to extensive differences of pharmaceutical importance, e.g., solubility, dissolution rate, and stability. It is claimed that most drug substances show polymorphism (Borka, 1991). As discussed previously, it is essential to determine which of the various forms should be used in a drug product to assure stable and reproducible formulation. A marked difference between the photostability of various crystal modifications of drug substances has been reported. This can be ascribed to differences in inter- and intramolecular binding, differences in diffusability (crystalline vs. amorphous structure), and differences in water content (crystal water, adsorbed water) (Hüttenrauch et al., 1986).

Exposure of solid samples to radiation almost invariably leads to discoloration even when the chemical transformation is moderate, or undetectable (Nyqvist and Wadsten, 1986; Matsuda et al., 1989; De Villiers et al., 1992). A distinctly different degree of photostability and also coloration is in many cases evident among the modifications (Lin et al., 1982; Matsuda and Tatsumi, 1990; Nord et al., 1997; Tønnesen et al., 1997). The possibility of conversion of one polymorphic form to another during manufacture and storage cannot be ruled out, resulting in a changed photosensitivity of the formulation. For example, milling or micronization of the raw material can make crystalline materials partly amorphous (Briggner et al., 1994; Ahmed et al., 1996). Grinding or compression of chloroquine diphosphate was demonstrated to result in an increased sensitivity to radiation (Nord et al., 1997) (Figure 16.4 and Figure 16.5).

The photostability of different crystal modifications can be further influenced by the presence of excipients, or a change in humidity or in surrounding oxygen concentration as demonstrated for cianidanol (Akimoto et al., 1985). In the case of chloroquine, a new and less photosensitive modification was formed at high humidity (Nord et al., 1997; Figure 16.4 and Figure 16.5). Water can dissolve in amorphous compounds because of the disordered state of the solid (Ahlneck and Zografi, 1990). This can further dissolve the drug substance converting the assumed solid state photochemistry into processes typical for solutions. However, only a minor effect on photostability could be detected in the example of a beta-lactamase inhibitor amorphous compound in spite of a gain in weight of 25% due to water absorption (Spilgies, 1998).

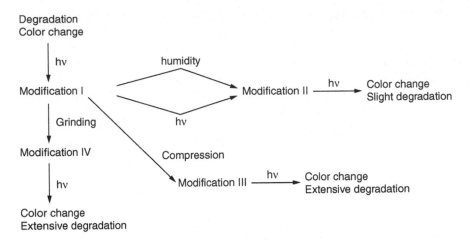

Figure 16.4   Formation of new crystal modifications during processing of chloroquine diphosphate.

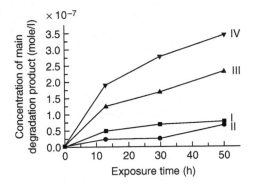

Figure 16.5   Formation of desethyl chloroquine, the main photodecomposition product of chloroquine, as a function of exposure time of the various polymorphic forms I through IV.

16.2.4 Reduction in Particle Size

The dissolution rate of a drug substance is directly proportional to the effective surface area of the solid available for dissolution according to the Noyes–Whitney equation:

$$dC/dt = AD(C_s - C)/h \tag{16.3}$$

where dC/dt is the rate of dissolution, A is the surface area available for dissolution, D is the diffusion coefficient of the compound in the actual medium, C_s is the solubility, C is the concentration at time t of the compound in the medium, and h is the thickness of the diffusion boundary layer adjacent to the surface of the dissolving compound.

The surface area available for dissolution can be increased by a decrease in particle size and/or by optimizing the wetting characteristics of the compound surface. The drug particle size can be reduced by micronization or nanosuspension approaches. Micronization of raw materials is done by milling techniques. The resulting particle size is typically in the range 0.1 to 25 μm. Nanosuspensions are basically produced by high-pressure homogenization, precipitation, or pearl milling to achieve particles of 1 to 2 μm. The major concern related to all particle size reducing methods is the eventual conversion of the high-energy polymorph to a low-energy crystalline form. Such a conversion may result in a change in photostability as discussed in Section 16.2.3

16.3 ARRANGED LIQUID MEDIA

Solid-state manipulation is often not advantageous or practical in attempting to obtain increased aqueous solubility of a drug substance. It is therefore necessary to alter the medium. For ionizable drugs, a change in pH of the solvent may be the simplest approach to increase solubility. This can be combined with formation of highly soluble salts. In the case of nonpolar drugs, the solubility can be increased by a modification of the solvent polarity through use of cosolvents. The influence of pH and medium polarity on the photostability of drugs in solution is also discussed in Chapter 14.

Complexation with cyclodextrins and solubilization by use of surfactants represent alternative approaches to increase drug solubility. From a photochemical point of view, such formulations introduce a nonhomogenous environment to the drug molecule. The medium can be described as "arranged," although it may be in the form of a liquid preparation. This nonhomogenous, arranged environment may alter the photochemistry of a substance compared to a homogenous solution or to the solid state.

16.3.1 Complex Formation with Cyclodextrins

Cyclodextrins are cyclic oligomers of glucose. The naturally occurring cyclodextrins consist of 6, 7, or 8 D-glucopyranose units known as α-, β-, and γ-cyclodextrins, respectively. A cyclodextrin molecule is hydrophilic on the outside. The presence of hydroxyl groups on the outer surface makes the molecules relatively soluble in water. In modified cyclodextrins, a number of these outer hydroxyl groups are modified, leading to a change (increase) in solubility. The inner core presents a hydrophobic environment due to the presence of electron-rich glycosidic oxygen atoms. The inner cavity is large enough to accept nonpolar molecules or parts of molecules, thus enabling complex formation with, for instance, poorly water-soluble drugs. When a guest molecule (drug) enters the host cavity (cyclodextrin), the contact between water and the nonpolar regions of both is reduced. The degree to which a solute molecule will be solubilized by a cyclodextrin molecule depends on the polarity, size, and shape of the guest, and the possibility for intermolecular interactions between the two molecules.

Solubility enhancement by use of cyclodextrins is achieved for a number of drug substances, an approach of interest in formulation of drugs for topical, parenteral, and oral use (Stella and Rajewski, 1997; Loftsson and Masson, 2001; Qi and Sikorski, 2001). The solubilizing effect can be extensive even in low concentrations of cyclodextrin. The use of 0.1 M sulfobutyl-ether-β-cyclodextrin increases the solubility of prednisolone acetate and testosterone by a factor of 426 and 2020, respectively (Myrdal and Yalkowsky, 2002). Cyclodextrin encapsulation of a molecule will affect many of its physicochemical properties (Loftsson, 1995). As a result of complexation, solubility, pK_a value, spectral properties, and the chemical reactivity of the included substance will change. The cyclodextrins are known to affect molecular orientation and to have an influence on rates and efficiency of electron transfer, excited state proton transfer, and rate of decomposition (Chattopadhyay, 1991; Fox, 1991; Sur et al., 2000). Cyclodextrins can also be used in combination with liposomes; a cyclodextrin–liposome entity represents an even more complex environment to the drug molecule (Loukas et al., 1995).

The solubilized molecules are in a dynamic equilibrium between complexed and uncomplexed forms. Therefore, a fraction of the molecules will react in the uncomplexed form in solution during exposure to radiation. Some molecules are likely to interact with the hydroxyl residues or substituents on the outer surface of the cyclodextrin unit and thereby experience a partly hindered degree of freedom. The remaining fraction will experience a less polar environment, protection from oxygen, and a decrease of the intramolecular rotational freedom inside the cavity during excitation. The inhomogenity of cyclodextrin systems can be demonstrated by a typically broad lifetime distribution and a red shift of the emission maximum when excited in the red edge of the absorption band (Bortolus and Monti, 1996).

Several factors determine photoreactivity changes of the included molecules. The microenvironment (e.g., polarity, steric constraints, specific interactions) will affect the excited state deactiviation pathways. Physical and chemical constraints can be imposed on the evolution of intermediates while the mode of complexation (stoichiometry, geometry) affects the ground state. This can further influence the distribution of stable photoproducts, which is demonstrated for tolmetin and suprofen (Sortino et al., 1998, 1999). It is also known that the various cyclodextrins can influence the photophysical properties of a host molecule differently. A substance may be photochemically stable in the inclusion complex of one cyclodextrin, but in the presence of another cyclodextrin the compound becomes photoactive. This behavior is attributed to a different conformation of the various complexes (Bortolus and Monti, 1996). In the case of nifedipine, nitardipine, promazine, phenothiazine, and trimeprazine, the photostability was clearly dependent on the type of cyclodextrin (Gyéresi et al., 1995, 1996; Mielcarek, 1996; Lutka, 1999a, 2000; Lutka and Koziara, 2000). The complexation method can also be important (Mielcarek, 1995; Lutka, 1999b).

A photostabilizing effect by complex formation with cyclodextrins has been reported for a number of biologically active compounds (Zejtli et al., 1980; Uekama et al., 1983; Ammar and El-Nahhas, 1995; Chen et al., 1996; Lin et al., 2000; Bayomi et al., 2002, Tønnesen et al., 2002). However, formation of an inclusion complex does not necessarily lead to a stabilization of the guest molecule. This is clearly demonstrated in the case of molsidomine, naproxen, and flutamide (Piel et al., 1996; Jiménez et al., 1997; Sortino et al., 2001). In the case of flutamide a 20-fold increase in photodecomposition quantum yield was observed in the presence of β-cyclodextrin.

The photoreactivity of a guest molecule is also influenced by the presence of other excipients in the formulation, particularly alcohols that might be used as cosolvents. Cyclodextrins frequently promote stereoselective photoreactions (Utsuki et al., 1993). Bimolecular reactions are also possible, for instance, in cases in which more than one guest molecule is incorporated in the cavity or when interactions between bound and free molecules take place. The cyclodextrin can take an active part in the process, e.g., by acting as H-donating partner for an excited drug molecule or by forming a photoinduced complex with the guest. This is demonstrated in the case of ketoprofen (Monti et al., 1998). Refer to Bortolus and Monti (1996) for a detailed description on photoreactions of organic molecules incorporated in cyclodextrins.

16.3.2 Micellar Solutions

A wide range of insoluble drugs has been formulated using the principle of solubilization by surface-active compounds. Micellar systems consist of submicroscopical agglomerates of amphiphilic (surface active) molecules surrounded by water. An amphiphilic molecule has a nonpolar hydrocarbon part and a polar end group, which can be neutral or charged, and interacts strongly with water. The hydrophobic groups form the core of the micelle and are thereby shielded from the water. The micelles are in dynamic equilibrium with free molecules (monomers) in the solution, i.e., the micelles are continuously breaking down and reforming. Therefore, the micellar solutions are polydisperse without a well-defined aggregation number. A mean aggregation number for each type of micellar system is often referred to in the literature. Due to the dynamic structure, water molecules are also found deep within the micelle and subtrates incorporated into the structure can therefore still form hydrogen bonds with water. The solubilized molecules are in a dynamic equilibrium between the micellar-bound and the unbound, or free, form, and like the cyclodextrins, the micellar systems represent an inhomogenous reaction medium. The hydrophobic core behaves like an organic phase and is capable of dissolving nonpolar molecules.

Water-insoluble compounds containing polar groups are orientated with the polar moiety at the surface of the micelle and the hydrophobic part buried in the micelle hydrocarbon core. Some compounds are even likely to be adsorbed on the micellar surface. The reaction pathways and rates in micellar systems depend on how deep the solubilized species are located within the amphiphilic moiety. Polarity and water content will vary through the structure. The surface layer of a micelle resembles that of a concentrated electrolyte solution with a dielectric constant lower than bulk water. Ionic micelles have a polarity near to that of pure ethanol (Tascioglu, 1996). Micelles can change the photoreactivity of a substrate due to a change in polarity, viscosity, molecular orientation, charge and redox properties, and to an up-concentration of reactants within a small volume. This can lead to stabilization or destabilization of the drug.

The compartmentalization effect can promote photoinduced proton or electron transfer, prohibit back-charge transfer, and permit controlled charge transfer across the interface (Tascioglu, 1996). Electron transfer through the micellar membrane is not favored compared to an aqueous medium because of the less polar environment in the aggregate. The efficiency of charge separations depends on the localization of the chromophore relative to the aqueous micellar interface. The quantum yield of bimolecular reactions like photoadditions is often greatly increased in micellar systems due to the high concentration of reactants within a defined area.

As described earlier, solute molecules with distinct polar and nonpolar parts will take up a specific orientation in the micellar structure, i.e., the polar part will be residing near the micellar surface. Photocycloaddition of such embedded molecules will therefore lead preferentially to the head-to-head dimer due to steric effects (Suppan, 1994). A charge on the micellar surface will facilitate or prevent interaction with ions. This can lead to a change in singlet–triplet intersystem crossing efficiencies in the presence of species like the bromidium ion. Although the number of photostability studies on drugs in micellar systems is rather limited, the existing reports clearly demonstrate that the photochemical properties of the drug molecules are altered in the presence of surfactants (Moore and Burt, 1981; Silva et al., 1991; Lin et al., 1992). Solubilization of drugs in the micelles can result in increase or decrease in photostability (Kostenbauder et al., 1965; Le et al., 1996; Mulinacci et al., 2001; Nazzal et al., 2002; Tønnesen, 2002).

Recently, novel micellar systems based on block copolymers have been developed for drug solubilization and delivery (Houlton, 2003). One example of such a system is poly-ethylene glycol–poly-aspartic acid block copolymer, which spontaneously forms into colloidal particles (micelles) in aqueous solution. In this type of micelle, the drug molecule will experience an environment characterized by the physicochemical properties of the polymer (see Section 16.2.2.3).

16.4 PHOTOSTABILIZATION OF DRUGS IN DOSAGE FORMS

As described earlier, the techniques applied to improve drug solubility may strongly influence photostability of the drug substance. Efforts should therefore be made to stabilize the formulation against photodecomposition when the final preparation has demonstrated light sensitivity. Most drug products are protected from radiation by a nontransparent market packaging. However, in many cases this outer container is removed during storage and use. The inner container can be made of transparent glass or plastic materials that offer little if any protection against UV and visible radiation (Tønnesen, 1989). When possible, the photostabilizing principles should therefore be incorporated in the formulation. Various approaches to photostabilization of drug products are outlined in Table 16.2.

Only radiation absorbed by the drug substance can cause photodegradation in a direct photochemical process. Thus, blocking of the wavelengths absorbed by the drug should prevent photodecomposition. The principle of spectral overlay therefore makes a useful tool in photoprotection (Thoma and Klimek, 1991). Stabilizers with absorption spectra similar to the drug can be incorporated in the formulation, e.g., as a component of a drug solution or incorporated in a film coating or capsule shell (Tønnesen and Karlsen, 1987; Thoma and Kerker, 1992; Thoma, 1996; Spilgies,

Table 16.2   Photostabilization of Drugs in Dosage Forms

Stabilizing Approach	Examples of Excipients
Coloring of tablet core; tablet coating; gelatin capsule shell	Food colorants; flavonoids; yellow-colored vitamins
Pigmentation of tablet core; tablet coating; gelatin capsule shell	Iron oxides; titanium dioxide
Addition of UV absorbers	Benzophenones; camphor derivatives; p-amino benzoic acid; vanillin
Increasing thickness of tablet coating; gelatin shell	
Absorbing, opaque or nontransparent immediate container	
Addition of scavengers; quenchers, antioxidants	Mannitol; glutathione; beta carotene; ascorbic acid
Remove oxygen	Inert atmosphere (nitrogen)

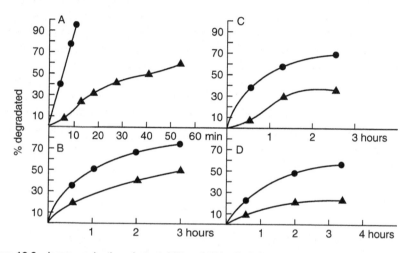

Figure 16.6   Increase in the photostability of (A) nifedipine; (B) chloramphenicol; (C) furosemide; and (D) clonazepam. ● = capsule shell without curcumin; ▲ = capsule shell colored with curcumin.

1998). Incorporation of 0.4% of the yellow compound curcumin in a soft gelatin capsule shell resulted in a threefold or higher increase in the half-life of the drugs nifedipine, chloramphenicol, furosemide, and clonazepam (Tønnesen and Karlsen, 1987) (Figure 16.6). When a photostabilizer is mixed with the drug formulation, the stabilizer must be pharmacologically and physicochemically inert, possess a solubility compatible with the formulation and route of administration, and have an absorption spectrum nearly identical to the drug substance. Such an excipient can be difficult to find and, if found, may gradually fade or even act as a photosensitizer (e.g., curcumin) (Tønnesen, 2002).

Alternatively, compounds that act by scattering incident radiation or that absorb over a wide wavelength range and have a very high optical density can be applied as stabilizers (Teraoka et al., 1988; Desai et al., 1994). In solid preparations of sorivudine and nifedipine, inclusion of 0.2% (w/w) of colored pigments into uncoated tablets provided a better stabilizing effect than coating the tablets with a white pigmented film (Desai et al., 1994). It is well known that the addition of pigments to polymeric films (e.g., tablet coatings) can change the mechanical properties of the coating. Titanium dioxide is a widely used pigment in pharmaceutical production. The pigment absorbs radiation below 405 nm (Anderson et al., 1997). In the visible region, TiO_2 is transparent and owes its whiteness to its high refractive index and small particle size — thus its ability to refract and scatter light. Titanium dioxide is photoreactive and can lead to the formation of all of the most common reactive oxygen species (ROS) (Skowronski et al., 1984).

The generation of ROS may be important with respect to photoinduced color changes sometimes observed in coated tablets. In the case of sulfisomidine tablets, a UV-absorbing film coating was sufficient to increase drug photoability (Matsuda et al., 1978). The film thickness may, however, play an important role (Teraoka et al., 1988; Béchard et al., 1992). In soft gelatin shells containing pigments, the protective effect was also a function of shell thickness. Variations in shell thickness may occur especially around the seal; in one standard type of capsule shell, the thickness was measured to decrease from 260 to 85 µm in the seal area. A strong correlation between the thickness of the gelatin shell and the photoprotective effect was found (Thoma, 1996). In cases in which the preparation cannot be sufficiently stabilized, it may be possible to use an absorbing, opaque, or nontransparent immediate container, e.g., a protective blister package (Narurkar et al., 1986; Thoma and Kerker, 1992; Thoma, 1996; Spilgies, 1998). Other approaches are to restrict the degradation process by addition of scavengers, quenchers, and antioxidants, or to use an inert atmosphere within the immediate container.

REFERENCES

Ahlneck, C. and Zografi, G. (1990) The molecular basis of moisture effects on the physical and chemical stability of drugs in the solid state, *Int. J. Pharm.*, 62, 87–95.

Ahmed, H., Buckton, G., and Rawlins, D.A. (1996) The use of isothermal microcalorimetry in the study of small degrees of amorphous content of a hydrophobic powder, *Int. J. Pharm.*, 130, 195–201.

Akimoto, K., Nakagawa, H., and Sugimoto, I. (1985) Photo-stability of cianidanol in aqueous solution and in the solid state, *Drug Dev. Ind. Pharm.*, 11, 865–889.

Ammar, H.O. and El-Nahhas, S.A. (1995) Improvement of some pharmaceutical properties of drugs by cyclodextrin complexation, *Pharmazie*, 50, 269–272.

Anderson, M.W., Hewitt, J.P., and Spruce, S.R. (1997) Broad-spectrum physical sunscreens: titanium dioxide and zinc oxide, in Lowe, N.J., Shaath, N.A., and Phatak, M.A. (Eds.) *Sunscreens — Development, Evaluation and Regulatory Aspects*, 2nd ed., New York: Marcel Dekker, pp. 365–367.

Bayomi, M.A., Abanumay, K.A., and Al-Angary, A.A. (2002) Effect of inclusion complexation with cyclodextrins on photostability of nifedipine in solid state, *Int. J. Pharm.*, 243, 107–117.

Béchard, S.R., Quraishi, O., and Kwong, E. (1992) Film coating: effect of titanium dioxide concentration and film thickness on the photostability of nifedipine, *Int. J. Pharm.*, 87, 133–139.

Borka, L. (1991) Review on crystal polymorphism of substances in the European pharmacopoeia, *Pharm. Acta Helv.*, 66, 16–22.

Bortolus, P. and Monti, S. (1996) Photochemistry in cyclodextrin cavities, *Adv. Photochem.*, 21, 1–133.

Briggner, L.-E., Buckton, G., Bystrom, K., and Darcy, P. (1994) The use of isothermal microcalorimetry in the study of changes in crystallinity induced during the processing of powders, *Int. J. Pharm.*, 105, 125–135.

Brittain, H.G. (2002) Polymorphism: pharmaceutical aspects, in Swarbrick, J. and Boylan, J.C. (Eds.) *Encyclopedia of Pharmaceutical Technology*, New York: Marcel Dekker, 2239–2249.

Carstensen, J.T. (1974) Stability of solids and solid dosage forms, *J. Pharm. Sci.*, 63, 1–14.

Chattopadhyay, N. (1991) Effect of cyclodextrin complexation on excited state proton transfer reactions, *J. Photochem. Photobiol. A Chem.*, 58, 31–36.

Chen, C.-Y., Chen, F.-A., Wu, A.-B., Hsu, H.-C., Kang, J.-J., and Cheng, H.-W. (1996) Effect of hydroxypropyl-β-cyclodextrin on the solubility, photostability and *in vitro* permeability of alkannin/shikonin enantiomers, *Int. J. Pharm.*, 141, 171–178.

Crowly, P. and Martini, L. (2001) Drug-excipient interactions, *Pharm. Technol. Eur.*, 13, 26–34.

Desai, D.S., Abdelnasser, M.A., Rubitski, B.A., and Varia, S.A. (1994) Photostabilization of uncoated tablets of sorivudine and nifedipine by incorporation of synthetic iron oxides, *Int. J. Pharm.*, 103, 69–76.

De Villiers, M.M., van der Watt, J.G., and Lötter, A.P. (1992) Kinetic study of the solid-state photolytic degradation of two polymorphic forms of furosemide, *Int. J. Pharm.*, 88, 275–283.

De Villiers, M.M., van der Watt, J.G., and Lötter, A.P. (1993) Influence of the cohesive behaviour of small particles on the solid-state photolytic degradation of furosemide, *Drug Dev. Ind. Pharm.*, 19, 383–394.

Ekiz–Gücer, N. and Reisch, J. (1991a) Photostabilität von Indometacin in kristallinem Zustand. II, *Pharm. Acta Helv.*, 66, 66–67.

Ekiz–Gücer, N. and Reisch, J. (1991b) Photostabilität von Digitoxin in kristallinem Zustand, *Liebigs Ann. Chem.*, 1105–1106.

Ekiz–Gücer, N., Zappel, J., and Reisch, J. (1991a) Photostabilität von kontrazeptiven Steroiden in kristallinem Zustand, *Pharm. Acta Helv.*, 66, 2–4.

Ekiz–Gücer, N., Reisch, J., and Nolte, G. (1991b) Photostabilität von kreuzkonjugierten Glucocorticoiden in kristallinem Zustand, *Eur. J. Pharm. Biopharm.*, 37, 234–237.

Ford, J.L. and Rubinstein, M.H. (1977) Phase equilibria and stability characteristics of chlorpropamide-urea solid dispersions, *J. Pharm. Pharmacol.*, 29, 209–211.

Fox, M.A. (1991) Photoinduced electron transfer in arranged media, *Topics Curr. Chem.*, 159, 67–101.

Giammona, G., Pitarresi, G., Tomarchio, V., De Guidi, G., and Giuffrida, S. (1998) Swellable microparticles containing suprofen: evaluation of the *in vitro* release and photochemical behavior, *J. Contr. Rel.*, 51, 249–257.

Gyéresi, Á., Tôkés, B., Regdon, G., and Kata, M. (1995) Increasing the solubility and photostability of nifedipine with cyclodextrins, *Proc.1st World Meeting APGI/APV*, Budapest, May 9–11.

Gyéresi, Á., Tôkés, B., Regdon, G., Kata, M., and Nagy, G. (1996) Study of the solubility and photostability of nifedipine and its derivatives with CDs, in Szejtli, J. and Szente, L. (Eds.) *Proc. 8th Int. Symp. Cyclodextrins*, 345–348, the Netherlands: Kluwer Academic Publishers.

Hadjoudis, E., Vitorakis, M., and Moustakali–Mavridis, I. (1986) Structural directing effects in solid state organic photochemistry, *Mol. Cryst. Liq. Cryst.*, 137, 1–15.

Hörter, D. and Dressman, J.B. (1997) Influence of physicochemical properties on dissolution of drugs in the gastrointestinal tract, *Adv. Drug Del. Rev.*, 25, 3–14.

Houlton, S. (2003) Blocking the way forward, *Manuf. Chem.*, 74, 27–30.

Hüttenrauch, R., Fricke, S., and Knop, M. (1986) Bedeutung des molekularen Ordnungszustands für die Lichtempfindlichkeit fester Wirkstoffe, *Pharmazie*, 41, 664–665.

Jiménez, M.C., Miranda, M.A., and Tormos, R. (1997) Photochemistry of naproxen in the presence of β-cyclodextrin, *J. Photochem. Photobiol. A Biol.*, 104, 119–121.

Kostenbauder, H.B., DeLuca, P.P., and Kowarski, C.R. (1965) Photobinding and photoreactivity of riboflavin in the presence of macromolecules, *J. Pharm. Sci.*, 54, 1243–1251.

Le, M.T., Coiffard, L.J.M., Peigné, F., and de Roeck–Holtzhauer, Y. (1996) Photodegradation of piroctone olamine in aqueous diluted solutions. Effect of the presence of anionic surfactants, *S.T.P. Pharm. Sci.*, 6, 455–458.

Lee, Y.J. and Lee, H.-K. (1990) Degradation kinetics of tolrestat, *J. Pharm. Sci.*, 79, 628–633.

Leuner, C. and Dressman, J. (2000) Improving drug solubility for oral delivery using solid dispersions, *Eur. J. Pharm. Biopharm.*, 50, 47–60.

Lin, C.-T., Perrier, P., Clay, G.G., Sutton, P.A., and Byrn, S.R. (1982) Solid state photooxidation of 21-cortisol *tert*-butylacetate to 21-cortisone *tert*-butylacetate, *J. Org. Chem.*, 47, 2978–2981.

Lin, C.T., Mertz, C.J., Bitting, H.C., and El-Sayed, M.A. (1992) Fluorescence anisotropy studies of dibucaine-HCl in micelles and bacteriorhodopsin, *J. Photochem. Photobiol. B Biol.*, 13, 169–185.

Lin, H.S., Chean, C.S., Ng, Y.Y., Chan, S.Y., and Ho, P.C. (2000) 2-Hydroxypropyl-β-cyclodextrin increases aqueous solubility and photostability of all-*trans*-retinoic acid, *J. Clin. Pharm. Ther.*, 25, 265–269.

Lipper, R.A. (1999) E pluribus product, *Modern Drug Discovery*, 2, 55–60.

Löbenberg, R. and Amidon, G.L. (2000) Modern bioavailability, bioequivalence and biopharmaceutics classification system. New scientific approaches to international regulatory standards, *Eur. J. Pharm. Biopharm.*, 50, 3–12.

Loftsson, T. (1995) Effects of cyclodextrins on the chemical stability of drugs in aqueous solution, *Drug Stab.*, 1, 22–33.

Loftsson, T. and Masson, M. (2001) Cyclodextrins in topical drug formulations: theory and practice, *Int. J. Pharm.*, 225, 15–30.

Loukas, Y.L., Jayasekera, P., and Gregoriadis, G. (1995) Novel liposome-based multicomponent systems for protection of photolabile agents, *Int. J. Pharm.*, 117, 85–94.

Lutka, A. (1999a) Effect of cyclodextrin complexation on the photostability of promazine in aqueous solution, *Pharmazie*, 54, 549–550.

Lutka, A. (1999b) Comparison of the effect of complexation method with cyclodextrins on photostability of thiordiazine in aqueous solution, *Acta Pol. Pharm. Drug Res.*, 56, 425–430.

Lutka, A. (2000) Effect of cyclodextrin complexation on aqueous solubility and photostability of phenothiazine, *Pharmazie*, 55, 120–123.

Lutka, A. and Koziara, J. (2000) Interaction of trimeprazine with cyclodextrins in aqueous solution, *Acta Pol. Pharm. Drug Res.*, 57, 369–374.

Marciniec, B. and Rychcik, W. (1994) Kinetic analysis of nifedipine photodegradation in the solid state, *Pharmazie*, 49, 894–897.

Marciniec, B., Bugaj, A., and Kedziora, W. (1997) Kinetic studies of the photodegradation of nitroimidazole derivatives in the solid state, *Pharmazie*, 52, 220–223.

Matsuda, Y., Inouye, H., and Nakanishi, R. (1978) Stabilization of sulfisomidine tablets by use of film coating UV absorber: protection of coloration and photolytic degradation from exaggerated light, *J. Pharm. Sci.*, 67, 196–201.

Matsuda, Y., Teraoka, R., and Sugimoto, I. (1989) Comparative evaluation of photostability of solid-state nifedipine under ordinary and intensive light irradiation conditions, *Int. J. Pharm.*, 54, 211–221.

Matsuda, Y. and Tatsumi, E. (1990) Physicochemical characterization of furosemide modifications, *Int. J. Pharm.*, 60, 11–26.

Mielcarek, J. (1995) Inclusion complexes of nifedipine and other 1,4-dihydropyridine derivatives with cyclodextrins. The IR and XRD-ray diffraction study on photochemical stability of the inclusion complexes formed by nimodipine, nisoldipine and nitrendipine with β-cyclodextrin, *Acta Pol. Pharm. Drug Res.*, 52, 465–470.

Mielcarek, J. (1996) Photochemical stability of the inclusion complexes of nicardipine with α-, γ-cyclodextrin, methyl-β-cyclodextrin and hydroxypropyl-β-cyclodextrin in the solid state and in solution, *Pharmazie*, 51, 477–479.

Monti, S., Sortino, S., De Guidi, G., and Marconi, G. (1998) Supramolecular photochemistry of 2-(3-benzoylphenyl)propionic acid (ketoprofen). A study in the β-cyclodextrin cavity, *New J. Chem.*, 599–604.

Moore, D.E. and Burt, C.D. (1981) Photosensitization by drugs in surfactant solutions, *Photochem. Photobiol.*, 34, 431–439.

Mulinacci, N., Romani, A., Pinelli, P., Gallori, S., Giaccherini, C., and Vincieri, F.F. (2001) Stabilization of natural anthocyanins by micellar systems, *Int. J. Pharm.*, 216, 23–31.

Myrdal, P.B. and Yalkowsky, S.H. (2002) Solubilization of drugs in aqueous media, in Swarbrick, J. and Boylan, J.C. (Eds.) *Encyclopedia of Pharmaceutical Technology*, New York: Marcel Dekker, pp. 2458–2480.

Narurkar, A.N., Sheen, P.-C., Bernstein, D.F., and Augustine, M.A. (1986) Studies on the light stability of flordipine tablets in amber blister packaging material, *Drug Dev. Ind. Pharm.*, 12, 1241–1247.

Nazzal, S., Guven, N., Reddy, I.K., and Khan, M.A. (2002) Preparation and characterization of coenzyme Q_{10}-Eudrasit® solid dispersion, *Drug Dev. Ind. Pharm.*, 28, 49–57.

Nord, K., Andersen, H., and Tønnesen, H.H. (1997) Photoreactivity of biologically active compounds. XII. Photostability of polymorphic modifications of chloroquine diphosphate, *Drug Stab.*, 1, 243–248.

Nyqvist, H. and Wadsten, T. (1986) Preformulation of solid dosage forms: light stability testing of polymorphs as a part of a preformulation program, *Acta Pharm. Technol.*, 32, 130–132.

Opel, M. and Venturini, F. (2002) Raman scattering in solids, *Eur. Pharm. Rev.*, 5, 76–82.

Piel, G., Pochet, L., Delattre, L., and Delarge, J. (1996) Study of the influence of γ-cyclodextrin on the molsidomine photostability, in Szejtli, J. and Szente, L. (Eds.) *Proc. 8th Int. Symp. Cyclodextrins*, 297–300, the Netherlands: Kluwer Academic Publishers, pp. 297–300.

Pignatello, R., Ferro, M., De Guidi, G., Salemi, G., Vandelli, M.A., Guccione, S., Geppi, M., Forte, C., and Puglisi, G. (2001) Preparation, characterization and photosensitivity studies of solid dispersions of diflunisal and Eudragit RS100 and RL100, *Int. J. Pharm.*, 218, 27–42.

Qi, Z.H. and Sikorski, C.T. (2001) Controlled drug delivery using cyclodextrin technology, *Pharm. Technol. Eur.*, 13, 17–27.

Rabek, J.F. (1996) *Photodegradation of Polymers. Physical Characterization and Applications*, Berlin: Springer–Verlag.

Reichmanis, E., Smith, B.C., and Gooden, R. (1985) *o*-Nitrobenzyl photochemistry: solution *vs.* solid state behavior, *J. Polym. Sci.*, 23, 1–8.

Reisch, J. and Müller, M. (1984) Über die Photostabilität offizineller Arznei- und Hilfsstoffe. I. Barbiturate, *Pharm. Acta Helv.*, 59, 56–61.

Reisch, J., Topaloglu, Y., and Henkel, G. (1986) Über die Photostabilität offizineller Arznei- und Hilfstoffe. II. Die gekreuzt-konjugierten Corticoide des europäischen Arzneibuches, *Arch. Pharm. Technol.*, 32, 115–121.

Reisch, J., Ekiz, N., and Takács, M. (1989a) Untersuchungen zur Photostabilität von Testosteron und Methyltestosteron im kristallinem Zustand, *Arch. Pharm.*, 322, 173–175.

Reisch, J., Ekiz–Gücer, N., Takács, M., and Gunaherath (1989b) Über die Photoisomerisierung des Azapropazons, *Arch. Pharm.*, 322, 295–296.

Reisch, J., Iranshahi, L., and Ekiz–Gücer, N. (1992a) Photostabilität von Glucocorticoiden in kristallinem Zustand, *Liebigs Ann. Chem.*, 1199–1200.

Reisch, J., Ekiz–Gücer, N., and Tewes, G. (1992b) Photostabilität einiger 1,4-Benzodiazepine in kristallinem Zustand, *Liebigs Ann. Chem.*, 69–70.

Reisch, J., Zappel, J., Henkel, G., and Ekiz–Gücer, N. (1993) Photochemische Studien, 65. Mitt [1]: Untersuchungen zur Photostabilität von Ethisteron und Norethisteron sowie deren Kristallstrukturen, *Monatshefte für Chemie*, 124, 1169–1175.

Reisch, J., Zappel, J., Rao, A.R.R., and Henkel, G. (1994a) Dimerization of levonorgestrel in solid state ultraviolet irradiation, *Pharm. Acta Helv.*, 69, 97–100.

Reisch, J., Zappel, J., Rao, A.R.R., and Henkel, G. (1994b) Photostability studies in solid state and crystal structure of mifepristone, *Arch. Pharm.*, 327, 809–811.

Sande, S.A. (1996) Mathematical models for studies of photochemical reactions, in Tønnesen, H.H. (Ed.) *Photostability of Drugs and Drug Formulations*, 1st ed., London: Taylor & Francis, pp. 323–340.

Silva, E., Rückert, V., Lissi, E., and Abuin, E. (1991) Effects of pH and ionic micelles on the riboflavin-sensitized photoprocess of tryptophan in aqueous solution, *J. Photochem. Photobiol. B Biol.*, 11, 57–68.

Skowronski, T.A., Rabek, J.F., and Rånby, B. (1984) The rôle of commercial pigments in the photodegradation of poly(vinyl chloride) (PVC), *Polym. Degr. Stab.*, 8, 37–53.

Sortino, S., De Guidi, G., Marconi, G., and Monti, S. (1998) Triplet photochemistry of suprofen in aqueous environment and in the β-cyclodextrin inclusion complex, *Photochem. Photobiol.*, 67, 603–611.

Sortino, S., Scaiano, J.C., De Guidi, G., and Monti, S. (1999) Effect of β-cyclodextrin complexation on the photochemical and photosensitizing properties of tolmetin: a steady-state and time-resolved study, *Photochem. Photobiol.*, 70, 549–556.

Sortino, S., Giuffrida, S., De Guidi, G., Chillemi, R., Petralia, S., Marconi, G., Condorelli, G., and Sciuto, S. (2001) The photochemistry of flutamide and its inclusion complex with β-cyclodextrin. Dramatic effect of the microenvironment on the nature and on the efficiency of the photodegradation pathways, *Photochem. Photobiol.*, 73, 6–13.

Spilgies, H. (1998) Investigations on the Photostability of Cephalosporins and Beta-Lactamase Inhibitors, Thesis, Munich, Ludwig–Maximilians–Universität.

Stella, V.J. and Rajewski, R.A. (1997) Cyclodextrins: their future in drug formulation and delivery, *Pharm. Res.*, 14, 556–567.

Suppan, P. (Ed.) (1994) *Chemistry and Light*, Cambridge: Royal Society of Chemistry.

Sur, D., Purkayastha, P., and Chattopadhyay, N. (2000) Kinetics of the photoconversion of diphenylamine in β-cyclodextrin environments, *J. Photochem. Photobiol. A Chem.*, 134, 17–21.

Takács, M. and Reisch, J. (1986) Versuche zur Prüfung der Photostabilität pharmaceutischer Festsubstanzen mit einer Kombination von Drehreaktor und Novasoltest–Prüfgerät, *Sci. Pharm.*, 54, 277.

Takács, M., Reisch, J., Gergely–Zobin, A., and Gücer–Ekiz, N. (1990) Zur Untersuchung der Lichtempfindlichkeit von Festsubstanzen, *Sci. Pharm.*, 58, 289–297.

Takács, M., Ekiz–Gücer, J., Reisch, J., and Gergely–Zobin, A. (1991) Lichtemfindlichkeit von Corticosteroiden in kristallinem Zustand, *Pharm. Acta Helv.,* 66, 137–140.

Tascioglu, S. (1996) Micellar solutions as reaction media, *Tetrahedron,* 52, 11113–11152.

Teraoka, R., Matsuda, Y., and Sugimoto, I. (1988) Quantitative design for photostabilization of nifedipine by using titanium dioxide and/or tartrazine as colourants in model film coating systems, *J. Pharm. Pharmacol.,* 41, 293–297.

Thoma, K. and Klimek, R. (1991) Photoinstabilität und Stabilisierung von Arzneistoffen, *Pharm. Ind.,* 53, 504–507.

Thoma, K. and Kerker, R. (1992) Photoinstabilität von Arzneistoffen. VI. Mitteilung: Untersuchungen zur Photoinstabilität von Molsidomin, *Pharm. Ind.,* 54, 630–638.

Thoma, K. (1996) Photodecomposition and stabilization of compounds in dosage forms, in Tønnesen, H.H. (Ed.) *Photostability of Drugs and Drug Formulations,* 1st ed., London: Taylor & Francis, pp. 111–140.

Tønnesen, H.H. and Karlsen, J. (1987) Studies on curcumin and curcuminoids. X. The use of curcumin as a formulation aid to protect light-sensitive drugs in soft gelatin capsules, *Int. J. Pharm.,* 38, 247–249.

Tønnesen, H.H. (1989) Emballasjens betydning ved formulering av fotokjemisk ustabile legemidler, *Norg. Apot. Tidsskr.,* 97, 79–85.

Tønnesen, H.H., Skrede, G., and Martinsen, B.K. (1997) Photoreactivity of biologically active compounds. XIII. Photostability of mefloquine hydrochloride in the solid state, *Drug Stab.,* 1, 249–253.

Tønnesen, H.H. (2002) Studies on curcumin and curcuminoids. XXVIII. Solubility, chemical and photochemical stability of curcumin in surfactant solution, *Pharmazie,* 57, 820–824.

Tønnesen, H.H., Másson, M., and Loftsson, T. (2002) Studies on curcumin and curcuminoids. XXVII. Cyclodextrin complexation; solubility, chemical and photochemical stability, *Int. J. Pharm.,* 244, 127–135.

Uekama, K., Oh, K., Otagiri, M., Seo, H., and Tsuruoka, M. (1983) Improvement of some pharmaceutical properties of clofibrate by cyclodextrin complexation, *Pharm. Acta Helv.,* 58, 338–342.

Utsuki, T., Imamura, K., Hirayama, F., and Uekama, K. (1993) Stoichiometry-dependent changes of solubility and photoreactivity of an antiulcer agent, 2′-carboxymethoxy-4,4′-bis(3-methyl-2-butenyloxy)chalcone, in cyclodextrin inclusion complexes, *Eur. J. Pharm. Sci.,* 1, 81–87.

Zejtli, J., Bolla–Pusztai, É., Szabó, P., and Ferenczy, T. (1980) Enhancement of stability and biological effect of cholecalciferol by β-cyclodextrin complexation, *Pharmazie,* 35, 779–787.

Zhan, X., Yin, G., and Liu, S. (1995) Kinetic study on the photostability of solid vesnarinone and the equivalent relationship between daylight and lamplight, *J. Pharm. Sci.,* 84, 624–626.

Useful Terms and Expressions in the Photoreactivity Testing of Drugs

(The list is not comprehensive.)

Actinometer A chemical system or physical device which determines the number of photons in a beam integrally or per unit time. This name is commonly applied to devices used in the ultraviolet and visible wavelength ranges. Solutions of iron(III) oxalate can be used as a chemical actinometer while thermopiles are examples of physical devices giving a reading that can be correlated to the number of photons detected.

Action spectrum A plot of the reciprocal of the number of incident photons required to produce a given effect compared with the wavelength of the radiation employed.

Black-body A body which completely absorbs radiation of any wavelength falling upon it at any angle.

Black-body radiator A black-body which emits, in every direction and at any wavelength, the maximum possible radiant energy, as compared with other temperature radiators of the same temperature, geometrical shape and dimension.

Black-light lamp Fluorescent lamp that emits ultraviolet radiation in a broad band from 320 to 380 nm.

Candle Equivalent to light produced by a spermaceti candle 7/8 inch in diameter burning at rate of 120 grams per hour.

Candlepower See luminous intensity.

Conversion spectrum A plot of a quantity related to the absorption (absorbance, cross-section, etc.) multiplied by the quantum yield for the considered process against a suitable measure of photon energy, such as frequency, ν, wavenumber, σ, or wavelength, λ.

Cut-off filter An optical device which only permits the transmission of radiation of wavelengths that are longer than or shorter than a specified wavelength.

Deactivation Any loss of energy by an excited molecular entity.

Dose The energy or amount of photons absorbed per unit area or unit volume by an irradiated object during a particular exposure time. Dose can also be used in the sense of the energy or amount of photons per unit area or unit volume received by an irradiated object during a particular exposure time.

Efficiency spectrum A plot of the biological or chemical change or response per absorbed photon *vs.* wavelength.

Electronically excited state A state of an atom or molecular entity which has greater electronic energy than the ground state of the same entity.

Emission Radiative deactivation of an excited state.

Footcandle Unit of intensity of illumination, obtained when a source of 1 candlepower illuminates a screen 1 ft away. 1 footcandle = 10.76 lux.

Illuminance (E_v) The quotient of the luminous flux elements ($d\Phi_v$) divided by the irradiated surface element (dA). Unit: (lm m^{-2}) = lux.

Inner filter effect During a light irradiation experiment the term refers to a sample with a high optical density, resulting in a significant reduction in light intensity at the center of the cuvette compared with an infinitely dilute solution.

Irradiance (E) The radiant flux or radiant power (P) *incident* on an infinitesimal element of surface containing the point under consideration divided by the area of the element (S) (dP/dS), simplified: E = P/S when the radiant power is constant over the surface area considered. The SI unit is W m^{-2}.

Isosbestic point A wavelength, wavenumber, or frequency at which absorption coefficients are equal, i.e., the total absorbance of a sample at this wavelength does not change during a chemical reaction or physical change of the sample.

Lumen Unit of luminous flux falling on a square centimeter at a distance of one centimeter from one international candle, cell cavity, passageway, or opening.

Luminance Luminous flux per unit solid angle leaving element of surface in a given direction, divided by the area of orthogonal projection on the plane perpendicular to this direction.

Luminosity factor Ratio of total luminous flux to total energy emitted by a light source at a given wavelength.

Luminous flux Total visible energy emitted by a source per unit time.

Luminous intensity Amount of luminous flux emitted by a point source of light per solid angle, compared with a standard candle.

Lux Measure of illumination of a surface, equal to 0.092 902 footcandle or 1.000 lumen per square meter.

Photoacoustic effect Generation of heat after absorption of radiation, due to radiationless deactivation or chemical reaction.

Photoallergy An acquired immunologic reactivity dependent on antibody or cell-mediated hypersensitivity.

Photochemical reaction A chemical reaction caused by absorption of ultraviolet, visible, or infrared radiation.

Photochemotherapy The combination of a photoactive chemical and light. The interaction of the light and the chemical produces a synergistic effect.

Photodynamic effect Photoinduced damage requiring the simultaneous presence of light, photosensitizer and molecular oxygen.

Photometry The measurement of quantities associated with light, i.e., based on the average apparent intensity of a light source as viewed by a normal light-adapted human eye. Photometric units report light intensity in terms of the illuminance, e.g., candlepower (lumen/ft^2) or lux (lumen/m^2). Photometric units are only appropriate for visible radiation.

Photo-oxidation Oxidation reactions induced by light, e.g., the loss of one or more electrons from a chemical species as a result of photoexcitation of that species or the reaction of a substance with oxygen under the influence of light.

Photooxygenation Incorporation of molecular oxygen into a molecular entity. There are three common mechanisms:

> Type I   The reaction of triplet molecular oxygen with radicals formed photochemically.
>
> Type II   The reaction of photochemically produced singlet molecular oxygen with molecular entities to give rise to oxygen containing molecular entities.
>
> Type III Mechanism proceeds by electron transfer producing superoxide anion as the reactive species.

Photophysical process Photoexcitation and subsequent events which lead from one to another state of a molecular entity through radiation and radiationless transitions. No chemical change results.

Photopolymerization Polymerization processes requiring a photon for the propagation step.

Photoreduction Reduction reactions induced by light, e.g., addition of one or more electrons to a photoexcited species or the photochemical hydrogenation of a substance.

Photosensitivity A broad term used to describe an adverse reaction to light, which may be phototoxic or photoallergic in nature.

Photosensitization The process by which a photochemical or photophysical alteration occurs in one molecular entity as a result of initial absorption of radiation by another molecular entity called a photosensitizer.

Photosensitized oxidation Two mechanisms named Type I and Type II (see *Photooxygenation*).

> Type I   Substrate or solvent reacts with the sensitizer excited state (either singlet or triplet sens*) to give radicals or radical ions, respectively, by hydrogen atom or electron transfer, leading to oxygenated products.
>
> Type II   The excited sensitizer reacts with oxygen to form singlet molecular oxygen which then reacts with substrate to form the products.

Photothermal effect An effect produced by photoexcitation resulting partially or totally in the production of heat.

Phototoxicity The conversion of an otherwise nontoxic chemical to one directly toxic to tissues after the absorption of electromagnetic radiation.

Quantum counter A medium emitting with a quantum yield independent of the excitation energy over a defined spectral range, e.g., concentrated rhodamine 6G solution between 300 and 600 nm.

Quantum yield (Φ) The number of defined events which occur per photon absorbed by the system.

Quencher A molecular entity that deactivates (quenches) an excited state of another molecular entity, either by energy transfer, electron transfer, or by a chemical mechanism.

Quenching The deactivation of an excited molecular entity intermolecularly by an external environmental influence (such as a quencher) or intramolecularly by a substituent through a nonradiactive process.

Radiance (L) The radiant power (P) leaving or passing through a surface element (S) in a given direction from the source, divided by the projection dS cos Q of the surface element, where Q is the angle between the direction of radiation and the normal to the surface. SI units W m^{-2} (parallel beam) or W m^{-2} sr^{-1} (divergent beam).

Radiant energy (Q) The total energy emitted, transferred or received as radiation in a defined period of time, i.e., the product of radiant power (P) and time (t) when the radiant power is constant over time considered. The SI unit is J.

Radiant (energy) flux Same as radiant power (P). Power emitted, transferred, or received as radiation. The SI unit is $J\ s^{-1} = W$.

Radiant exposure (H) The irradiance, E, integrated over the time of irradiation. SI unit is $J\ m^{-2}$.

Radiant intensity (I) Radiant (energy) flux or radiant power, P, per unit solid angle, ω. The SI unit is $W\ sr^{-1}$.

Radiant power (P) See radiant (energy) flux.

Radiometry The measurement of quantities associated with radiant energy. The radiometric unit of intensity is irradiance.

Self-quenching Quenching of an excited atom or molecular entity by interaction with another atom or molecular entity of the same species in the ground state.

Singlet molecular oxygen The oxygen molecule (dioxygen), O_2, in an excited singlet state. The ground state of O_2 is a triplet.

Singlet state A state having a total electron spin quantum number equal to 0.

Temperature radiation Every body which has a temperature higher than 0 K emits radiation due to its own temperature.

Thermopile Radiation-measuring instrument consisting of a number of thermocouples connected together in series. Measures the incident total radiant flux (calibrated in microwatts). The irradiance is obtained by dividing the measured value by the effective sensitive receiving surface area.

Triplet state A state having a total electron spin quantum number of 1.

Relevant Literature on the Photostability of Specific Drug Substances and Drug Formulations

(The list is not comprehensive. See individual chapters for further information.)

DRUG SUBSTANCES/DRUG FORMULATIONS

Adenosine

Arce, R., Martinez, L., and Danielsen, E. (1993) The photochemistry of adenosine: intermediates contributing to its photodegradation mechanism in aqueous solution at 298 K and characterization of the major product, *Photochem. Photobiol.*, 58, 318–328.

Adrenaline

Brustugun, J., Tønnesen, H.H., Klem, W., and Kjønniksen, I. (2000) Photodestabilization of epinephrine by sodium metabisulphite, *PDA J. Pharm. Sci. Technol.*, 54, 136–143.

Chafetz, L. and Chow, L.H. (1988) Photochemical hydroxylation of phenylephrine to epinephrine (adrenaline), *J. Pharm. Biomed. Anal.*, 6, 511–514.

De Mol, N.J., Beijersbergen van Henegouwen, G.M.J., and Gerritsma, K.W. (1979) Photodecomposition of catecholamines. Photoproducts, quantum yields and action spectrum of adrenaline, *Photochem. Photobiol.*, 29, 7–12.

De Mol, N.J., Beijersbergen van Henegouwen, G.M.J., and Gerritsma, K.W. (1979) Photochemical decomposition of catecholamines. II. The extent of aminochrome formation from adrenaline, isoprenaline and noradrenaline induced by ultraviolet light, *Photochem. Photobiol.*, 29, 479–482.

Hoevenaars, P.C.M. (1965) Stabiliteit van adrenaline in injectievloeistoffen, *Pharm. Weekbl.*, 100, 1151–1162.

Larson, T.A. and Schilling, C.G. (1996) Stability of epinephrine hydrochloride in an extemporaneously compounded topical anaesthetic solution of lidocaine, racepinephrine and tetracaine, *Am. J. Health-Syst. Pharm.*, 53, 659–662.

Newton, D.W., Yin Yee Fung, E., and Williams, D.A. (1981) Stability of five catecholamines and terbutaline sulfate in 5% dextrose injection in the absence and presence of aminophylline, *Am. J. Hosp. Pharm.*, 38, 1314–1319.

Wollmann, H. and Grünert, R. (1984) Einfluss des sichtbaren Lichtes auf die Haltbarkeit von Isoprenalin-, Epinephrin- und Levarterenollösungen in unterschiedlichen Behältnissen, *Pharmazie*, 39, 161–163.

Adriamycin

Alegria, A.E. and Riesz, P. (1988) Photochemistry of aqueous adriamycin and daunomycin. A spin trapping study with ^{17}O enriched oxygen and water, *Photochem. Photobiol.*, 48, 147–152.

Bosanquet, A.G. (1986) Stability of solutions of antineoplastic agents during preparation and storage for *in vitro* assays. II. Assay methods, adriamycin and the other antitumor antibiotics, *Cancer Chemother. Pharmacol.*, 17, 1–10.

Tavoloni, N., Guarino, A.M., and Berk, P.D. (1980) Photolytic degradation of adriamycin, *J. Pharm. Pharmacol.*, 32, 860–862.

Amidopyrin

Reisch, J. and Fitzek, A. (1967) Über die zersetzung von wässrigen amidopyrinlösungen unter dem einfluss von licht und γ-strahlen, *Dtsch. Apoth.-Ztg.*, 107, 1358–1359.

Amidinohydrazones

Schleuder, M., Richter, P.H., Keckeis, A., and Jira, T.H. (1993) Antiarrythmisch wirksame Amidinohydrazone substituierter Benzophenone, *Pharmazie*, 48, 33–37.

Amiloride

Hamoudi, H.I., Hellis, P.F., Jones, R.A., Navaratnam, J.S., Parsons, B.J., Philipps, G.O., Vandenburg, M.J., and Currie, W.J.C. (1984) A laser flash photolysis and pulse radiolysis study of amiloride in aqueous and alcoholic solution, *Photochem. Photobiol.*, 40, 35–39.

4-Aminobenzoic acid

Langford, S.A., Sugden, J.K., and Fitzpatric, R.W. (1996) Detection and determination of the hydrazo and azo products of 4-aminobenzoic acid by high-performance liquid chromatography, *J. Pharm. Biomed. Anal.*, 14, 1615–1623.

Aminophenazone

Reisch, J. and Abdel–Khalek, M. (1979) Zur fotooxidation von kristallinem aminophenazon, *Pharmazie*, 34, 408–410.

Aminophylline

Boak, L.R. (1987) Aminophylline stability, *Can. J. Hosp. Pharm.*, 40, 155.

Aminosalicylic acid

Jensen, J., Cornett, C., Olsen, C.E., Tjørnelund, J., and Hansen, S.H. (1992) Identification of major degradation products of 5-aminosalicylic acid formed in aqueous solutions and in pharmaceuticals, *Int. J. Pharm.*, 88, 177–187.

Amiodarone

Li, A.S.W. and Chignell, C.F. (1987) A spin trapping study of the photolysis of amiodarone and desethylamiodarone, *Photochem. Photobiol.*, 45, 191–197.

Paillous, N. and Verrier, M. (1988) Photolysis of amiodarone, an antiarrhythmic drug, *Photochem. Photobiol.*, 47, 337–343.

Amodiaquine

Owoyale, J.A. (1989) Amodiaquine less sensitive than chloroquine to photochemical reactions, *Int. J. Pharm.*, 56, 213–215.

Amonafide

Sánchez, M.A.C., Suáres, A.I.T., and Sanz, M.P. (1989) Estabilidad de disoluciones de amonafide frente a la luz y la temperatura, *Cienc. Ind. Farm.*, 8, 104–109.

Amphotericin B

Block, E.R. and Bennett, J.E. (1973) Stability of amphotericin B in infusion bottles, *Antimicrob. Agents Chemother.*, 4, 648–649.

Gallelli, J.F. (1967) Assay and stability of amphotericin B in aqueous solution, *Drug Intel.*, 1, 103–105.

Lee, M.D., Hess, M.M., Boucher, B.A., and Apple, A.M. (1994) Stability of amphotericin B in 5% dextrose injection stored at 4 or 25°C for 120 hours, *Am. J. Hosp. Pharm.*, 51, 394–396.

Shadomy, S., Brummer, D.L., and Ingroff, A.V. (1973) Light sensitivity of prepared solutions of amphotericin B, *Am. Rev. Respir. Dis.*, 107, 303–304.

Amsacrine

Cartwright–Shamoon, J.M., McElnay, J.C., and D'Arcy, P.F. (1988) Examination of sorption and photodegradation of amsacrine during storage in intravenous burette administration sets, *Int. J. Pharm.*, 42, 41–46.

Antimycotic agents

Thoma, K. and Kübler, N. (1996) Untersuchung der Photostabilität von Antimykotika. I. Mitt.: Photostabilität von Azolantimykotika, *Pharmazie*, 51, 885–893.

Thoma, K. and Kübler, N. (1997) Untersuchung der Photostabilität von Antimykotika. II. Mitt.: Photostabilität von Polyenantibiotika, *Pharmazie*, 52, 294–302.

Thoma, K., Kübler, N., and Reimann, E. (1996) Untersuchung der Photostabilität von Antimykotika. III. Mitt.: Photostabilität lokal wirksamer Antimykotika, *Pharmazie*, 52, 362–373.

Thoma, K., Kübler, N., and Reimann, E. (1997) Untersuchung der Photostabilität von Antimykotika. IV. Mitt.: Photostabilität von Flucytocin und Griseofulvin, *Pharmazie*, 52, 455–463.

L-Ascorbic acid (see also vitamin C)

Ho, A.H.L., Puri, A., and Sugden, J.K. (1994) Effect of sweetening agents on the light stability of aqueous solutions of L-ascorbic acid, *Int. J. Pharm.*, 107, 199–203.

Sidhu, D.S. and Sugden, J.K. (1992) Effect of food dyes on the photostability of aqueous solutions of L-ascorbic acid, *Int. J. Pharm.*, 83, 263–266.

Azapropazon

Reisch, J., Ekiz–Gücer, N., Takàcs, M., Gunaherath, G.M., and Kamal, B. (1989) Photochemische studien, LIII. Mitt. Über die photoisomerisierung des azapropazons, *Arch. Pharm.*, 322, 295–296.

Azathioprine

Hemmens, V.J. and Moore, D.E. (1986) Photochemical sensitization by azathioprine and its metabolites. I. 6-Mercaptopurine, *Photochem. Photobiol.*, 43, 247–255.
Hemmens, V.J. and Moore, D.E. (1986) Photochemical sensitization by azathioprine and its metabolites. II. Azathioprine and nitroimidazoale metabolites, *Photochem. Photobiol.*, 43, 257–262.
Moore, D.E., Chignell, C.F., Sik, R.H., and Motten, A.G. (1986) Generation of radical anions from metronidazole, misonidazole and azathioprine by photoreduction in the presence of EDTA, *Int. J. Radiat. Biol.*, 50, 885–891.

Aztreonam

Fabre, H., Ibrok, H., and Lerner, D.A. (1992) Photodegradation kinetics under UV light of aztreonam solutions, *J. Pharm. Biomed. Anal.*, 10, 645–650.

Barbituric acid derivatives

Asker, A.F. and Islam, M.S. (1994) Effect of sodium thiosulphate on the photolysis of phenobarbital: evidence of complex formation, *PDA J. Pharm. Sci. Technol.*, 48, 205–210.
Barton, H., Mokrosz, J., Bojarski, J., and Klimczak, M. (1980) Photochemical degradation of barbituric acid derivatives. I. Products of photolysis and hydrolysis of pentobarbital, *Pharmazie*, 35, 155–158.
Barton, H.J., Bojarski, J., and Zurowska, A. (1986) Stereospecificity of the photoinduced conversion of methylphenobarbital to mephenytoin, *Arch. Pharm.*, 319, 457–461.
Jochym, K., Barton, H., and Bojarski, J. (1988) Photochemical degradation of barbituric acid derivatives. VIII. Photolysis of sodium salts of barbiturates in solid state, *Pharmazie*, 43, 623–624.
Mokrosz, J., Klimczak, M., Barton, H., and Bojarski, J. (1980) Photochemical degradation of barbituric acid derivatives. II. Kinetics of pentobarbital photolysis, *Pharmazie*, 35, 205–208.
Mokrosz, J. and Bojarski, J. (1980) Photochemical degradation of barbituric acid derivatives. III. Rate constants of photolysis of barbituric and thiobarbituric acid derivatives, *Pharmazie*, 35, 768–773.
Mokrosz, J., Zurowska, A., and Bojarski, J. (1982) Photochemical degradation of barbituric acid derivatives. IV. Kinetics and TLC investigations of photolysis of proxibarbal, *Pharmazie*, 37, 832–835.
Paluchowska, M.H. and Bojarski, J. (1988) Photochemical formation of primidone from 2-thiophenobarbital, *Arch. Pharm.*, 321, 343–344.
Reisch, J., Müller, M., and Münster (1984) Über die photostabilität offizineller Arnei- und hilfstoffe. I. Barbiturate, *Pharm. Acta Helv.*, 59, 56–61.

Benorylate

Castell, J.V., Gomez-L., M.J., Mirabet, V., Miranda, M.A., and Morera, I.M. (1987) Photolytic degradation of benorylate: effects of the photoproducts on cultured hepatocytes, *J. Pharm. Sci.,* 76, 374–378.

Benoxaprofen

Kochevar, I.E., Hoover, K.W., and Gawienowski, M. (1984) Benoxaprofen photosensitization of cell membrane disruption, *J. Invest. Dermatol.,* 82, 214–218.

Moore, D.W. and Chappuis, P.P. (1988) A comparative study of the photochemistry of the non-steroidal anti-inflammatory drugs, naproxen, benoxaprofen and indomethacin, *Photochem. Photobiol.,* 47, 173–180.

Navaratnam, S., Hughes, J.L., Parsons, B.J., and Phillips, G.O. (1985) Laser flash and steady-state photolysis of benoxaprofen in aqueous solution, *Photochem. Photobiol.,* 41, 375–380.

Reszka, K. and Chighell, C.F. (1983) Spectroscopic studies of cutaneous photosensitizing agents. IV. The photolysis of benoxaprofen, an anti-inflammatory drug with photo-toxic properties, *Photochem. Photobiol.,* 38, 281–291.

Benzamide

Nyqvist, H. and Wadsten, T. (1986) Preformulation of solid dosage forms: light stability testing of polymorphs as a part of a preformulation program, *Acta Pharm. Technol.,* 32, 103–112.

Benzocaine

Chingpaisal, P., Fletcher, G., and Davis, D.J.G. (1977) The effect of CTAB on the radiation sensitivity of benzocaine to hydrated electrons and hydroxyl radicals in aqueous solution, *J. Pharm. Pharmacol.,* 29, 47P.

Fletcher, G. and Davis, D.J.G. (1974) The effect of surfactants on the radiation sensitivity of benzocaine in aqueous solution, *J. Pharm. Pharmacol.,* 26, 82P.

Benzodiazepine

Reisch, J., Ekiz–Gücer, N., and Tewes, G. (1992) Photochemische studien. LXIII. Photosta-bilität einiger 1,4-Benzodiazepine in kristallinem Zustand, *Liebigs Ann. Chem.,* 69–70.

Sur, D., Purkayastha, P., and Chattopadhyay, N. (2000) Photophysics of 1*H*-1,5-benzodiaz-epine in aqueous cyclodextrin environments, *Indian J. Chem.,* 39A, 389–391.

Benzoquinones

Kallmayer, H.J. and Fritzen, W. (1987) 2-Amino-3,5,6-tribromo-1,4-benzochinone und ihre labilität am tageslicht, *Arch. Pharm.,* 320, 769–775.

Kallmayer, H.J. and Fritzen, W. (1992) Photoreaktivität einiger 2-amino-3,5,6-tribromo-1,4-benzochinone, *Pharm. Acta Helv.,* 67, 210–213.

Benzydamine

Vargas, F., Rivas, C., Machado, R., and Sarabia, Z. (1993) Photodegradation of benzydamine: phototoxicity of an isolated photoproduct on erythrocytes, *J. Pharm. Sci.,* 82, 371–372.

Betamethasone

Thoma, K., Kerker, R., and Weissbach, C. (1987) Untersuchung des Einflusses von Bestrahlungsmethoden auf die Photostabilität von Betamethason, *Pharm. Ind.*, 49, 961–963.
Thoma, K. and Kerker, R. (1992) Photoinstabilität von Arzneimitteln. V. Mitteilung: Untersuchungen zur Photostabilisierung von Glukokortikoiden, *Pharm. Ind.*, 54, 551–554.

Bleomycin

Douglas, K.T., Ratwatte, H.A.M., and Thakrar, N. (1983) Photoreactivity of bleomycin and its implications, *Bull. Cancer*, 70, 372–380.
Douglas, K.T. (1983) Photoactivity of bleomycin, *Biomed. Pharmacother.*, 37, 191–193.
Thakrar, N. and Douglas, K.T. (1981) Photolability of bleomycin and its complexes, *Cancer Lett.*, 13, 265–268.

Bupivacaine

Tu, Y.-H., Stiles, M.L., and Allen, L.V. (1990) Stability of fentanyl citrate and bupivacaine hydrochloride in portable pump reservoirs, *Am. J. Hosp. Pharm.*, 47, 2037–2040.

Butibufen

Castell, J.V. and Gómez–Lechón, M.J. (1992) Phototoxicity of non-steroidal anti-inflammatory drugs: *in vitro* testing of the photoproducts of butibufen and flurbiprofen, *J. Photochem. Photobiol. B Biol.*, 13, 71–81.

Carbamazepine

Matsuda, Y., Akazawa, R., Teraoka, R., and Otsuka, M. (1994) Pharmaceutical evaluation of carbamazepine modifications: comparative study for photostability of carbamazepine polymorphs by using Fourier-transformed reflection–absorption infrared spectroscopy and colorimetric measurement, *J. Pharm. Pharmacol.*, 46, 162–167.

Carbisocaine

Bezakova, Z., Bachratà, M., Blesovà, M., and Borovansky, A. (1986) Studium lokalnych anestetik. LXXXIII. Stabilita karbizokainumchloridu a pentakainumchloridu, *Farm. Obz.*, 55, 195–203.

Carboplatin

Torres, F., Girona, V., Puiol, M., Prat, J., and de Bolós, J. (1996) Stability of carboplatin in 5% glucose solution exposed to light, *Int. J. Pharm.*, 129, 275–277.

Carmustine

Fredriksson, K., Lundgren, P., and Landersjø, L. (1986) Stability of carmustine — kinetics and compatability during administration, *Acta Pharm. Suec.*, 23, 115–124.

Cefotaxime

Lerner, D.A., Bonnefond, G., Fabre, H., Mandrou, B., and de Buochberg, S.M. (1988) Photodegradation paths of cefotaxime, *J. Pharm. Sci.*, 77, 699–703.

Cefuroxime axetil

Fabre, H., Ibork, H., and Lerner, D.A. (1994) Photoisomerization kinetics of cefuroxime axetil and related compounds, *J. Pharm. Sci.*, 83, 553–558.

Cephaeline

Teshima, D., Otsubo, K., Higuchi, S., Hirayama, F., Uekama, K., and Aoyama, T. (1989) Effects of cyclodextrins on degradations of emetine and cephaeline in aqueous solution, *Chem. Pharm. Bull.*, 37, 1591–1594.

Cephalexine

Oliveira, A.G. and Petrovick, P.R. (1984) Photochemical stability of cephalexine monohydrated suspension in mineral oil. Degradation kinetics and half-life period, *Rev. Cienc. Farm.*, 6, 63–66.

Cephradine

Signoretti, E.C., Onori, S., Valvo, L., Fattibene, P., Savella, A.L., De Sena, C., and Alimonti, S. (1993) Ionizing radiation induced effects on cephradine. Influence on sample moisture content, irradiation dose and storage conditions, *Drug. Dev. Ind. Pharm.*, 19, 1693–1708.

Chalcone

Utsuki, T., Imamura, K., Hirayama, F., and Uekama, K. (1993) Stoichiometry-dependent changes of solubility and photoreactivity of an antiulcer agent, 2-carboxymethoxy-4,4-bis(3-methyl-2-butenyloxy)chalcone, in cyclodextrin inclusion complexes, *Eur. J. Pharm. Sci.*, 1, 81–87.

Chloramphenicol

Beijersbergen van Henegouwen, G.M.J. (1991) New trends in photobiology, *J. Photochem. Photobiol. B Biol.*, 10, 183–210.

DeVries, H., Beijersbergen van Henegouwen, G.M.J., and Huf, F.A. (1984) Photochemical decomposition of chloramphenicol in a 0.25% eyedrop and in a therapeutic intraocular concentration, *Int. J. Pharm.*, 20, 265–271.

Mubarak, S.I.M., Stanford, J.B., and Sugden, J.K. (1982) Some aspects of the photochemical degradation of chloramphenicol, *Pharm. Acta Helv.*, 57, 226–230.

Mubarak, S.I.M., Stanford, J.B., and Sugden, J.K. (1983) Photochemical reactions of chloramphenicol with diols, *Pharm. Acta Helv.*, 58, 343–347.

Reisch, J. and Weidmann, K.G. (1971) Photo und Radiolysis des Chloramphenicols, *Arch. Pharm.*, 304, 911–919.

Shih, I.K. (1971) Photodegradation products of chloramphenicol in aqueous solution, *J. Pharm. Sci.*, 60, 1889–1890.

Thoma, K. and Kerker, R. (1992) Photoinstabilität von Arzneimitteln. I. Mitteilung über das Verhalten von nur im UV-bereich Absorbierenden Substanzen bei der Tageslichtsimulation, *Pharm. Ind.*, 54, 169–177.

Chlordiazepoxide

Beijersbergen van Henegouwen, G.M.J. (1991) New trends in photobiology, *J. Photochem. Photobiol. B Biol.*, 10, 183–210.

Cornelissen, P.J.G., Beijersbergen van Henegouwen, G.M.J., and Gerritsma, K.W. (1979) Photochemical decomposition of 1,4-benzodiazepines. Chlordiazepoxide, *Int. J. Pharm.*, 3, 205–220.

Moore, D.E. and Tamat, S.R. (1980) Photosensitization by drugs: photolysis of some chloride-containing drugs, *J. Pharm. Pharmacol.*, 32, 172–177.

Chloroquine

Aaron, J.J. and Fidanza, J. (1982) Photochemical analysis studies. III. A fluorimetric and ultraviolet spectrophotometric study of the photolysis of chloroquine on silica-gel thin layers, and its analytical application, *Talanta*, 29, 383–389.

Nord, K., Karlsen, J., and Tønnesen, H.H. (1991) Photochemical stability of biologically active compounds. IV. Photochemical degradation of chloroquine, *Int. J. Pharm.*, 72, 11–18.

Nord, K., Karlsen, J., and Tønnesen, H.H. (1994) Photochemical stability of biologically active compounds. IX. Characterization of the spectroscopic properties of the 4-aminoquinolines chloroquine and hydroxychloroquine, and of selected metabolites by absorption, fluorescence and phorphorescence measurements, *Photochem. Photobiol.*, 60, 427–431.

Nord, K., Andersen, H., and Tønnesen, H.H. (1997) Photoreactivity of biologically active compounds. XII. Photostability of polymorphic modifications of chloroquine diphosphate, *Drug Stab.*, 1, 243–248.

Nord, K., Orsteen, A.-L., Karlsen, J., and Tønnesen, H.H. (1997) Photoreactivity of biologically active compounds. X. Photoreactivity of chloroquine in aqueous solution, *Pharmazie*, 52, 598–603.

Nord, K. (1997) Photoreactivity of Chloroquine, thesis, University of Oslo, Norway.

Odusote, M.O. and Nasipuri, R.N. (1988) Effect of pH and storage conditions on the stability of a novel chloroquine phosphate syrup formulation, *Pharm. Ind.*, 50, 367–369.

Owoyale, J.A. (1989) Decomposition profile of ultraviolet-irradiated chloroquine, *Int. J. Pharm.*, 52, 179–181.

Chlorotiazide

Revelle, L.K., Musser, S.M., Rowe, B.J., and Feldman, I.C. (1997) Identification of chlorothiazide and hydrochlorothiazide UV-A photolytic decomposition products, *J. Pharm. Sci.*, 86, 631–634.

Ulvi, V. and Tammilehto, S. (1989) Photodecomposition studies on chlorothiazide and hydrochlorothiazide, *Acta Pharm. Nord.*, 1, 195–200.

Chlorpromazine

Beijersbergen van Henegouwen, G.M.J. (1991) New trends in photobiology, *J. Photochem. Photobiol. B Biol.*, 10, 183–210.

Davies, A.K., Navaratnam, S., and Phillips, G.O. (1976) Photochemistry of chlorpromazine (2-chloro-*N*-(3-dimethyl-aminopropyl)phenothiazine) in propan-2-ol solution, *J. Chem. Soc. Perkin Trans.*, 2, 25–29.

Hall, R.D., Buettner, G.R., Motten, A.G., and Chignell, C.F. (1987) Near infrared detection of singlet molecular oxygen produced by photosensitization with chlorpromazine, *Photochem. Photobiol.*, 46, 295–300.

Huang, C.L. and Sands, F.L. (1964) The effect of UV irradiation on chlorpromazine, *J. Chromatogr.*, 13, 246–249.

Huang, C.L. and Sands, F.L. (1967) Effect of ultra-violet irradiation on chlorpromazine. II. Anaerobic condition, *J. Pharm. Sci.,* 56, 259–264.

Matsuda, Y. and Masahara, R. (1980) Comparative evaluation of coloration of photosensitive solid drugs under various light sources, *Yakugaku Zasshi,* 100, 953–957.

Moore, D.E. and Tamat, S.R. (1980) Photosensitization by drugs: photolysis of some chloride-containing drugs, *J. Pharm. Pharmacol.,* 32, 172–177.

Motten, A.G., Buettner, G.R., and Chignell, C.F. (1985) Spectroscopic studies of cutaneous photosensitizing agents. VIII. A spin-trapping study of light induced free radicals from chlorpromazine and promazine, *Photochem. Photobiol.,* 42, 9–15.

Rosenthal, I., Ben-Hur, E., Prager, A., and Riklis, E. (1978) Photochemical reactions of chlorpromazine: chemical and biochemical implications, *Photochem. Photobiol.,* 28, 591–594.

Chlortetracycline

Moore, D.E. and Tamat, S.R. (1980) Photosensitization by drugs: photolysis of some chloride-containing drugs, *J. Pharm. Pharmacol.,* 32, 172–177.

Cianidanol

Akimoto, K., Inoue, K., and Sugimoto, I. (1985) Photostability of several crystal forms of cianidanol, *Chem. Pharm. Bull.,* 33, 4050–4053.

Akimoto, K., Nakagawa, H., and Sugimoto, I. (1985) Photostability of cianidanol in aqueous solution and in the solid state, *Drug Dev. Ind. Pharm.,* 2, 865–889.

Cinoxacin

Vargas, F., Rivas, C., and Canudas, N. (1994) Photosensitized lipid peroxidation by cinoxacin and its photoproducts. Involvement of a derived peroxide in its phototoxicity, *Pharmazie,* 49, 742–745.

Ciprofloxacin

Tiefenbacher, E.-M., Haen, E., Przybilla, B., and Kurz, H. (1994) Photodegradation of some quinolones used as antimicrobial therapeutics, *J. Pharm. Sci.,* 83, 463–467.

Torniainen, K., Tammiletho, S., and Ulvi, V. (1996) The effect of pH, buffer type and drug concentration on the photodegradation of ciprofloxacin, *Int. J. Pharm.,* 132, 53–61.

Cisplatin

Cubells, M.P., Jane, M.T., Antunez, X.D., and De Bolos Capdevila, J. (1992) Degradación fotoquimica del cisplatino, *Circ. Farm.,* 314, 149–154.

Greene, R.F., Challeji, D.C., Hiranaka, P.K., and Gallelli, J.F. (1979) Stability of cisplatin in aqueous solution, *Am. J. Hosp. Pharm.,* 36, 38–43.

Honda, D.H., Jansen, J.R., Minor, D.R., and Good, J.W. (1979) Preprinted physician's order form for intravenous cisplatin therapy, *Am. J. Hosp. Pharm.,* 36, 742–743.

Macka, M., Boràk, J., Semènkovà, L., and Kiss, F. (1994) Decomposition of cisplatin in aqueous solutions containing chlorides by ultrasonic energy and light, *J. Pharm. Sci.,* 83, 815–818.

Stewart, C.F. and Fleming, R.A. (1990) Compatibility of cisplatin and fluorouracil in 0.9% sodium chloride injection, *Am. J. Hosp. Pharm.,* 47, 1373–1377.

Villarejo, A.D., Revilla, P., Alonso, P., and Doadrio, A. (1986) Estudio de la estabilidad fotoquimica del cis-dicloro diamino Pt(II) por cromatografia liquida, *Ann. R. Acad. Farm.*, 52, 265–278.

Zieske, P.A., Koberda, M., Hines, J.L., Knight, C.C., Sriram, R., Raghavan, N.V., and Rabinow, B.E. (1991) Characterization of cisplatin degradation as affected by pH and light, *AJHP*, 48, 1500–1506.

Clofazimine

Singh, S., Khanna, M., and Sarin, J.P.S. (1987) *In vivo* evaluation of tablet containing clofazimine and dapsone, *Indian J. Pharm. Sci.*, 49, 236–238.

Clonazepam

Bebawy, L.I. and ElKousy, N. (1997) Stability-indicating method for the determination of hydrochlorothiazide, benzydamine hydrochloride and clonazepam in the presence of their degradation products, *Anal. Lett.*, 30, 1379–1397.

Wad, N. (1986) Degradation of clonazepam in serum by light confirmed by means of a high performance liquid chromatographic method, *Ther. Drug Monit.*, 8, 358–360.

Colchicine

Ammar, H.O. and El-Nahhas, S.A. (1995) Improvement of some pharmaceutical properties by cyclodextrin complexation. II. Colchicine, *Pharmazie*, 50, 269–272.

Habib, M.J. and Asker, A.F. (1989) Influence of certain additives on the photostability of colchicine solutions, *Drug Dev. Ind. Pharm.*, 15, 845–849.

Habib, M.J. and Asker, A.F. (1989) Influence of certain additives on the photostability of colchicine solutions, *Drug Dev. Ind. Pharm.*, 15, 1904–1909.

Contraceptives (see also steroids)

Beijersbergen van Henegouwen, G.M.J. (1991) New trends in photobiology, *J. Photochem. Photobiol. B Biol.*, 10, 183–210.

Ekiz–Gücer, N., Zappel, J., and Reisch, J. (1991) Photostabilität von kontrazeptiven Steroiden in kristallinem Zustand, *Pharm. Acta Helv.*, 66, 2–4.

Corticoids

Reisch, J., Topaloglu, Y., and Henkel, G. (1986) Über die Photostabilität offizineller Arznei- und Hilfsstoffe. II. Die gekreuzt-konjugierten Corticoide des europäischen Arzneibuches, *Acta Pharm. Technol.*, 32, 115–121.

Takács, M., Ekiz–Gücer, N., Reisch, J., and Gergely–Zobin, A. (1991) Lichtempfindlichkeit von Corticosteroiden in kristallinem Zustand, *Pharm. Acta Helv.*, 66, 137–140.

Cytarabine

Chevrier, R., Sautou, V., Pinon, V., Demeocq, F., and Chopineau, J. (1995) Stability and compatibility of a mixture of the anti-cancer drugs etoposinde, cytarabine and daunorubicine for infusion, *Pharm. Acta Helv.*, 70, 141–148.

Dacarbazine

Baird, G.M. and Willoughby, M.L.N. (1978) Photodegradation of dacarbazine, *Lancet*, 2, 681.

Horton, J.K. and Stevens, M.F.G. (1981) A new light on the photodecomposition of the antitumor drug DTIC, *J. Pharm. Pharmacol.*, 33, 808–811.

Islam, M.S. and Asker, A.F. (1994) Photostabilization of dacarbazine with reduced glutathione, *J. Pharm. Sci. Technol.*, 48, 38–40.

Kirk, B. (1987) The evaluation of a light-protective giving set. The photosensitivity of intravenous dacarbazine solutions, *Intensive Ther. Clin. Monit.*, 8, 78–86.

Sauer, H. (1990) Aufbewarung von Zytostatika–Lösungen, *Krankenhauspharmazie*, 11, 373–375.

Dapsone

Singh, S., Khanna, M., and Sarin, J.P.S. (1987) *In vivo* evaluation of tablet containing clofazimine and dapsone, *Indian J. Pharm. Sci.*, 49, 236–238.

Daunorubicin

Chevrier, R., Sautou, V., Pinon, V., Demeocq, F., and Chopineau, J. (1995) Stability and compatibility of a mixture of the anti-cancer drugs etoposide, cytarabine and daunorubicine for infusion, *Pharm. Acta Helv.*, 70, 141–148.

Islam, M.S. and Asker, A.F. (1995) Photoprotection of daunorubicin hydrochloride with sodium sulfite, *PDA J. Pharm. Sci. Technol.*, 49, 122–126.

Maniez–Devos, D.M., Baurain, R., Lesne, M., and Trouet, A. (1986) Degradation of doxorubicin and daunorobicin in human and rabbit biological fluids, *J. Pharm. Biomed. Anal.*, 4, 353–365.

Thoma, K. and Klimek, R. (1991) Photostabilization of drugs in dosage forms without protection from packaging materials, *Int. J. Pharm.*, 67, 169–175.

Wood, M.J., Irwin, W.J., and Scott, D.K. (1990) Photodegradation of doxorubicin, daunorubicin and epirubicin measured by high-performance liquid chromatography, *J. Clin. Pharm. Ther.*, 15, 291–300.

Demeclocycline

Ferdous, A.J. and Asker, A. (1996) Photostabilization of demeclocycline hydrochloride with reduced glutathione, *Drug. Dev. Ind. Pharm.*, 22, 119–124.

Moore, D.E. and Tamat, S.R. (1980) Photosensitization by drugs: photolysis of some chloride-containing drugs, *J. Pharm. Pharmacol.*, 32, 172–177.

Dichloracetamids

Reisch, J. and Weidmann, K.G. (1971) Notize zur Photochemie des Dichloracetamids, *Arch. Pharm.*, 304, 920–922.

Diclofenac

Moore, D.E., Roberts–Thomson, S., Zhen, D., and Duke, C.C. (1990) Photochemical studies on the anti-inflammatory drug diclofenac, *Photochem. Photobiol.*, 52, 685–690.

Diflunisal

Giammona, G., Cavallaro, G., Fontana, G., de Guidi, G., and Guiffrida, S. (1996) Macromolecular prodrug of diflunisal. II. Investigations of *in vitro* release and of photochemical behavior, *Eur. J. Pharm. Sci.*, 4, 273–282.

Digitoxin

Ekiz–Gücer, N. and Reisch, J. (1991) Photochemische Studien. LXI. Photoisomerisierung von Digitoxin in kristallinem Zustand, *Liebs. Ann. Chem.*, 1105–1106.

Digoxin

Reisch, J., Zappel, J., and Rao, R.R. (1994) Photostability studies of ouabain, α-acetyldigoxin and digoxin in solid state, *Pharm. Acta Helv.*, 69, 47–50.

Suleiman, M.S., Abdulhameed, M.E., Najib, N.M., and Muti, H.Y. (1989) Effect of ultraviolet radiation on the stability of diltiazem, *Int. J. Pharm.*, 50, 71–73.

Dihydroergotamin

Thoma, K. and Klimek, R. (1991) Photostabilization of drugs in dosage forms without protection from packaging materials, *Int. J. Pharm.*, 67, 169–175.

Dihydropyridines

Barbato, F., Grumetto, L., and Morrica, P. (1994) Analysis of calcium channel blocking dihydropyridines and their degradation products by gas chromatography, *Farmaco*, 49, 461–466.

Diltiazem

Suleiman, M.S., Abdulhameed, M.E., Najib, N.M., and Muti, H.Y. (1989) Effect of ultraviolet radiation on the stability of diltiazem, *Int. J. Pharm.*, 50, 71–73.

Diosgenin

Joshi, S. and Dhar, D.N. (1986) A novel photo-oxygenation reaction of diosgenin, *Indian J. Chem.*, 25B, 432.

Diphenhydramine

Beijersbergen van Henegouwen, G.M.J., van de Zijde, H.J., van de Griend, J., and de Vries, H. (1987) Photochemical decomposition of diphenhydramine in water, *Int. J. Pharm.*, 35, 259–262.

Beijersbergen van Henegouwen, G.M.J. (1991) New trends in photobiology, *J. Photochem. Photobiol. B Biol.*, 10, 183–210.

Dipyridamole

Ameer, B., Callahan, R.J., and Dragotakes, S.C. (1989) Preparation and stability of an oral suspension of dipyridamole, *J. Pharm. Technol.*, 5, 202–205.

Dithranol

Müller, K., Wiegrebe, W., and Younes, M. (1987) Formation of active oxygen species by dithranol. III. Dithranol, active oxygen species and lipid peroxidation *in vivo*, *Arch. Pharm.*, 320, 59–66.

Dobutamin

Pramar, Y., Das Gupta, V., Gardner, S.N., and Yau, B. (1991) Stabilities of dobutamine, dopamine, nitroglycerin and sodium nitroprusside in disposable plastic syringes, *J. Clin. Pharm. Ther.*, 16, 203–207.

Dopamine

Dandurand, K.R. and Stennett, D.J. (1985) Stability of dopamine hydrochloride exposed to blue-light phototherapy, *Am. J. Hosp. Pharm.*, 42, 595–597.

Newton, D.W., Fung, Y.Y.E., and Williams, D.A. (1981) Stability of five catecholamines and terbutaline sulfate in 5% dextrose injection in the abscence and presence of aminophylline, *Am. J. Hosp. Pharm.*, 38, 1314–1319.

Pramar, Y., Das Gupta, V., Gardner, S.N., and Yau, B. (1991) Stabilities of dobutamine, dopamine, nitroglycerin and sodium nitroprusside in disposable plastic syringes, *J. Clin. Pharm. Ther.*, 16, 203–207.

Dothiepin

Tammilehto, S. and Torniainen, K. (1989) Photochemical stability of dothiepin in aqueous solutions, *Int. J. Pharm.*, 52, 123–128.

Doxorubicin

Asker, A.F. and Habib, M.J. (1988) Effect of glutathione on photolytic degradation of doxorubicin hydrochloride, *J. Parent. Sci. Technol.*, 42, 153–156.

Habib, J.M. and Asker, A.F. (1989) Photostabilization of doxorubicin hydrochloride with radioprotective and photoprotective agents: potential mechanism for enhancing chemotherapy during radiotherapy, *J. Parent. Sci. Technol.*, 43, 259–261.

Maniez–Devos, D.M., Baurain, R., Lesne, M., and Trouet, A. (1986) Degradation of doxorubicin and daunorobicin in human and rabbit biological fluids, *J. Pharm. Biomed. Anal.*, 4, 353–365.

Wood, M.J., Irwin, W.J., and Scott, D.K. (1990) Photodegradation of doxorubicin, daunorubicin and epirubicin measured by high-performance liquid chromatography, *J. Clin. Pharm. Ther.*, 15, 291–300.

DTIC

Horton, J.K. and Stevens, M.F.G. (1981) A new light on the photo-decomposition of the antitumor drug DTIC, *J. Pharm. Pharmacol.*, 33, 808–811.

Epinephrine (see Adrenaline)

Epirubicin

Wood, M.J., Irwin, W.J., and Scott, D.K. (1990) Photodegradation of doxorubicin, daunorubicin and epirubicin measured by high-performance liquid chromatography, *J. Clin. Pharm. Ther.*, 15, 291–300.

Ergotamine

Thoma, K. and Klimek, R. (1991) Photostabilization of drugs in dosage forms without protection from packing materials, *Int. J. Pharm.*, 67, 169–175.

Etoposide

Chevrier, R., Sautou, V., Pinon, V., Demeocq, F., and Chopineau, J. (1995) Stability and compatibility of a mixture of the anti-cancer drugs etoposide, cytarabine and daunorubicine for infusion, *Pharm. Acta Helv.*, 70, 141–148.
McLeod, H.L. and Retting, M.V. (1992) Stability of etoposide solution for oral use, *Am. J. Hosp. Pharm.*, 49, 2784–2785.

Famotidine

Underberg, W.J.M., Koomen, J.M., and Beijner, J.H. (1988) Stability of famotidine in commonly used nutritional infusion fluids, *J. Parent. Sci. Tehnol.*, 42, 94–97.

Fenofibrate

Vargas, F., Rivas, C., and Canudas, N. (1993) Formation of perbenzoic acid derivative in the photodegradation of fenofibrate: phototoxicity studies on erythrocytes, *J. Pharm. Sci.*, 82, 590–591.
Vargas, F. and Canudas, N. (1993) Photostability and photohemolytic studies of fibrates, *Pharmazie*, 48, 900–904.

Fentanyl

Allen, L.V., Stiles, M.L., and Tu, Y.-H. (1990) Stability of fentanyl citrate in 0.9% sodium chloride solution in portable infusion pumps, *Am. J. Hosp. Pharm.*, 47, 1572–1574.
Tu, Y.-H., Stiles, M.L., and Allen, L.V. (1990) Stability of fentanyl citrate and bupivacaine hydrochloride in portable pump reservoirs, *Am. J. Hosp. Pharm.*, 47, 2037–2040.

Flordipine

Narurkar, A.N., Sheen, P.-C., Bernstein, D.F., and Augustine, M.A. (1986) Studies on the light stability of flordipine tablets in amber blister packaging material, *Drug. Dev. Ind. Pharm.*, 12, 1241–1247.

Flucytosine

Biondi, L. and Nairn, J.G. (1986) Stability of 5-fluorouracil and flucytosine in parenteral solutions, *Can. J. Hosp. Pharm.*, 39, 60–64.
Thoma, K., Kübler, N., and Reimann, E. (1997) Untersuchung der Photostabilität von Antimykotika. IV. Mitt.: Photostabilität von Flucytocin und Griseofulvin, *Pharmazie*, 52, 455–463.

Flunitrazepam

Busker, R.W., Beijersbergen van Henegouwen, G.M.J., Kwee, B.M.C., and Winkens, J.H.M. (1987) Photobinding of flunitrazepam and its major photo-decomposition product N-desmethylflunitrazepam, *Int. J. Pharm.*, 36, 113–120.

Givens, R.S., Gingrich, J., and Mecklenburg, S. (1986) Photochemistry of flunitrazepam: a product and model study, *Int. J. Pharm.*, 29, 67–72.

5-Fluorouracil

Biondi, L. and Nairn, J.G. (1986) Stability of 5-fluorouracil and flucytosine in parenteral solutions, *Can. J. Hosp. Pharm.*, 39, 60–64.

Lorillon, P., Corbel, J.C., Mordelet, M.-F., Basle, B., and Guesnier, L.R. (1992) Photosensibilité du 5-fluorouracile et du méthotrexate dans des perfuseurs translucides ou opaques, *J. Pharm. Clin.*, 11, 285–295.

Milovanovic, D. and Nairn, J.G. (1980) Stability of fluorouracil in amber glass bottles, *Am. J. Hosp. Pharm.*, 37, 164–165.

Stewart, C.F. and Fleming, R.A. (1990) Compatibility of cisplatin and fluorouracil in 0.9% sodium chloride injection, *Am. J. Hosp. Pharm.*, 47, 1373–1377.

Flurbiprofen

Castell, J.V. and Gómez–Lechón, M.J. (1992) Phototoxicity of non-steroidal anti-inflammatory drugs: *in vitro* testing of the photoproducts of butibufen and flurbiprofen, *J. Photochem. Photobiol. B Biol.*, 13, 71–81.

Flutamide

Vargas, F., Rivas, C., Mendez, H., Fuentes. A., Fraile, G., and Velasquez, M. (2000) Photochemistry and phototoxicity studies of flutamide, a photoxic anti-cancer drug, *J. Photochem. Photobiol. B Biol.*, 58, 108–114.

Furnidipine

Nunez–Vergara, L.J., Sunkel, C., and Squella, J.A. (1994) Photodecomposition of a new 1,4-dihydropyridine: furnidipine, *J. Pharm. Sci.*, 83, 502–507.

Furosemide

Asker, A.F. and Ferdous, A.J. Photodegradation of furosemide solution, *PDA J. Pharm. Sci. Technol.*, 50, 158–162.

Bundgaard, H., Nørgaard, T., and Nielsen, T.M. (1988) Photodegradation and hydrolysis of furosemide and furosemide esters in aqueous solutions, *Int. J. Pharm.*, 42, 217–224.

De Villiers, M.M., van der Watt, J.G., and Lötter, A.P. (1992) Kinetic study of the solid-state photolytic degradation of two polymorphic forms of furosemide, *Int. J. Pharm.*, 88, 275–283.

Kerremans, A.L.M., Tan, Y., van Ginneken, C.A.M., and Gribnau, F.W.J. (1982) Specimen handling and high-performance liquid chromatographic determination of furosemide, *J. Chrom.*, 229, 129–139.

Matsuda, Y. and Masahara, R. (1980) Comparative evaluation of coloration of photosensitive solid drugs under various light sources, *Yakugaku Zasshi*, 100, 953–957.

Moore, D.E. and Tamat, S.R. (1980) Photosensitization by drugs: photolysis of some chlorine-containing drugs, *J. Pharm. Pharmacol.*, 32, 172–177.

Moore, D.E. and Sithipitaks, V. (1983) Photolytic degradation of frusemide, *J. Pharm. Pharmacol.*, 35, 489–493.

Rowbotham, P.C., Stanford, J.B., and Sugden, J.K. (1976) Some aspects of the photochemical degradation of frusemide, *Pharm. Acta Helv.*, 51, 304–307.

Thoma, K. and Klimek, R. (1991) Photostabilization of drugs in dosage forms without protection from packaging materials, *Int. J. Pharm.*, 67, 169–175.

Ulvi, V. and Keski–Hynnilä, H. (1994) First-derivative UV spectrophotometric and high-performance liquid chromatographic analysis of some thiazide diuretics in the presence of their photodecomposition products, *J. Pharm. Biomed. Anal.*, 12, 917–922.

Yahya, A.M., McElnay, J.C., and D'Arcy, P.F. (1986) Photodegradation of frusemide during storage in burette administration sets, *Int. J. Pharm.*, 31, 65–68.

Griseofulvin

Thoma, K., Kübler, N., and Reimann, E. (1997) Untersuchung der Photostabilität von Antimykotika. IV. Mitt.: Photostabilität von Flucytocin und Griseofulvin, *Pharmazie*, 52, 455–463.

Haloperidol

Thoma, K. and Klimek, R. (1991) Photostabilization of drugs in dosage forms without protection from packaging materials, *Int. J. Pharm.*, 67, 169–175.

Heptacaine

Cizmarik, J., Bezakova, Z., van Lau, T., and Soviar, K. (1991) Study of local anaesthetics. XCIV. Study of heptacaine and pentacaine bases stability, *Acta Fac. Pharm. Univ. Comen.*, 45, 117–136.

Hexachlorophane

Moore, D.E. and Tamat, S.R. (1980) Photosensitization by drugs: photolysis of some chloride-containing drugs, *J. Pharm. Pharmacol.*, 32, 172–177.

Hydralazine

Halasi, S. and Nairn, J.G. (1990) Stability of hydralazine hydrochloride in parenteral solutions, *Can. J. Hosp. Pharm.*, 43, 237–241.

Hydrochlorothiazide

Moore, D.E. and Tamat, S.R. (1980) Photosensitization by drugs: photolysis of some chloride-containing drugs, *J. Pharm. Pharmacol.*, 32, 172–177.

Moore, D.E. and Mallesch, J.L. (1991) Photochemical interaction between triamterene and hydrochlorthiazide, *Int. J. Pharm.*, 76, 187–190.

Revelle, L.K., Musser, S.M., Rowe, B.J., and Feldman, I.C. (1997) Identification of chlorothiazide and hydrochlorothiazide UV-A photolytic decomposition products, *J. Pharm. Sci.*, 86, 631–634.

Ulvi, V. and Tammilehto, S. (1989) Photodecomposition studies on chlorothiazide and hydrochlorothiazide, *Acta Pharm. Nord.*, 1, 195–200.

Hydrocortisone

Hamlin, W.E., Chulski, T., Johnson, R.H., and Wagner, J.G. (1960) A note on the photolytic degradation of anti-inflammatory steroids, *J. Am. Pharm. Assn.*, 49, 253–255.

Hydroxychloroquine

Tønnesen, H.H., Grislingaas, A.-L., Woo, S.O., and Karlsen, J. (1988) Photochemical stability of antimalarials. I. Hydroxychloroquine, *Int. J. Pharm.*, 43, 215–219.

Hypochloride

Vincent–Ballereau, F., Merville, C., and Lafleuriel, M.T. (1989) Sodium hypochloride as a disinfectant for injection materials in third world rural dispensaries, *Int. J. Pharm.*, 50, 87–88.

Ibuprofen

Castell, J.V., Gòmez, L.M.J., Miranda, M.A., and Morea, I.M. (1987) Photolytic degradation of ibuprofen, *Photochem. Photobiol.*, 46, 991–996.

Pandit, J.K., Pal, R.N., and Mishra, B. (1989) Effect of formulation variables and storage conditions on the release rate of ibuprofen solid dosage forms, *East. Pharm.*, 32, 133–137.

Imipramin

Matsuda, Y. and Masahara, R. (1980) Comparative evaluation of coloration of photosensitive solid drugs under various light sources, *Yakugaku Zasshi*, 100, 953–957.

Indapamide

Davis, R., Wells, C.H.J., and Taylor, A.R. (1979) Photolytic decomposition of indapamide, *J. Pharm. Sci.*, 68, 1063–1064.

Indomethacin

Dabestani, R., Sik, R.H., Davis, D.G., Dubay, G., and Chignell, C.F. (1993) Spectroscopic studies of cutaneous photosensitizing agents. XVIII. Indomethacin, *Photochem. Photobiol.*, 58, 367–373.

Ekiz–Gücer, N. and Reisch, J. (1991) Photostabilität von Indometacin in kristallinem Zustand (2). Photochemische Studien. LVIII. Mitt., *Pharma Acta Helv.*, 66, 66–67.

Matsuda, Y., Itooka, T., and Mitsuhashi, Y. (1980) Photostability of indomethacin in model gelatin capsules: effects of film thickness and concentration of titanium dioxide on the coloration and photolytic degradation, *Chem. Pharm. Bull.*, 28, 2665–2671.

Matsuda, Y. and Masahara, R. (1980) Comparative evaluation of coloration of photosensitive solid drugs under various light sources, *Yakugaku Zasshi*, 100, 953–957.

Moore, D.W. and Chappuis, P.P. (1988) A comparative study of the photochemistry of the non-steroidal anti-inflammatory drugs, naproxen, benoxaprofen and indomethacin, *Photochem. Photobiol.*, 47, 173–180.

Thoma, K. and Kerker, R. (1992) Photoinstabilität von Arzneimitteln. I. Mitteilung über das Verhalten von nur im UV-bereich absorbierenden Substanzen bie der Tageslichtsimulation, *Pharm. Ind.*, 54, 169–177.

Indoprofen

Lhiaubet–Vallet, V., Trzcionka, J., Encinas, S., Miranda, M.A., and Chouini–Lalanne, N. (2003) Photochemical and photophysical properties of indoprofen, *Photochem. Photobiol.*, 77, 487–491.

Isoprenaline

De Mol, N.J., Beijersbergen van Henegouwen, G.M.J., and Gerritsma, K.W. (1979) Photochemical decomposition of catecholamines. II. The extent of aminochrome formation from adrenaline, isoprenaline and noradrenaline induced by ultraviolet light, *Photochem. Photobiol.*, 29, 479–482.

Wollmann, H. and Grünert, R. (1984) Einfluss des sichtbaren Lichtes auf die Haltbarkeit von Isoprenalin-, Epinephrin- und levarterenollösungen in unterschiedlichen Behältnissen, *Pharmazie*, 39, 161–163.

Isopropylaminophenazone

Reisch, J., Ekiz, N., and Güneri, T. (1986) Photo- und Strahlenchemische Studien. XLVI. Mitt. Zur Photochemie des Isopropylaminophenazones im kristallinen Zustand und wässeriger Lösung, *Arch. Pharm.*, 319, 973–977.

Isoproterenol

Newton, D.W., Fung, Y.Y.E., and Williams, D.A. (1981) Stability of five catecholamines and terbutaline sulfate in 5% dextrose injection in the absence and presence of aminophylline, *Am. J. Hosp. Pharm.*, 38, 1314–1319.

Ketoconazole

Kumer, K.P., Okonomah, A.D., and Bradshaw, W.G. (1991) Stability of ketoconazole in ethanolic solutions, *Drug Dev. Ind. Pharm.*, 17, 577–580.

Ketoprofen

Boscá, F., Miranda, M.A., Carganico, G., and Mauleón, D. (1994) Photochemical and photobiological properties of ketoprofen associated with the benzophenone chromophore, *Photochem. Photobiol.*, 60, 96–101.

Ketorolac tromethamine

Gu, L., Chiang, H.-S., and Johnson, D. (1988) Light degradation of ketorolac tromethamine, *Int. J. Pharm.*, 41, 105–113.

Levomepromazine

Vargas, F., Carbonell, K., and Camacho, M. (2003) Photochemistry and *in vitro* phototoxicity studies of levomepromazine (methotrimeprazine), a phototoxic neuroleptic drug, *Pharmazie*, 58, 315–319.

Levonorgestrel

Reisch, J., Zappel, J., Rao, A.R.R., and Henkel, G. (1994) Dimerization of levonorgestrel in solid state ultraviolet light irradiation, *Pharm. Acta Helv.*, 69, 97–100.

Levothyroxine

Kato, Y., Saito, M., Koizumi, H., and Toyoda, Y. (1986) Determination and stability of levothyroxine sodium injections prepared in hospital, *Jpn. J. Hosp. Pharm.*, 12, 253–256.

Mebendazole

Karim, E.F.I.A., Ahmed, M.H., and Salama, R.B. (1996) Studies on the thermal and photochemical decomposition of mebendazole, *Int. J. Pharm.*, 142, 251–255.

Meclofenamic acid

Philip, J. and Szulczewski, D.H. (1973) Photolytic decomposition of *N*-(2,6-Dichloro-m-tolyl) anthranilic acid (meclofenamic acid), *J. Pharm. Sci.*, 62, 1479–1482.

Medazepam

Topaloglu, Y. and Yener, G. (1995) Photoinstability and degradation kinetics of medazepam, *Pharmazie*, 50, 119–121.

Mefloquine

Navaratnam, S., Hamblett, I., and Tønnesen, H.H. (2000) Photoreactivity of biologically active compounds. XVI. Formation and reactivity of free radicals in mefloquine, *J. Photochem. Photobiol. B Biol.*, 56, 25–38.

Tønnesen, H.H. and Grislingaas, A.-L. (1990) Photochemical stability of biologically active compounds. II. Photochemical stability of mefloquine in water, *Int. J. Pharm.*, 60, 157–162.

Tønnesen, H.H. and Moore, D.E. (1991) Photochemical stability of biologically active compounds. III. Mefloquine as a photosensitizer, *Int. J. Pharm.*, 70, 95–101.

Tønnesen, H.H., Skrede, G., and Martinsen, B.K. (1997) Photoreactivity of biologically active compounds. XIII. Photostability of mefloquine hydrochloride in the solid state, *Drug Stab.*, 1, 249–253.

Tønnesen, H.H. (1999) Photoreactivity of biologically active compounds. XV. Photochemical behavior of mefloquine in aqueous solution, *Pharmazie*, 54, 590–594.

Menadione

Asker, A.F. and Habib, M.J. (1989) Photostabilization of menadione sodium bisulfite by glutathione, *J. Parent. Sci. Technol.*, 43, 204–207.

Menatetrenone (see Vitamin K₂)

6-Mercaptopurine

Hemmes, V.J. and Moore, D.E. (1984) Photo-oxidation of 6-mercaptopurine in aqueous solution, *J. Chem. Soc. Perkin Trans.*, 2, 209–210.

Methadone hydrochloride

Denson, D.D., Crews, J.C., Grummich, K.W., Stirm, E.J., and Sue, C.A. (1991) Stability of methadone hydrochloride in 0.9% sodium chloride injection in single-dose plastic containers, *Am. J. Hosp. Pharm.*, 48, 515–517.

Methaqualone-1-oxide

Pöhlmann, H., Theil, F.P., and Otto, H. (1986) Nachweis eines Oxaziridins als Zwischenprodukt der fotochemischer Umsetzung von Methaqualon-1-oxid, *Pharmazie*, 41, 390–394.

Theil, F.P., Pöhlmann, H., Pfeifer, S., and Franke, P. (1985) Fotochemische Reaktivität von Methaqualon-1-oxid, *Pharmazie*, 40, 328–331.

Methotrexate

Chatterji, D.C. and Gallelli, J.F. (1978) Thermal and photolytic decomposition of methotrexate in aqueous solutions, *J. Pharm. Sci.*, 67, 526–531.

Dyvik, O., Grislingaas, A.-L., Tønnesen, H.H., and Karlsen, J. (1986) Methotrexate in infusion solutions — a stability test for the hospital pharmacy, *J. Clin. Hosp. Pharm.*, 11, 343–348.

Lorillon, P., Corbel, J.C., Mordelet, M.-F., Basle, B., and Guesnier, L.R. (1992) Photosensibilité du 5-fluorouracile et du méthotrexate dans des perfuseurs translucides ou opaques, *J. Phar. Clin.*, 11, 285–295.

Methylprednisolone

Hamlin, W.E., Chulski, T., Johnson, R.H., and Wagner, J.G. (1960) A note on the photolytic degradation of anti-inflammatory steroids, *J. Am. Pharm. Assoc.*, 49, 253–255.

Reisch, J., Zappel, J., and Rao, A.R.R. (1995) Solid state photochemical investigations of methylprednisolone, prednisolone and triamcinolone acetonide, *Acta Pharm. Turc.*, 37, 13–17.

Metronidazole

Cano, S.B. and Glogiewicz, F.L. (1986) Storage requirements for metronidazole injection, *Am. J. Hosp. Pharm.*, 43, 2983–2985.

Elfaith, I.A., Ibrahim, K.E., and Adam, M.E. (1991) Studies on the photochemical decomposition of metronidazole, *Int. J. Pharm.*, 76, 261–264.

Godfrey, R. and Edwards, R. (1991) A chromatographic and spectroscopic study of photodegraded metronidazole in aqueous solution, *J. Pharm. Sci.*, 80, 212–218.

Habib, M.J. and Asker, A.F. (1989) Complex formation between metronidazole and sodium urate: effect on photodegradation of metronidazole, *Pharm. Res.*, 6, 58–61.

Karim, E.I.A., Ibrahim, K.E., and Adam, M.E. (1991) Studies on the photochemical decomposition of metronidazole, *Int. J. Pharm.*, 76, 261–264.

Moore, D.E., Chignell, C.F., Sik, R., and Motton, A.G. (1986) Generation of radical anions from metronidazole and azathioprine by photoreduction in the presence of EDTA, *Int. J. Radiat. Biol.*, 50, 885–891.

Moore, D.E. and Wilkins, B.J. (1990) Common products from gamma-radiolysis and ultraviolet photolysis of metronidazole, *Radiat. Phys. Chem.*, 36, 547–550.

Midazolam

Andersin, R. and Tammilehto, S. (1989) Photochemical decomposition of midazolam. II. Kinetics in ethanol, *Int. J. Pharm.*, 56, 175–179.

Andersin, R., Ovaskainen, J., and Kaltia, S. (1994) Photochemical decomposition of midazolam. III. Isolation and identification of products in aqueous solutions, *J. Pharm. Biomed. Anal.*, 12, 165–172.

Andersin, R. and Mesilaasko, M. (1995) Structure elucidation of 6-chloro-2-methyl-4(1H)-quinazolinone, a photodecomposition product of midazolam, *J. Pharm. Biomed. Anal.*, 13, 667–670.

Andersin, R. and Tammilehto, S. (1995) Photochemical decomposition of midazolam. IV. Study of pH-dependent stability by high performance liquid chromatography, *Int. J. Pharm.*, 123, 223–235.

Bleasel, M.D., Peterson, G.M., and Jestrimsky, K.W. (1993) Stability of midazolam in sodium chloride infusion packs, *Aust. J. Hosp. Pharm.*, 23, 260–262.

McMullan, S.T., Schaiff, R.A.B., and Dietzen, D.J. (1995) Stability of midazolam hydrochloride in polyvinyl chloride bags under fluorescent light, *Am. J. Health-Syst. Pharm,* 52, 2018–2020,

Reisch, J., Ekiz–Gücer, N., and Tewes, G. (1992) Photochemical studies. LXIII. Photostability of some 1,4-benzodiazepines in the crystalline state, *Liebigs Ann. Chem.*, 69–70.

Selkämaa, R. and Tammilehto, S. (1989) Photochemical decomposition of midazolam. I. Isolation and identification of products, *Int. J. Pharm.*, 49, 83–89.

Mifepristone

Reisch, J., Zappel, J., Rao, A.R.R., and Henkel, G. (1994) Photostability studies in solid state and crystal structure of mifepristone, *Arch. Pharm.*, 327, 809–811.

Minoxidil

Chinnan, D. and Asker, A. (1996) Photostability of minoxidil solutions, *PDA J. Pharm. Sci. Technol.*, 50, 94–98.

Mitomycin C

Beijnen, J.H., van Gijn, R., and Underberg, W.J.M. (1990) Chemical stability of the antitumor drug mitomycin C in solutions for intravenical instillation, *J. Parent. Sci. Technol.*, 44, 332–335.

Mitonafide

Suárez, A.I.T. and Camacho, M.A. (1994) Photolability evaluation of the new cytostatic drug mitonafide, *Drug Res.*, 44, 81–83.

Mitoxantrone

Bosanquet, A.G. (1986) Stability of solutions of antineoplastic agents during preparation and storage for *in vitro* assays. II. Assay methods, adriamycin and the other antitumor antibiotics, *Cancer Chemother. Pharmacol.*, 17, 1–10.

Molsidomine

Thoma, K. and Kerker, R. (1992) Photoinstabilität von Arzneimitteln. VI. Mitt.: Untersuchungen zur Photostabilität von Molsidomin, *Pharm. Ind.*, 54, 630–638.

Aman, W. and Thoma, K. (2002) The influence of formulation and manufacturing process on the photostability of tablets, *Int. J. Pharm.*, 243, 33–41.

Morphine

Grassby, P.F. and Hutchings, L. (1993) Factors affecting the physical and chemical stability of morphine sulphate solutions stored in syringes, *Int. J. Pharm.*, 2, 39–43.

Poggi, G.L. (1991) Compatibility of morphine tartratea admixtures in propylene syringes, *Austral. J. Hosp. Pharm.*, 21, 316.

Nabumetone

Valero, M. and Costa, S.M.B. (2003) Photodegradation of nabumetone in aqueous solutions, *J. Photochem. Photobiol. A Chem.*, 157, 93–101.

Nalidixic acid

Fernandez, E., Peña, W., Vinet, R., and Hidalgo, M.E. (1986) Kinetics of photodegradation of nalidixic acid, *Ann. Quim., Ser. A*, 82, 96–98.

Moore, D.E., Hemmens, V.S., and Yip, H. (1984) Photosensitization by drugs: nalidixic and oxolinic acids, *Photochem. Photobiol.*, 39, 57–61.

Tate, T.J., Diffey, B.L., and Davis, A. (1980) An ultraviolet radiation dosimeter based on the photosensitizing drug, nalidixic acid, *Photochem. Photobiol.*, 31, 27–30.

Thoma, K. and Kübler, N. (1997) Untersuchungen zur Photostabilität von Gyrasehemmern, *Pharmazie*, 52, 519–529.

Naproxen

Boscà, F., Miranda, M.A., Vano, L., and Vargas, F. (1990) New photodegradation for naproxen, a phototoxic non-steroidal anti-inflammatory drug, *J. Photochem. Photobiol. A Chem.*, 54, 131–134.

Moore, D.E. (1987) Principles and practice of drug photodegradation studies, *J. Pharm. Biomed. Anal.*, 5, 441–453.

Moore, D.W. and Chappuis, P.P. (1988) A comparative study of the photochemistry of the non-steroidal anti-inflammatory drugs, naproxen, benoxaprofen and indomethacin, *Photochem. Photobiol.*, 47, 173–180.

Neocarzinostatin

Burger, R.M., Peisach, J., and Horwitz, S.B. (1978) Effect of light and oxygen on neocarzino-statin stability and DNA-clearing activity, *J. Biol. Chem.*, 253, 4830–4832.

Nifedipine

Al-Turk, W.A., Majeed, I.A., Murray, W.J., Newton, D.W., and Othman, S. (1988) Some factors affecting the photodecomposition of nifedipine, *Int. J. Pharm.*, 41, 227–230.

Aman, W. and Thoma, K. (2002) The influence of formulation and manufacturing process on the photostability of tablets, *Int. J. Pharm.*, 243, 33–41.

Béchard, S.R., Quraishi, O., and Kwong, E. (1992) Film coating: effect of titanium dioxide concentration and film thickness on the photostability of nifedipine, *Int. J. Pharm.*, 87, 133–139.

Beijersbergen van Henegouwen, G.M.J. (1991) New trends in photobiology, *J. Photochem. Photobiol. B Biol.*, 10, 183–210.

Desai, D.S., Abdelnasser, M.A., Rubitski, B.A., and Varia, S.A. (1994) Photostabilization of uncoated tablets of sorivudine and nifedipine by incorporation of synthetic iron oxides, *Int. J. Pharm.*, 103, 69–76.

Feltkamp, H., Meyle, E., and Toppel, G. (1989) Nifedipin — Qualität und Chargenkonformität retardierter Nifedipin-Generika, *Dtsch. Apoth. Ztg.*, 129, 2697–2704.

Gibbs, N.K., Traynor, N.J., Johnson, B.E., and Ferguson, J. (1992) *In vitro* phototoxicity of nifedipine: sequential induction of toxic and non-toxic photoproducts with VVA radiation, *J. Photochem. Photobiol. B Biol.*, 13, 275–288.

Grundy, J.S., Kherani, R., and Foster, R.T. (1994) Photostability determination of commercially available nifedipine oral dosage formulations, *J. Pharm. Biomed. Anal.*, 12, 1529–1535.

Hayase, N., Itagaki, Y.-I., Ogawa, S., Akutsu, S., Inagaki, S.-I., and Abiko, Y. (1994) Newly discovered photodegradation products of nifedipine in hospital prescriptions, *J. Pharm. Sci.*, 83, 532–538.

Hiller, H. and Meyle, E. (1994) Nifedipin-Retardpräparate zur einmal täglichen Applikation, *Dtsch. Apoth. Ztg.*, 134, 23–26.

Klimek, R. (1978) Untersuchungen zur Stabilitätskinetik und Stabilisierung photoinstabiler Arzneistoffe. Inaugwal. Dissertation, Frankfurt A.M., Germany.

Logan, B.K. and Patrick, K.S. (1990) Photodegradation of nifedipine and nitrendipine evaluated by liquid and gas chromatography, *J. Chromatogr.*, 529, 175–181.

Majeed, I.A., Murray, W.J., Newton, D.W., Othman, S., and Al-Turk, W.A. (1987) Spectrophotometric study of the photodecomposition kinetics of nifedipine, *J. Pharm. Pharmacol.*, 39, 1044–1046.

Marciniec, B. and Rychcik, W. (1994) Kinetic analysis of nifedipine photodegradation in the solid state, *Pharmazie*, 49, 894–897.

Matsuda, Y., Teraoka, R., and Sugimoto, I. (1989) Comparative evaluation of photostability of solid-state nifedipine under ordinary and intensive light irradiation conditions, *Int. J. Pharm.*, 54, 211–221.

Müller, B.W. and Albers, E. (1992) Complexation of dihydropyridine derivatives with cyclodextrins and 2-hydroxypropyl-β-cyclodextrin in solution, *Int. J. Pharm.*, 79, 273–288.

Ogawa, S., Itagaki, Y., Hayase, N., Takemoto, I., Kasahara, N., Akutsu, S., and Inagaki, S. (1990) Photostability of nifedipine in powder, obtained by crushing tablet, granule or fine granule, *Jpn. J. Hosp. Pharm.*, 16, 189–197.

Pietta, P., Rava, A., and Biondi, P. (1981) High-performance liquid chromatography of nifedipine, its metabolites and photochemical degradation products, *J. Chromatogr.*, 210, 516–521.

Sadana, G.S. and Ghogare, A.B. (1991) Quantitative proton magnetic resonance spectroscopic determination of nifedipine and its photodecomposition products from pharmaceutical preparations, *J. Pharm. Sci.*, 80, 895–898.

Teraoka, R., Matsuda, Y., and Sugimoto, I. (1988) Quantitative design for photostabilization of nifedipine by using titanium dioxide and/or tartrazine as colorants in model film coating systems, *J. Pharm. Pharmacol.*, 41, 293–297.

Thoma, K. and Klimek, R. (1985a) Untersuchungen zur Photoinstabilität von Nifedipin. I. Mitt.: Zersetzungskinetik und Reaktionsmechanismus, *Pharm. Ind.*, 47, 207–215.

Thoma, K. and Klimek, R. (1985b) Untersuchungen zur Photoinstabilität von Nifedipin. II. Mitt.: Einfluss von Milieubedingungen, *Pharm. Ind.*, 47, 319–327.

Thoma, K. and Klimek, R. (1991a) Untersuchungen zur Photoinstabilität von Nifedipin. III. Mitt.: Photoinstabilität und Stabilisierung von Nifedipin in Arzneizubereitungen, *Pharm. Ind.*, 53, 388–396.

Thoma, K. and Klimek, R. (1991b) Photostabilization of drugs in dosage forms without protection from packaging materials, *Int. J. Pharm.*, 67, 169–175.

Thoma, K. and Kerker, R. (1992a) Photostabilität von Arzneimitteln. III. Mitt.: Photoinstabilität und Stabilisierung von Nifedipin in Arzneiformen, *Pharm. Ind.*, 54, 359–365.

Thoma, K. and Kerker, R. (1992b) Photostabilität von Arzneimitteln. IV. Mitt.: Untersuchung zu den Zersetzungprodukten von Nifedipin, *Pharm. Ind.*, 54, 465–468.
Tucker, F.A., Minty, P.S.B., and MacGregor, G.A. (1985) Study of nifedipine photodecomposition in plasma and whole blood using capillary gas–liquid chromatography, *J. Chromatogr.*, 342, 193–198.
Vargas, F., Rivas, C., and Machado, R. (1992) Photodegradation of nifedipine under aerobic conditions: evidence of formation of singlet oxygen and radical intermediate, *J. Pharm. Sci.*, 81, 399–400.

Nimodipine

Rango, G., Veronico, M., and Vetuschi, C. (1995) Analysis of nimodipine and its photodegradation product by derivative spectrophotometry and gas chromatography, *Int. J. Pharm.*, 119, 115–119.
Zanocco, A.L., Díaz, L., López, M., Nunez–Vergara, L.J., and Squella, J.A. (1992) Polarographic study of the photodecompostion of nimodipine, *J. Pharm. Sci.*, 81, 920–924.

Nisoldipine

Müller, B.W. and Albers, E. (1992) Complexation of dihydropyridine derivatives with cyclodextrins and 2-hydroxypropyl-β-cyclodextrin in solution, *Int. J. Pharm.*, 79, 273–288.

Nitrazepam

Roth, H.J. and Adomeit, M. (1973) Photochemie des Nitrazepams, *Acta Pharm.*, 306, 889–897.

Nitrendipine

Logan, B.K. and Patrick, K.S. (1990) Photodegradation of nifedipine relative to nitrendipine evaluated by liquid and gas chromatography, *J. Chromatogr.*, 529, 175–181.
Müller, B.W. and Albers, E. (1992) Complexation of dihydropyridine derivatives with cyclodextrins and 2-hydroxypropyl-β-cyclodextrin in solution, *Int. J. Pharm.*, 79, 273–288.
Squella, J.A., Zanocco, A., Perna, S., and Nuñez–Vergara, L.J. (1990) A polarographic study of the photodegradation of nitrendipine, *J. Pharm. Biomed. Anal.*, 8, 43–47.

Nitrobenzaldehydes

Reisch, J. and Weidmann, K.G. (1971) Photochemie des p-nitrobenzaldehydes, *Arch. Pharm.*, 304, 906–910.

Nitrofurantoin

Beijersbergen van Henegouwen, G.M.J. (1991) New trends in photobiology, *J. Photochem. Photobiol. B Biol.*, 10, 183–210.

Nitrofurazone

Quilliam, M.A., McCarry, B.E., Hoo, K.H., McCalla, D.R., and Viatekunas, S. (1987) Identification of the photolysis products of nitrofurazone irradiated with laboratory illumination, *Can. J. Chem.*, 65, 1128–1132.
Shahjahan, M. and Enever, R.P. (1996) Photolability of nitrofurazone in aqueous solution. I. Quantum yield studies, *Int. J. Pharm.*, 143, 75–82.

Shahjahan, M. and Enever, R.P. (1996) Photolability of nitrofurazone in aqueous solution. II. Kinetic studies, *Int. J. Pharm.*, 143, 83–92.

Thoma, K. and Klimek, R. (1991) Photostabilization of drugs in dosage forms without protection from packaging materials, *Int. J. Pharm.*, 67, 169–175.

Thoma, K. and Kerker, R. (1992) Photostabilität von Arzneimitteln. II. Mitteilung über das Verhalten von im Sichtbaren bereich Absorbierenden Substanzen bei der Tageslicht-simulation, *Pharm. Ind.*, 54, 287–293.

Nitroglycerin

Ludwig, D.J. and Ueda, C.T. (1978) Apparent stability of nitroglycerin in dextrose 5% in water, *Am. J. Hosp. Pharm.*, 35, 541–544.

Pramar, Y., Das Gupta, V., Gardner, S.N., and Yau, B. (1991) Stabilities of dobutamine, dopamine, nitroglycerin and sodium nitroprusside in disposable plastic syringes, *J. Clin. Pharm. Ther.*, 16, 203–207.

Sturek, J.K., Sokoloski, T.D., Winsley, W.T., and Stach, P.E. (1978) Stability of nitroglycerin injection determined by gas chromatography, *Am. J. Hosp. Pharm.*, 35, 537–541.

Nitroimidazole

Marciniec, B., Bugaj, A., and Kedziora, W. (1997) Kinetic studies of the photodegradation of nitroimidazole derivatives in the solid state, *Pharmazie*, 52, 220–223.

Nitroprusside

Asker, A.F. and Gragg, R. (1983) Dimethyl sulfoxide as a photoprotective agent for sodium nitroprusside solutions, *Drug Dev. Ind. Pharm.*, 9, 837–848.

Asker, A.F. and Canady, D. (1984) Influence of certain additives on the photostabilizing effect of dimethyl sulfoxide for sodium nitroprusside solutions, *Drug Dev. Ind. Pharm.*, 10, 1025–1039.

Davidson, S.W. and Lyall, D. (1987) Sodium nitroprusside stability in light-protective administration sets, *Pharm. J.*, 239, 599–601.

Edwards, R. (1986) Stability and light sensitivity of sodium nitroprusside infusions, *Aust. J. Hosp. Pharm.*, 16, 145.

Frank, M.J., Johnson, J.B., and Rubin, S.H. (1976) Spectrophotometric determination of sodium nitroprusside and its photodegradation products, *J. Pharm. Sci.*, 65, 44–48.

Leeuwenkamp, O.R., van Bennekom, W.P., van der Mark, E.J., and Bult, A. (1984) Nitro-prusside, antihypertensive drug and analytical reagent. Review of (photo) stability, pharmacology and analytical properties, *Pharm. Weekbl., Sci. Ed.*, 6, 129–140.

Leeuwenkamp, O.R., van der Mark, E.J., van Bennekom, W.P., and Bult, A. (1985) Investigation of the photochemical and thermal degradation of aqueous nitroprusside solutions using liquid chromatography, *Int. J. Pharm.*, 24, 27–41.

Mahony, C., Brown, J.E., Stargel, W.W., Verghese, C.P., and Bjornsson, T.D. (1984) *In vitro* stability of sodium nitroprusside solutions for intravenous administration, *J. Pharm. Sci.*, 73, 838–839.

Pramar, Y., Das Gupta, V., Gardner, S.N., and Yau, B. (1991) Stabilities of dobutamine, dopamine, nitroglycerin and sodium nitroprusside in disposable plastic syringes, *J. Clin. Pharm. Ther.*, 16, 203–207.

Saunders, A. (1986) Stability and light sensitivity of sodium nitroprusside infusions, *Aust. J. Hosp. Pharm.*, 16, 55–56.

Vessey, C.J. and Batistoni, G.A. (1977) The determination and stability of sodium nitroprusside in aqueous solutions (determination and stability of SNP), *J. Clin. Pharm.*, 2, 105–117.

Noradrenaline (norepinephrine)

De Mol, N.J., Beijersbergen van Henegouwen, G.M.J., and Gerritsma, K.W. (1979) Photochemical decomposition of catecholamines. II. The extent of aminochrome formation from adrenaline, isoprenaline and noradrenaline induced by ultraviolet light, *Photochem. Photobiol.*, 29, 479–482.

Newton, D.W., Fung, Y.Y.E., and Williams, D.A. (1981) Stability of five catecholamines and terbutaline sulfate in 5% dextrose injection in the absence and presence of aminophylline, *Am. J. Hosp. Pharm.*, 38, 1314–1319.

Norfloxacin

Nangia, A., Lam, F., and Hung, C.T. (1991) A stability study of aqueous solution of norfloxacin, *Drug Dev. Ind. Pharm.*, 17, 681–694.

Olaquindox

Beijersbergen van Henegouwen, G.M.J. (1991) New trends in photobiology, *J. Photochem. Photobiol. B Biol.*, 10, 183–210.

Oxolinic acid

Moore, D.E., Hemmens, V.S., and Yip, H. (1984) Photosensitization by drugs: nalidixic and oxolinic acids, *Photochem. Photobiol.*, 39, 57–61.

Oxytocine

De Groot, A.N.J.A., Hekser, Y.A., Vree, T.B., and van Dongen, P.W.J. (1995) Oxytocin and desamino-oxytocin tablets are not stable under simulated tropical conditions, *J. Clin. Pharm. Ther.*, 20, 115–119.

Pentacaine

Cizmarik, J., Bezakova, Z., van Lau, T., and Soviar, K. (1991) Study of local anaesthetics XCIV. Study of heptacaine and pentacaine bases stability, *Acta Fac. Pharm. Univ. Comenianae*, 45, 117–136.

van Lau, T., Cizmarik, J., and Soviar, K. (1991) Decomposition of the pentacaine base, *Pharmazie*, 46, 356.

Perazine

Pawelczyk, E. and Marciniek, B. (1977) Kinetics of drug decomposition. XLVII. Effect of substituents on photochemical stability of perazine derivatives, *Pol. J. Pharmacol. Pharm.*, 29, 143–149.

Perphenazine

Matsuda, Y. and Masahara, R. (1980) Comparative evaluation of coloration of photosensitive solid drugs under various light sources, *Yakugaku Zasshi*, 100, 953–957.

Phenazone

Marciniec, B. (1983) Photochemical decomposition of phenazone derivatives. II. Kinetics of decomposition in the solid state, *Pharmazie*, 38, 848–850.

Marciniec, B. (1985) Photochemical decomposition of phenazone derivatives. IV. Kinetics of photolysis and photo-oxidation in solutions, *Acta Polon. Pharm.*, 42, 448–454.

Reisch, J., Ekiz, N., and Güneri, T. (1986) Fotoabbau von 4-Isopropyl-2,3-dimethyl-1-phenyl-3-pyrazolin-5-on (propyphenazon), *Pharmazie*, 41, 287–288.

Phenothiazines

Beijersbergen van Henegouwen, G.M.J. (1991) New trends in photobiology, *J. Photochem. Photobiol. B Biol.*, 10, 183–210.

Phenylbutazone

Fabre, H., Hussam–Eddine, N., Lerner, D., and Mandrou, B. (1984) Autoxidation and hydrolysis kinetics of the sodium salt of phenylbutazone in aqueous solution, *J. Pharm. Sci.*, 73, 1709–1713.

Phenylephrine

Chafetz, L. and Chow, L.H. (1988) Photochemical hydroxylation of phenylephrine to epinephrine (adrenaline), *J. Pharm. Biomed. Anal.*, 6, 511–514.

Phenytoin

Matsuda, Y. and Masahara, R. (1980) Comparative evaluation of coloration of photosensitive solid drugs under various light sources, *Yakugaku Zasshi*, 100, 953–957.

Physostigmine

Asker, A.F. and Harris, C.W. (1988) Influence of certain additives on the photostability of physostigmine sulfate solutions, *Drug Dev. Ind. Pharm.*, 14, 733–746.

Piroxicam

Bartsch, H., Eiper, A., and Kopelent–Frank, H. (1999) Stability indicating assays for the determination of piroxicam — comparison of methods, *J. Pharm. Biomed. Anal.*, 20, 531–541.

Pralidoxime

Holcombe, D.G. (1986) Stability of pralidoxime mesylate injections, *Anal. Proc.*, 23, 320–321.

Prednisolone

Hamlin, W.E., Chulski, T., Johnson, R.H., and Wagner, J.G. (1960) A note on the photolytic degradation of anti-inflammatory steroids, *J. Am. Pharm. Assn.*, 49, 253–255.

Reisch, J., Zappel, J., and Rao, A.R.R. (1995) Solid state photochemical investigations of methylprednisolone, prednisolone and triamcinolone acetonide, *Acta Pharm. Turc.*, 37, 13–17.

Primaquine

Brossi, A., Gessner, W., Hufford, C.D., Baker, J.K., Homo, F., Millet, P., and Landau, I. (1987) Photo-oxidation products of primaquine. Structure, antimalarial activity and hemolytic effects, *FEBS Lett.*, 223, 77–81.

Kristensen, S., Grislingaas, A.-L., Greenhill, J.V., Skjetne, T., Karlsen, J., and Tønnesen, H.H. (1993) Photochemical stability of biologically active compounds. V. Photochemical degradation of primaquine in aqueous medium, *Int. J. Pharm.,* 100, 15–23.
Kristensen, S., Nord, K., Orsteen, A.-L., and Tønnesen, H.H. (1998) Photochemical stability of biologically active compounds. XIV. Influence of oxygen on light induced reactions of primaquine, *Pharmazie,* 53, 98–103.
Kristensen, S. (1997) Photoreactivity of Antimalarial Drugs, Thesis, University of Oslo, Norway.

Proguanil

Owoyale, J.A. and Elmarakby, Z.S. (1989) Effect of sunlight, ultraviolet irradiation and heat on proguanil, *Int. J. Pharm.,* 50, 219–221.
Taylor, R.B., Moody, R.R., Ochekpe, N.A., Low, A.S., and Harper, M.I.A. (1990) A chemical stability study of proguanil hydrochloride, *Int. J. Pharm.,* 60, 185–190.

Promazine

Motton, A.G., Buettner, G.R., and Chignell, C.F. (1985) Spectroscopic studies of cutaneous photosensitizing agents. VIII. A spin-trapping study of light induced free radicals from chlorpromazine and promazine, *Photochem. Photobiol.,* 42, 9–15.

Promethazine

Matsuda, Y. and Masahara, R. (1980) Comparative evaluation of coloration of photosensitive solid drugs under various light sources, *Yakugaku Zasshi,* 100, 953–957.

Proxibarbital

Mokrosz, J., Zurowska, A., and Bojarski, J. (1982) Photochemical degradation of barbituric acid derivatives. IV. Kinetics and TLC investigations of photolysis of proxibarbital, *Pharmazie,* 37, 768–773.

Psoralen

Hearst, J.E. (1982) Psoralen photochemistry and nucleic acid structure, *Biorg. Chem.,* 8, 945–953.

Pyrazinamide

Cain, M.J., Spencer, N.R., Dusci, L., and Ong, R. (1986) Stability of pyrazinamide infusion, *Aust. J. Hosp. Pharm.,* 16, 217–218.

Quinidine

Pöhlmann, H., Theil, F.-P., and Pfeifer, S. (1986) *In-vitro* und *in-vivo*-Untersuchungen zur fotochemischer Reaktivität der Chinin- und Chinidin-N–Oxide, *Pharmazie,* 41, 859–862.

Quinine

Laurie, W.A., McHale, D., and Saag, K. (1986) Photoreactions of quinine in citric acid solutions, *Tetrahedron,* 42, 3711–3714.

Laurie, W.A., McHale, D., Saag, K., and Sheridan, J.B. (1988) Photoreactions of quinine in citric acid solutions. II. Some end products, *Tetrahedron,* 44, 5905–5910.

McHale, D., Laurie, W.A., Saag, K., and Sheridan, J.B. (1989) Photoreactions of quinine in citric acid solutions. III. Products formed in aqueous 2-hydroxy-2-methylpropionic acid, *Tetrahedron,* 45, 2127–2130.

Pöhlmann, H., Theil, F.-P., and Pfeifer, S. (1986) *In-vitro* und *in-vivo*-Untersuchungen zur fotochemischer Reaktivität der Chinin- und Chinidin-N–Oxide, *Pharmazie,* 41, 859–862.

Quinolones

Tiefenbacher, E.-M., Haen, E., Przybilla, B., and Kurz, H. (1994) Photodegradation of some quinolones used as antimicrobial therapeutics, *J. Pharm. Sci.,* 83, 463–467.

Ranitidine

Caufield, W. (2001) HPLC separation of zidovudine and selected pharmaceuticals using a hexadecylsilane amide column, *Chromatographia,* 54, 561–568.

Inagaki, G.M., Okamoto, M.P., and Takagi, J. (1992) Stability of ranitidine hydrochloride with aztreonam, ceftazidime, or piperacillin sodium during simulated Y-site administration, *Am. J. Hosp. Pharm.,* 49, 2769–2772.

Inagaki, G.M., Okamoto, M.P., and Takagi, J. (1993) Chemical compatibility of cefmetazole sodium with ranitidine hydrochloride during simulated Y-site administration, *J. Parent. Sci. Technol.,* 47, 35–39.

Kondo, S., Kagaya, M., Yamada, Y., Matsusaka, H., and Jimbow, K. (2000) UVB Photosensitivity due to ranitidine, *Dermatology,* 201, 71–73.

Schrijvers, D., Tai–Apin, C., de Smet, M.C., Cornil, P., Vermorken, J.B., and Bruyneel, P. (1998) Determination of compatibility and stability of drugs used in palliative care, *J. Clin. Pharm. Ther.,* 23, 311–314.

Todd, P., Norris, P., Hawk, J.L.M., and Du Vivier, A.W.P. (1995) Ranitidine-induced photosensitivity, *Clin. Exp. Dermatol.,* 20, 146–148.

Williams, M.F., Hak, L.J., and Dukes, G. (1990) *In vitro* evaluation of the stability of rantidine hydrochloride in total parenteral nutrient mixtures, *Am. J. Hosp. Pharm.,* 47, 1574–1579.

Retinoic acid

Bempong, D.K., Honigberg, I.L., and Meltzer, N.M. (1993) Separation of 13-*cis* and all *trans* retinoic acid and their photodegradation products using capillary zone electrophoresis and micellar electronic chromatography (MEC), *J. Pharm. Biomed. Anal.,* 11, 829–833.

Riboflavine

Ahmad, I., David, H., and Rapson, C. (1990) Multicomponent spectrophotometric assay of riboflavine and photoproducts, *J. Pharm. Biomed. Anal.,* 8, 217–223.

Asker, A.F. and Habib, M.J. (1990) Effect of certain stabilizers on photobleaching of riboflavin solutions, *Drug Dev. Ind. Pharm.,* 16, 149–156.

Habib, M.J. and Asker, F. (1991) Photostabilization of riboflavin by incorporation into liposomes, *J. Parent. Sci. Technol.* 45, 124–127.

Rubidazone

Bosanquet, A.G. (1986) Stability of solutions of antineoplastic agents during preparation and storage for *in vitro* assays. II. Assay methods, adriamycin and the other antitumor antibiotics, *Cancer Chemother. Pharmacol.*, 17, 1–10.

Salicylanilides

Davies, A.K., Hilal, N.S., McKellar, J.F., and Phillips, G.O. (1975) Photodegradation of salicylanilides, *Br. J. Derm.*, 92, 143–147.

Salicylic acid

Mills, A., Holland, C.E., Davies, R.H., and Worsley, D. (1994) Photomineralization of salicylic acid: a kinetic study, *J. Photochem. Photobiol. A Chem.*, 83, 257–263.

Sorivudine

Desai, D.S., Abdelnasser, M.A., Rubitski, B.A., and Varia, S.A. (1994) Photostabilization of uncoated tablets of sorivudine and nifedipine by incorporation of synthetic iron oxides, *Int. J. Pharm.*, 103, 69–76.

Spironolactone

Mathur, L.K. and Wickman, A. (1989) Stability of extemporaneously compounded spirono-lactone suspensions, *Am. J. Hosp. Pharm.*, 46, 2040–2042.

Steroids

Ekiz–Gücer, N., Reisch, J., and Nolte, G. (1991) Photostabilität von kreuzkonjugierten Glu-cocorticoiden in kristallinem Zustand. Photochemische studien. LX. Mitt, *Eur. J. Pharm. Biopharm.*, 37, 234–237.

Ekiz–Gücer, N., Zappel, J., and Reisch, J. (1991) Photostabilität von Kontrazeptiven Steroiden in kristallinem Zustand, *Pharm. Acta Helv.*, 66, 2–4.

Hamlin, W.E., Chulski, T., Johnson, R.H., and Wagner, J.G. (1960) A note on the photolytic degradation of anti-inflammatory steroids, *J. Am. Pharm. Assn.*, 49, 253–257.

Lin, C.-T., Perrier, P., Clay, G.G., Sutton, P.A., and Byrn, S.R. (1982) Solid-state photo-oxidation of 21-cortisol tert-butylacetate to 21-cortisone tert-butylacetate, *J. Org. Chem.*, 47, 2978–2981.

Reisch, J., Ekiz, N., and Takàcs, M. (1989) Photochemische Studien. LII. Mitt.: Untersuchun-gen zur Photostabilität von Testosteron und Methyltestosteron in kristallinem Zustand, *Arch. Pharm.*, 322, 173–175.

Reisch, J., Ekiz–Gücer, N., Takàcs, M., and Henkel, G. (1989) Photochemische Studien. LIV. Photodimerizierung von Testosteronpropionat in kristallinem Zustand, *Liebigs Ann. Chem.*, 595–597.

Reisch, J., Henkel, G., Ekiz–Gücer, N., and Nolte, G. (1992) Photochemische Studien. LXII. Untersuchungen zur Photostabilität von Halometason und Prednicarbat in kristallinem Zustand sowie Kristallstruktur von Halometason, *Liebigs Ann. Chem.*, 63–67.

Reisch, J., Iranshahi, L., and Ekiz–Gücer, N. (1992) Photochemische Studien. LXIV. Photo-stabilität von Glucocorticoiden in kristallinem Zustand, *Liebigs Ann. Chem.*, 1199–1200.

Reisch, J., Topaloglu, Y., and Henkel, G. (1986) Über die Photostabilität offizineller Arznei- und Hilfstoffe. II. Die gekreuzt-konjugierten Corticoide des Europäischen Arzneibuches, *Acta Pharm. Technol.*, 32, 115–121.

Reisch, J., Zappel, J., Henkel, G., and Ekiz–Gücer, N. (1993) Photochemische Studien. LXV. Mitt.: Untersuchungen zur Photostabilität von Ethisteron und Norethisteron sowie deren Kristallstrukturen, *Monatshefte für Chemie*, 124, 1169–1175.

Reisch, J., Zappel, J., Raghu, A., and Rao, R. (1995) Photochemical studies. LXVIII. Solid state photochemical investigation of methylprednisolone, prednisolone and triamcinolone, *Acta Pharm. Turc.*, 37, 13–17.

Reisch, J., Zappel, J., Rao, A.R.R., and Henkel, G. (1994) Dimerization of levonorgestrel in solid state ultraviolet irradiation, *Pharm. Acta Helv.*, 69, 97–100.

Scdee, A.G.J., Beijersbergen van Henegouwen, G.M.J., and Blaauw–Geers, H.J.A. (1983) Isolation, identification and densitometric determination of norethisterone 4β,5β-epoxide after photochemical decomposition of norethisterone, *Int. J. Pharm.*, 15, 149–158.

Sedee, A.G.J., Beijersbergen van Henegouwen, G.M.J., de Vries, H., Guit, W., and Haasnoot, C.A.G. (1985) The photochemical decomposition of the progestronic 19-nor-steroid, norithestrone, in aqueous medium, *Pharm. Weekbl. Sci. Ed.*, 7, 194–201.

Stiles, L.M., Allen, L.V., Jr., Resztak, K.E., and Prince, S.J. (1993) Stability of octreotide acetate in polypropylene syringes, *Am. J. Hosp. Pharm.*, 50, 2356–2358.

Takàcs, M., Ekiz–Gücer, N., Reisch, J., and Gergely–Zobin, A. (1991) Lichtempfindlichkeit von Corticosteroiden in kristallinem Zustand. Photochemische Studien. LIX. Mitt., *Pharm. Acta Helv.*, 66, 137–140.

Thoma, K. and Kerker, R. (1992) Photoinstabilität von Arzneimitteln. I. Mitteilung über das Verhalten von nur im UV-Bereich absorbierenden Substanzen bei der Tageslichtsimulation, *Pharm. Ind.*, 54, 169–177.

Thoma, K. and Kerker, R. (1992) Photoinstabilität von Arzneimitteln. V. Mitteilung: Untersuchungen zur Photostabilisierung von Glucocorticoiden, *Pharm. Ind.*, 54, 551–554.

Succinylcholine

Schmutz, C.W. and Mühlebach, S.F. (1991) Stability of succinylcholine chloride injection, *Am. J. Hosp. Pharm.*, 48, 501–506.

Sulfacetamide

Land, E.J., Navaratnam, S., Parsons, B.J., and Phillips, G.O. (1982) Primary processes in the photochemistry of aqueous sulfacetamide: a laser flash photolysis study, *Photochem. Photobiol.*, 35, 637–642.

Reisch, J. and Niemeyer, D.H. (1972) Photochemische Studien am Sulfanilamid, Sulfacetamid und 4-Aminobenzoesäure-Äthylester, *Arch. Pharm.*, 305, 135–140.

Tammilehto, S. (1989) Photodecomposition of sulphacetamide in water-ethanol solutions, *Acta Pharm. Fenn.*, 98, 225–230.

Sulfametoxazole

Zhou, W. and Moore, D.E. (1994) Photochemical decomposition of sulfamethoxazole, *Int. J. Pharm.*, 110, 55–63.

Sulfanilamide

Pawlaczyk, J. and Turowska, W. (1976) Photochemistry of photodynamic compounds. VII. Identification of some products of photolysis of sulphanilamide aqueous solutions, *Acta Polon. Pharm.*, 33, 505–509.

Reisch, J. and Niemeyer, D.H. (1972) Photochemische Studien am Sulfanilamid, Sulfacetamid und 4-Aminobenzoesäure-äthylester, *Arch. Pharm.*, 305, 135–140.

Sulfathiazole

Asker, A.F. and Larose, M. (1987) Influence of uric acid on photostability of sulfathiazole sodium solutions, *Drug Dev. Ind. Pharm.*, 13, 2239–2248.

Sulfisomidine

Matsuda, Y., Inouye, H., and Nakanishi, R. (1978) Stabilization of sulfisomidine tablets by use of film coating containing UV absorber: protection of coloration and photolytic degradation from exaggerated light, *J. Pharm. Sci.*, 67, 196–201.

Sulpyrine

Matsuda, Y. and Masahara, R. (1980) Comparative evaluation of coloration of photosensitive solid drugs under various light sources, *Yakugaku Zasshi*, 100, 953–957.

Suprofen

Castell, J.V., Gomez–Lechon, M.J., Grassa, C., Martinez, L.A., Miranda, M.A., and Tarrega, P. (1994) Photodynamic lipid peroxidation by the photosensitizing non-steroidal anti-inflammatory drugs suprofen and tiaprofenic acid, *Photochem. Photobiol.*, 59, 35–39.

Suramin

Beiner, J.H., van Gijn, R., Horenblas, S., and Underberg, W.J.M. (1990) Chemical stability of suramin commonly used infusion solutions, DICP, *Ann. Pharmacother.*, 24, 1056–1058.

Tauromustine (TCNU)

Betteridge, R.F., Culverwell, A.L., and Bosanquet, A.G. (1989) Stability of tauromustine (TCNU) in aqueous solutions during preparation and storage, *Int. J. Pharm.*, 56, 37–41.

Terbutaline

Glascock, J.C., diPiro, J.T., Cadwallader, D.E., and Pern, M. (1987) Stability of terbutaline sulfate repackaged in disposable plastic syringes, *Am. J. Hosp. Pharm.*, 44, 2291–2293.

Newton, D.W., Fung, Y.Y.E., and Williams, D.A. (1981) Stability of five catecholamines and terbutaline sulfate in 5% dextrose injection in the abscence and prescence of aminophylline, *Am. J. Hosp. Pharm.*, 38, 1314–1319.

Terfenadine

Chen, T.-M., Sill, A.D., and Housmyer, A.L. (1986) Solution stability study of terfenadine by high-performance liquid chromatography, *J. Pharm. Biomed. Anal.*, 4, 533–539.

Tetracyclines

Asker, A.F. and Habib, M.J. (1991) Effect of certain additives on photodegradation of tetra-cycline hydrochloride solutions, *J. Parent. Sci. Technol.*, 45, 113–115.
Wiebe, J.A. and Moore, D.E. (1977) Oxidation photosensitized by tetracyclines, *J. Pharm. Sci.*, 66, 186–189.

Thiazide

Moore, D.E. and Mallesch, J.L. (1991) Photochemical interaction between triamterene and hydrochlorthiazide, *Int. J. Pharm.*, 76, 187–190.
Moore, D.E. and Tamat, S.R. (1980) Photosensitization by drugs: photolysis of some chlorine-containing drugs, *J. Pharm. Pharmacol.*, 32, 172–177.
Tamat, S.R. and Moore, D.E. (1983) Photolytic decomposition of hydrochlorthiazide, *J. Pharm. Sci.*, 72, 180–183.
Ulvi, V. and Tammilehto, S. (1989) Photodecomposition studies on chlorothiazide and hydro-chlorothiazide, *Acta Pharm. Nord.*, 1, 195–200.
Ulvi, V. and Tammilehto, S. (1990) High-performance liquid chromatographic method for studies on the photodecomposition kinetics of chlorothiazide, *J. Chromatogr.*, 507, 151–156.

Thiocolchicoside

Vargas, F., Mendez, H., Fuentes, A., Sequera, J., Fraile, G., Velasques, G., Caceres, G., and Cuello, K. (2001) Photosensitizing activity of thiocolchicoside: photochemical and *in vitro* phototoxicity studies, *Pharmazie*, 56, 83–88.

Thioridazine

Eap, C.B., Koeb, L., and Baumann, P. (1993) Artifacts in the analysis of thioridazine and other neuroleptics, *J. Pharm. Biomed. Anal.*, 11, 451–457.
Pawelczyk, E. and Marciniec, B. (1974) Chemical characterization of decomposition products of drugs, *Pharmazie*, 29, 585–589.

Thiorphan

Gimenez, F., Postaire, E., Prognon, P., Le Hoang, M.D., Lecomte, J.M., Pradeau, D., and Hazebroucq, G. (1988) Study of thiorphan degradation, *Int. J. Pharm.*, 43, 23–30.

Thiothixene

Thoma, K. and Klimek, R. (1991) Photostabilization of drugs in dosage forms without protection from packaging materials, *Int. J. Pharm.*, 67, 169–175.

Tiaprofenic acid

Boscá, F., Miranda, M.A., and Vargas, F. (1992) Photochemistry of tiaprofenic acid, a non-steroidal anti-inflammatory drug with phototoxic side effects, *J. Pharm. Sci.*, 81, 181–182.
Castell, J.V., Gomez–Lechon, M.J., Grassa, C., Martinez, L.A., Miranda, M.A., and Tarrega, P. (1994) Photodynamic lipid peroxidation by the photosensitizing non-steroidal anti-inflammatory drugs suprofen and tiaprofenic acid, *Photochem. Photobiol.*, 59, 35–39.
Castell, J.V., Gomez–Lechon, M.J., Herandez, D., Martinez, L.A., and Miranda, M.A. (1994) Molecular basis of drug phototoxicity: photosensitized cell damage by the major photoproduct of tiaprofenic acid, *Photochem. Photobiol.*, 60, 586–590.

Tinidazole

Salomies, H. (1991) Structure elucidation of the photolysis and hydrolysis products of tinidazole, *Acta Pharm. Nord.*, 3, 211–214.

Tolmetin

Giuffrida, S., De Guidi, G., Sortino, S., Chillemi, R., Costanzo, L.L., and Condorelli, G. (1995) Molecular mechanism of drug photosensitization. VI. Photohaemolysis sensitized by tolmetin, *J. Photochem. Photobiol. B Biol.*, 29, 125–133.

Tretinoin

Brisaert, M.G., Everaerts, G.R., Bultinck, G.R., and Plaizier–Vercammen, J.A. (1992) Onderzoek naar de chemische stabiliteit van tretinoine in dermatologische preparaten, *Farm. Tijdschr. Belg.*, 69, 194–199.

Triamcinolone

Reisch, J., Zappel, J., and Rao, A.R.R. (1995) Solid state photochemical investigations of methylprednisolone, prednisolone and triamcinolone acetonide, *Acta Pharm. Turc.*, 37, 13–17.

Triamterene

Moore, D.E. and Mallesch, J.L. (1991) Photochemical interaction between triamterene and hydrochlorthiazide, *Int. J. Pharm.*, 76, 187–190.

Trifluoperazine

Abdel–Moety, E.M., Al-Rashood, K.A., Rauf, A., and Khattab, N.A. (1996) Photostability-indicating HPLC method for determination of trifluoperazine in bulk form and pharmaceutical formulations, *J. Pharm. Biomed. Anal.*, 14, 1639–1644.

Trimethoprim

Bergh, J.J., Breytenbach, J.C., and Wessels, P.L. (1989) Degradation of trimethoprim, *J. Pharm. Sci.*, 78, 348–350.

Ubidecarenone

Matsuda, Y. and Masahara, R. (1983) Photostability of solid-state ubidecarenone at ordinary and elevated temperatures under exaggerated UV irradiation, *J. Pharm. Sci.*, 72, 1198–1203.
Matsuda, Y. and Teraoka, R. (1985) Improvement of the photostability of ubidecarenone microcapsules by incorporating fat-soluble vitamins, *Int. J. Pharm.*, 26, 289–301.
Takeuchi, H., Sasaki, H., Niwa, T., Hino, T., Kawashima, Y., Uesugi, K., and Ozawa, H. (1992) Improvement of the photostability of ubidecarenone in the formulation of a novel powdered dosage form termed redispersible dry emulsion, *Int. J. Pharm.*, 86, 25–33.

Vancomycin

Khalfi, F., Dine, T., Gressier, B., Luyckx, M., Brunet, C., Ballester, L., Goudaliez, F., Cazin, M., and Cazin, J.C. (1996) Comparability and stability of vancomycin hydrochloride with PVC infusion material in various conditions using stability-indicating high-performance liquid chromatography assay, *Int. J. Pharm.*, 139, 243–247.

Vinblastine

Black, J., Buechter, D.D., and Thurston, D.F. (1988) Stability of vinblastine sulphate when exposed to light, *Drug Intell. Clin. Pharm.*, 22, 634–635.

Black, J., Buechter, D.D., Chinn, J.W., Gard, J., and Thurston, D.E. (1988) Studies on the stability of vinblastine sulfate in aqueous solution, *J. Pharm., Sci.*, 77, 630–634.

Vitamin A

Allwood, M.C. and Plane, J.H. (1986) The wavelength-dependent degradation of vitamin A exposed to ultraviolet radiation, *Int. J. Pharm.*, 31, 1–7.

Bempong, D.K., Honigberg, I.L., and Meltzer, N.M. (1993) Separation of 13-*cis* and all-*trans* retinoic acid and their photodegradation products using capillary zone electrophoresis and micellar electrokinetic chromatography (MEC), *J. Pharm. Biomed. Anal.*, 11, 829–833.

Bluhm, D.P., Summers, R.S., Lowes, M.M.J., and Dürheim, H.H. (1991) Lipid emulsion content and vitamin A stability in TPN admixtures, *Int. J. Pharm.*, 68, 277–288.

Bluhm, D.P., Summers, R.S., Lowes, M.M.J., and Dürheim, H.H. (1991) Influence of container on vitamin A stability in TPN admixtures, *Int. J. Pharm.*, 68, 281–283.

Smith, J.L., Canham, J.E., and Wells, P.A. (1988) Effect of phototherapy light, sodium bisulfite and pH on vitamin stability in total parenteral nutrition admixtures, *J. Parent. Enter. Nutrit.*, 12, 394–402.

Vitamin B (see also riboflavine)

Ahmad, I., David, H., and Rapson, C. (1990) Multicomponent spectrophotometric assay of riboflavine and photoproducts, *J. Pharm. Biomed. Anal.*, 8, 217–223.

Asker, A.F. and Habib, M.J. (1990) Effect of certain stabilizers on photobleaching of riboflavine solutions, *Drug Dev. Ind. Pharm.*, 16, 149–156.

Chen, M.F., Boyce, H.W., and Triplett, L. (1983) Stability of the B vitamins in mixed parenteral nutrition solution, *J. Parent. Enter. Nutr.*, 7, 462–464.

Habib, M.J. and Asker, F. (1991) Photostabilization of riboflavine by incorporation into liposomes, *J. Parent. Sci. Technol.*, 45, 124–127.

Martens von Harrie, J.M. (1989) Stabilität wasserlöslicher Vitamine in verschiedenen Infusionsbeuteln, *Krankenhauspharmazie*, 10, 359–361.

Ostrea, E.M., Green, C.D., and Balun, J.E. (1982) Decomposition of TPN solutions exposed to phototherapy, *J. Pediatr.*, 100, 669–670.

Smith, J.L., Canham, J.E., and Wells, P.A. (1988) Effect of phototherapy light, sodium bisulfite and pH on vitamin stability in total parenteral nutrition admixtures, *J. Parent. Enter. Nutr.*, 12, 394–402.

Vitamin C

Ho, A.H.L., Puri, A., and Sugden, J.K. (1994) Effect of sweetening agents on the light stability of aqueous solutions of L-ascorbic acid, *Int. J. Pharm.*, 107, 199–203.

Martens von Harrie, J.M. (1989) Stabilität wasserlöslicher Vitamine in verschiedenen Infusionsbeuteln, *Krankenhauspharmazie*, 10, 359–361.

Sidhu, D.S. and Sugden, J.K. (1992) Effect of food dyes on the photostability of aqueous solutions of L-ascorbic acid, *Int. J. Pharm.*, 83, 263–266.

Smith, J.L., Canham, J.E., and Wells, P.A. (1988) Effect of phototherapy light, sodium bisulfite and pH on vitamin stability in total parenteral nutrition admixtures, *J. Parent. Enter. Nutr.*, 12, 394–402.

Vitamin D

Hüttenrauch, R., Fricke, S., and Matthey, K. (1985) Zur Notwendigkeit des Lichtschutzes für wässrige Vitamin-D-Lösungen, *Pharmazie*, 41, 742.

Vitamin E

Smith, J.L., Canham, J.E., and Wells, P.A. (1988) Effect of phototherapy light, sodium bisulfite and pH on vitamin stability in total parenteral nutrition admixtures, *J. Parent. Enter. Nutr.*, 12, 394–402.

Vitamin K$_1$

Hangarter, M.-A., Hörmann, A., Kamdzhilov, Y., and Wirz, J. (2003) Primary photoreactions of phylloquinone (vitamin K$_1$) and plastoquinone-1 in solution, *Photochem. Photobiol. Sci.*, 2, 524–535.

Vitamin K$_2$

Martinelli, E. and Mühlebach, S. (1995) Kunststoffumbeutel als Lichtschutz für Infusionen, *Krankenhauspharmazie*, 16, 286–289.
Nishikawa, M., Suzuki, K., Morita, H., Kawakage, I., and Fujii, K. (1986) Stability of vitamin K$_2$ syrup after dispensing, *Jpn. J. Hosp. Pharm.*, 12, 257–260.
Teraoka, R. and Matsuda, Y. (1993) Stabilization-oriented preformulation study of photolabile menatetrenone (vitamin K$_2$), *Int. J. Pharm.*, 93, 85–90.

Warfarin

Reisch, J. and Zappel, J. (1993) Photostabilitätsuntersuchungen an Natrium-Warfarin in kristallinem Zustand, *Sci. Pharm.*, 61, 283–286.

Reviews

Greenhill, J.V. (1994) Photodecomposition of drugs, in Swarbrick, J. and Boylan, J.C. (Eds.) *Encyclopedia of Pharmaceutical Technology*, New York: Marcel Dekker, Vol. 12, 105–135.
Moore, D.E. (1987) Techniques for analysis of photodegradation pathways of medicinal compounds, *Trends Anal. Chem.*, 6, 234–238.
Moore, D.E. (1987) Principles and practice of drug photodegradation studies, *J. Pharm. Biomed. Anal.*, 5, 441–453.
Moore, D.E. (2002) Drug-induced cutaneous photosensitivity. Incidence, mechanism, prevention and management, *Drug Saf.*, 25, 345–372.
Nema, S., Washkuhn, R.J., and Beussink, D.R. (1995) Photostability testing: an overview, *Pharm. Technol.*, 19, 170–185.
Oppenländer, T. (1988) A comprehensive photochemical and photophysical essay exploring the photoreactivity of drugs, *Chimia*, 42, 331–342.
Shahjahan, M. (1989) Ultraviolet light absorbers and their use in pharmacy, *East. Pharm.*, 32, 37–41.
Stewart, P.J. and Tucker, I.G. (1985) Prediction of drug stability. III. Oxidation and photolytic degradation, *Aust. J. Hosp. Pharm.*, 15, 111–117.

Thatcher, S., Mansfield, R.K., Miller, R.B., Davis, C.W., and Baertschi, S.W. (2001a) Pharmaceutical photostability: a technical guide and practical interpretation of the ICH guideline and its application to pharmaceutical stability. I, *Pharm. Technol. U.S.,* 25(3), 98–110.

Thatcher, S., Mansfield, R.K., Miller, R.B., Davis, C.W., and Baertschi, S.W. (2001b) Pharmaceutical photostability: a technical guide and practical interpretation of the ICH guideline and its application to pharmaceutical stability. II, *Pharm.Technol. U.S.,* 25(4), 50–62.

Thoma, K. and Struve, M. (1986) Einfluss hydrophiler Lösungsmittel auf die Instabilität von Arzneistoffen, *Pharm. Ind.,* 48, 179–183.

Thoma, K. and Kerker, R. (1992) Photoinstabilität von Arzneimitteln. I. Mitteilung über das Verhalten von im UV-Bereich absorbierenden Substanzen bei der Tageslichtsimulation, *Pharm. Ind.,* 54, 169–177.

Thoma, K. and Kerker, R. (1992) Photoinstabilität von Arzneimitteln. II. Mitteilung über das Verhalten von im sichtbaren Bereich absorbierenden Substanzen bei der Tageslichtsimulation, *Pharm. Ind.,* 54, 287–293.

Thoma, K. and Kübler, N. (1996) Einfluss der Wellenlänge auf die Photozersetzung von Arzneistoffen, *Pharmazie,* 51, 660–664.

Thoma, K. and Kübler, N. (1996) Photoinstabilität von Arzneistoffen und ihre analytischen Nachweismöglichkeiten, *Pharmazie,* 51, 919–923.

Thoma, K. and Kübler, N. (1996) Anwendung der HPLC-MS-Kopplung für die Photostabilitätsprüfung, *Pharmazie,* 51, 940–946.

Thoma, K. and Kübler, N. (1997) Einfluss von Hilfsstoffen auf die Photozersetzung von Arzneistoffen, *Pharmazie,* 52, 122–129.

Tønnesen, H.H. (2002) Photodecomposition of drugs, in Swarbrick, J., and Boylan, J.C. (Eds.) *Encyclopedia of Pharmaceutical Technology,* New York: Marcel Dekker, pp. 2197–2203.

Index

A